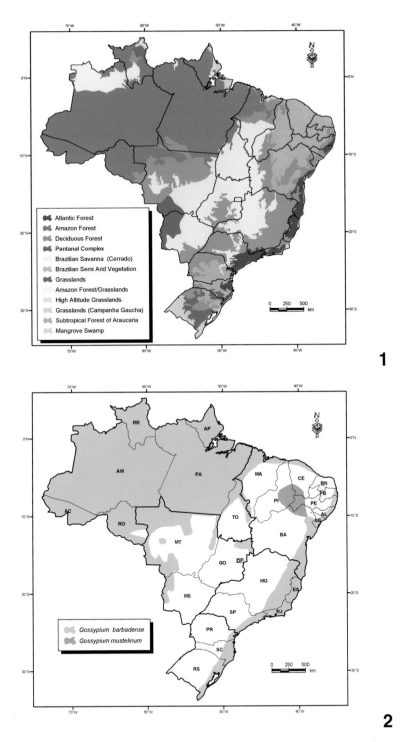

1

2

Plate 1. Natural vegetation zones (biomes) in Brazil (see Chapter 2). Source: Unidades de Conservação Federais do Brasil (IBGE, 1994).

Plate 2. Distribution of the native cotton species *Gossypium mustelinum* and the naturalized species *Gossypium barbadense*.

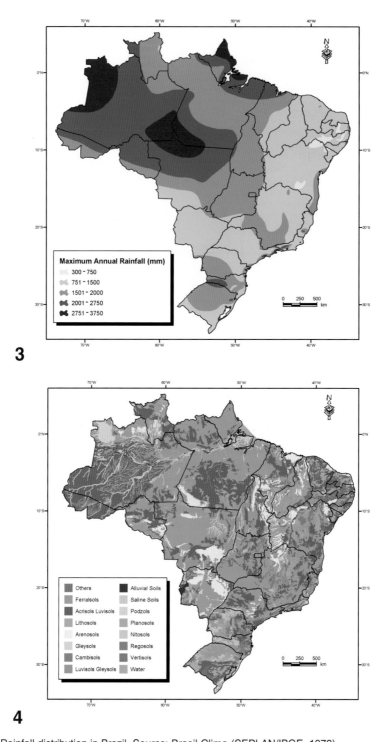

3

4

Plate 3. Rainfall distribution in Brazil. Source: Brasil Clima (SEPLAN/IBGE, 1978).

Plate 4. Soil types in Brazil. Source: Mapa de Solos do Brasil (Embrapa Solos, 1981).

ENVIRONMENTAL RISK ASSESSMENT OF GENETICALLY MODIFIED ORGANISMS SERIES

Volume 2. Methodologies for Assessing Bt Cotton in Brazil

ENVIRONMENTAL RISK ASSESSMENT OF GENETICALLY MODIFIED ORGANISMS SERIES

Titles available

Volume 1. A Case Study of Bt Maize in Kenya

Edited by A. Hilbeck and D.A. Andow

Volume 2. Methodologies for Assessing Bt Cotton in Brazil

Edited by A. Hilbeck, D.A. Andow and E.M.G. Fontes

ENVIRONMENTAL RISK ASSESSMENT OF GENETICALLY MODIFIED ORGANISMS

Volume 2. Methodologies for Assessing Bt Cotton in Brazil

Edited by

Angelika Hilbeck

Geobotanical Institute
Swiss Federal Institute of Technology
Zurich, Switzerland

David A. Andow

Department of Entomology
University of Minnesota
Minnesota, USA

and

Eliana M.G. Fontes

Embrapa Cenargen
Brasília, Brazil

Series Editors

A.R. Kapuscinski and P.J. Schei

CABI Publishing

CABI Publishing is a division of CAB International

CABI Publishing
CAB International
Wallingford
Oxfordshire OX10 8DE
UK

Tel: +44 (0)1491 832111
Fax: +44 (0)1491 833508
E-mail: cabi@cabi.org
Website: www.cabi-publishing.org

CABI Publishing
875 Massachusetts
Avenue, 7th Floor
Cambridge, MA 02139
USA

Tel: +1 617 395 4056
Fax: +1 617 354 6875
E-mail: cabi-nao@cabi.org

A catalogue record for this book is available from the British Library, London, UK.

A catalogue record for this book is available from the Library of Congress, Washington, DC.

ISBN-10: 1-84593-000-2
ISBN-13: 978-1-84593-000-4

Typeset by SPI Publisher Services, Pondicherry, India.
Printed and bound in the UK by Biddles Ltd, King's Lynn.

Contents

Contributors

R.P. de Almeida, *Embrapa Cotton, Rua Osvaldo Cruz 1143, Centenário, Caixa Postal 174, Campina Grande, 58107 720 PB, Brazil. e-mail: raul@cnpa.embrapa.br*

N.O. Amugune, *Plant Biotechnologist/Geneticist, Department of Botany, University of Nairobi, Riverside Drive, Chiromo Campus, PO Box 30197 GPO, Nairobi, Kenya. e-mail: noamugune@yahoo.co.uk*

D.A. Andow, *Professor of Insect Ecology, Department of Entomology, University of Minnesota, 219 Hodson Hall, 1980 Folwell Avenue, St Paul, MN 55108, USA. e-mail: dandow@umn.edu*

B.M. Anyango, *Senior Lecturer, Microbiology/Molecular Biology, Department of Botany, University of Nairobi, Riverside Drive, Chiromo Campus, PO Box 30197-00100 GPO, Nairobi, Kenya. e-mail: banyango@ uonbi.ac.ke*

F.J.L. Aragão, *Embrapa Genetic Resources and Biotechnology (Cenargen), Parque Estacão Biológica, Av. W5 Norte – Final, Brasília, 70770-900 DF, Brazil. e-mail: aragao@cenargen.embrapa.br*

S. Arpaia, *Researcher, Department of Biotechnology/Health and Environment Protection, ENEA – Italian National Agency for New Technologies, Energy and Environment, Research Centre Trisaia, SS 106 Ionica km 419+500, Rotondella, I-75026 MT, Italy. e-mail: salvatore.arpaia@trisaia. enea.it*

P.A.V. Barroso, *Embrapa Cotton, Rua Osvaldo Cruz 1143, Centenário, Caixa Postal 174, Campina Grande, 58107 720 PB, Brazil. e-mail: pbarroso@cnpa.embrapa.br*

N.E. de M. Beltrão, *Embrapa Cotton, Embrapa, Rua Osvaldo Cruz 1143, Centenário, Caixa Postal 174, Campina Grande, 58107 720 PB, Brazil. e-mail: nbeltrao@cnpa.embrapa.br*

A.N.E. Birch, *Research Entomologist, Host Parasite Coevolution, Scottish Crop Research Institute (SCRI), Invergowrie, Dundee DD2 5DA, Scotland, UK. e-mail: N.Birch@scri.sari.ac.uk*

C.L. Brubaker, *Senior Research Scientist, Centre for Plant Biodiversity Research, CSIRO Plant Industry, GPO Box 1600, Canberra, ACT 2601, Australia. e-mail: Curt.Brubaker@csiro.au*

D.M.F. Capalbo, *Embrapa Environment – Meio Ambiente (CNPMA), Rodovia SP 340, Km 127.5, CP 69, Tanquinho Velho, Jaguariúna, CEP 13820-000 SP, Brazil. e-mail: deise@cnpma.embrapa.br*

M. Caprio, *Professor, Insect Genetics, Entomology and Plant Pathology, Mississippi State University, Box 9775, MS 39762, USA. e-mail: mcaprio@entomology.msstate.edu*

L.P. de Carvalho, *Assistant Director of Research and Development, Embrapa Cotton, Rua Osvaldo Cruz 1143, Centenário, Caixa Postal 174, Campina Grande, 58107 720 PB, Brazil. e-mail: lpaulo@cnpa.embrapa.br*

E. Cia, *Centro Exp. de Algodão, Instituto Agronomico de Sao Paolo, Av. Barão Itapura 1481, CP 28, Campinas, 13020-902 SP, Brazil. e-mail: cia@iac.sp.gov.br*

A.Y. Ciampi, *Embrapa Genetic Resources and Biotechnology (Cenargen), Parque Estacão Biológica, Av. W5 Norte – Final, Brasília, 70770-900 DF, Brazil. e-mail: aciampi@cenargen.embrapa.br*

V.M. Cirino, *Researcher, Instituto Agronomico do Paraná, Rod. Celso Garcia Cid – KM 375, C.P. 481, Londrina, 86001-970 PR, Brazil. e-mail: vamoci@iapar.br*

L. Coradin, *Department of the National Programme for Biodiversity Conservation, IBAMA, Esplanada dos Ministérios, Bl B sala 819, Brasília, 70068-900 DF, Brazil. e-mail: lidio.coradin@mma.gov.br*

E.P.F. Diads, *Assessora da Comissão Nacional de Ética em Pesquisa (Consultant for the National Commission for Ethics in Research), Conselho Nacional da Sáude, Brazilian Ministry of Health, Brasília, DF, Brazil. e-mail: ednilza@terra.com.br*

J. de O. Duarte, *Researcher, Embrapa Maize and Sorghum, Rod. MG 424 km 65, Cx. Postal 151, Sete Lagoas, 35701-970 MG, Brazil. e-mail: jason@cnpms.embrapa.br*

G.S.J. Dubois, *Coordinator of GMO Environmental Licensing, Coordenação Geral de Licenciamento Ambiental, IBAMA (Brazilian Institute of Environment and Natural Renewable Resources), SCEN Trecho 02 – Ed. Sede do Ibama, Bloco C, Brasília, 70818-900 DF, Brazil. e-mail: gaetan.dubois@ibama.gov.br*

M.R. de Faria, *Entomologist, Embrapa Genetic Resources and Biotechnology (Cenargen), Parque Estacão Biológica, Av. W5 Norte – Final, Brasília, 70770-900 DF, Brazil. e-mail: mrf39@cornell.edu*

M.G. Fernandes, *Professor of Entomology and Ecology, Departamento de Ciências Agrárias, Universidade Federal de Mato Grosso do Sul, CP 533, Dourados, 79804-970, Mato Grosso do Sul, Brazil. e-mail: mgfernan@ceud.ufms.br*

O.A. Fernandes, *UNESP – Universidade Estadual Paulista (UNESP) Depto. Fitossanidade/FCAV, Rod. Prof. Paulo D. Castellane, km 5, Jaboticabal, 14884-900 SP, Brazil. e-mail: oafernan@fcav.unesp.br*

G.P. Fitt, *Strategy Director, CSIRO Entomology, Long Pocket Laboratories, 120 Meiers Road, Indooroopilly, QLD 4068, Australia. e-mail: Gary.Fitt@csiro.au*

V.L.I. Fonseca, *Ecology Department, Universidade de Sao Paolo, Rua do Matão-Travessa 14, 321, Butantã/Cidade Universitária, Sao Paolo, 05508-900 SP, Brazil. e-mail: vlifonse@ib.usp.br*

E.M.G. Fontes, *Leading Scientist, Embrapa Genetic Resources and Biotechnology (Cenargen), Parque Estacão Biológica, Av. W5 Norte – Final, Brasília, 70770-900 DF, Brazil. e-mail: eliana@cenargen.embrapa.br*

E.C. Freire, *Embrapa Cotton, Rua Osvaldo Cruz 1143, Centenário, Caixa Postal 174, Campina Grande, 58107 720 PB, Brazil. e-mail: eleusio@cnpa.embrapa.br*

M.G. Fuzatto, *Centro Exper. Algodão, Instituto Agronomico de Campinas, Av. Barão Itapura 1481, CP 28, Campinas, 13020-902 SP, Brazil. e-mail: mfuzatto@iac.sp.gov.br*

F.B. Gandara, *Departmento de Ciências Biológicas, Universidade de São Paolo, Escola Superior de Agricultura "Luiz de Queiroz", Av. Pádua Dais, 11, Piracicaba, 13418-900 SP, Brazil. e-mail: fgandara@esalq. usp.br*

M.F. Grossi De Sa, *Embrapa Genetic Resources and Biotechnology (Cenargen), Parque Estacão Biológica (SAIN Parque Rural), Av. W5 Norte – Final, Brasília, 70770-900 DF, Brazil. e-mail: fatimasa@cenargen. embrapa.br*

Truong Nam Hai, *Vietnam Institute of Biotechnology (IBT), Vietnamese Academy of Science and Technology (VAST), 18 Hoang Quoc Viet Road, Cau Giay, Hanoi, Vietnam. e-mail: tnhai@hn.vnn.vn*

A. Hilbeck, *Geobotanical Institute, Swiss Federal Institute of Technology Zurich, Zürichbergstrasse 38, Zürich, CH-8044, Switzerland. e-mail: angelika.hilbeck@env.ethz.ch*

Tran Thi Cuc Hoa, *Head of Department, Department of Biotechnology, Cuu Long Delta Rice Research Institute (CLRRI), Thoi Thanh, O Mon, Can Tho, Vietnam. e-mail: cuchoa@hcm.vnn.vn*

Nguyen Huu Huan, *Deputy Director General, Plant Protection Department, Vietnam Ministry of Agriculture and Rural Development (MARD), 28 Mac Dinh Chi Street, 1st District, Ho Chi Minh City, Vietnam. e-mail: ppdsouth@hcm.fpt.vn*

P.Y. Kageyama, *Director of National Program for the Conservation of Biological Diversity, Brazilian Environment Ministry, Esplanada dos Ministérios, B1 B sala 819, Brasília, 70068-900 DF, Brazil. e-mail: paulo.kageyama@mma.gov.br*

G.L. Lövei, *Senior Scientist, Department of Crop Protection, Danish Institute of Agricultural Science, Flakkebjerg Research Centre, Slagelse, DK-4200, Denmark. e-mail: gabor.lovei@agrsci.dk*

W.A. Lucena, *Researcher, Embrapa Cotton, Rua Oswaldo Cruz 1143, Centenário, Caixa Postal 174, Campina Grande, 58107 720 PB, Brazil. e-mail: wagner@cnpa.embrapa.br*

J. Lundgren, *Northern Grains Insect Research Laboratory, US Department of Agriculture (USDA) – Agricultural Research Service (ARS), 2923 Medary Avenue, Brookings, SD 57006, USA. e-mail: jlundgren@ngirl.ars. usda.gov*

A. de H.N. Maia, *Embrapa Environment – Meio Ambiente (CNPMA), Rod. SP 340, Km 127,5, Tanquinho Velho – C.P. 69, Jaguariúna, 13820-000SP, Brazil. e-mail: ahmaia@cnpma.embrapa.br*

C. Mallory-Smith, *Professor, Department of Crop and Soil Science, Oregon State University, Crop Science 107, Corvallis, OR 97331, USA. e-mail: carol.mallory-smith@oregonstate.edu*

I.S. Melo, *Senior Researcher, Embrapa Environment – Meio Ambiente (CNPMA), Rodovia SP 340, Km 127,5, Caixa Postal 69, Tanquinho Velho, Jaguariúna, 13820-000 SP, Brazil. e-mail: itamar@cnpma.embrapa.br*

L.C.S. Mendonça Hagler, *Professor, Institute of Microbiology, Universidade Federal Rio de Janeiro, Ilha do Fundão – CCS – BLI, CP 68028, Rio de Janeiro, 21944-970 RJ, Brazil. e-mail: leda@ibpinet.com.br*

J.E. Miranda, *Embrapa Cotton, Rua Osvaldo Cruz 1143, Centenário, Caixa Postal 174, Campina Grande, 58107 720 PB, Brazil. e-mail: miranda@cnpa.embrapa.br*

R.G. Monnerat, *Embrapa Genetic Resources and Biotechnology (Cenargen), Parque Estacão Biológica, Av. W5 Norte – Final, Brasília, 70770-900 DF, Brazil. e-mail: rose@cenargen.embrapa.br*

K.C. Nelson, *Forest Resources/Fisheries, Wildlife & Conservation Biology, University of Minnesota, 115 Green Hall, 1530 Cleveland Avenue North, St Paul, MN 55108, USA. e-mail: kcn@umn.edu*

A.L. Nepomuceno, *Embrapa Soja, Rodovia Carlos João Strass (Londrina/Warta), Accesso Orlando Amaral, Londrina, 86001-970 PR, Brazil. e-mail: nepo@cnpso.embrapa.br*

R.O. Nodari, *Gerência de Recursos Genéticos, CTNBio, Secretaria de Biodiversidade e Florestas, Projeto de Recursos Genéticos, Brazilian Environment Ministry, Sala 715, Esplanada dos Ministérios, Bloco B, Brasília, 70068-900 DF, Brazil. e-mail: rubens.nodari@mma.gov.br*

M.A. Okech, *Postdoctorate Fellow, Molecular Biology and Biochemistry Research, International Centre of Insect Physiology and Ecology (ICIPE), PO Box 30772-00100 GPO, Nairobi, Kenya. e-mail: maokech@icipe.org*

C. Omoto, *Faculty Professor of Entomology, Universidade de São Paulo, Escola Superior de Agricultura "Luiz de Queiroz", Avenida Pádua Dias 11, Piracicaba, 13418-900 SP, Brazil. e-mail: celomoto@esalq.usp.br*

E.O. Osir, *Head of MBB Department, Principal Scientist, Molecular Biology and Biotechnology (MBB), International Centre of Insect Physiology and Ecology (ICIPE), PO Box 30772-00100 GPO, Nairobi, Kenya. e-mail: asro@idrc.org.sg*

A. Pallini, *Departmento de Biologia Animal & Entomologia, Universidade Federal de Viçosa, Campus Universitário s/n, Viçosa, 36570-000 MG, Brazil. e-mail: pallini@ufv.br*

C.S.S. Pires, *Researcher, Embrapa Genetic Resources and Biotechnology (Cenargen), Parque Estacão Biológica, Av. W5 Norte – Final, Brasília, 70770-900 DF, Brazil. e-mail: cpires@cenargen.embrapa.br*

Vu Duc Quang, *Head of Molecular Biotechnology Lab, Institute of Agricultural Genetics (IAG), Vietnam Ministry of Agriculture and Rural Development (MARD), Co Nhue, Tu Liem, Hanoi, Vietnam. e-mail: vdquang@hn.vnn.vn*

Le Quang Quyen, *Director, Vietnam Institute for Cotton Research and Development, Nha Ho, Ninh Son, Ninh Thuan, Vietnam. e-mail: nhahocrc @hcm.vnn.vn*

F.S. de Ramalho, *Cotton Entomologist, Embrapa Cotton, Rua Osvaldo Cruz 1143, Centenário, Caixa Postal 174, Campina Grande, 58107 720 PB, Brazil. e-mail: ramalhohvv@globo.com*

E. Romano, *Embrapa Genetic Resources and Biotechnology (Cenargen), SAIN Parque Rural W5 Norte, Brasília, 70770-900DF, Brazil. e-mail: romano@cenargen.embrapa.br*

R.F. dos Santos, *Economia Rural, Embrapa Cotton, Rua Osvaldo Cruz 1143, Centenário, Caixa Postal 174, Campina Grande, 58107 720 PB, Brazil. e-mail: roberio@cnpa.embrapa.br*

F.G.V. Schmidt, *Embrapa Genetic Resources and Biotechnology (Cenargen), Parque Estacão Biológica, Av. W5 Norte – Final, Brasília, 70770-900 DF, Brazil. e-mail: schmidt@cenargen.embrapa.br*

M. Sétamou, *Research Scientist, Beneficial Insects Research Unit, USDA Agricultural Research Service (ARS), 2413 E Highway 83 Building 200, Weslaco, TX 78596, USA. e-mail: msetamou@weslaco.ars.usda.gov*

F.A. Silveira, *Instituto Ciências Biológicas Departamento de Zoologia, Universidade Federal de Minas Gerais, Av. Antonio Carlos, 6627, Pampulha, 31270-901 MG, Brazil. e-mail: fernando@mono.icb. ufmg.br*

P. Silvie, *CIRAD-CA, SHIS QI 15, Conjunto 15 casa 03, Brasília, CEP 71 635-350 DF, Brazil. e-mail: psilvie@terra.com.br*

M.F. Simon, *Secretaria de Gestão e Estratégia, Embrapa, Parque Estação Biológica S/N, Brasília, 70770-901 DF, Brazil. e-mail: marcelo.simon@ embrapa.br*

J.O. Siqueira, *Dep. Ciência do Solo, Universidade Federal de Lavras, Campus Universitário, CP37, Lavras, cep 37.200-000MG, Brazil. e-mail: siqueira@ufla.br*

D.A. Somers, *At time of writing: Professor, Agronomy and Plant Genetics, University of Minnesota, 411 Borlaug Hall, 1991 Upper Buford Circle, St Paul, MN 55108, USA. At time of publication: e-mail: david.a.somers@ monsanto.com*

J.M. Songa, *Entomologist, Biotechnology Centre, Kenya Agricultural Research Institute (KARI), P.O. Box 14733-00800, Nairobi, Kenya. e-mail: jmsonga@africaonline.co.ke*

M.L. de Souza, *Molecular Virologist, Embrapa Genetic Resources and Biotechnology (Cenargen), Parque Estacão Biológica, Av. W5 Norte – Final, Brasília, 70770-900 DF, Brazil. e-mail: marlinda@cenargen.embrapa.br*

E.R. Sujii, *Researcher Biological Control, Embrapa Genetic Resources and Biotechnology (Cenargen), Parque Estacão Biológica, Av. W5 Norte – Final, Brasília, 70770-900 DF, Brazil. e-mail: sujii@cenargen.embrapa.br*

Pham Van Toan, *Head of Division, Agricultural Microbiology Division, Vietnam Agricultural Science Institute (VASI), Thanh Tri, Hanoi, Vietnam. e-mail: pvtoan@hn.vnn.vn*

Nguyen Van Tuat, *Director, National Institute for Plant Protection (NIPP), Vietnam Ministry of Agriculture and Rural Development (MARD), Dong Ngac, Tu Liem, Hanoi, Vietnam. e-mail: tuat@hn.vnn.vn*

M.C. Valadares-Inglis, *Embrapa Genetic Resources and Biotechnology (Cenargen), Parque Estacão Biológica, Av. W5 Norte – Final, Brasília, 70770-900 DF, Brazil. e-mail: cleria@cenargen.embrapa.br*

S. Valle, *Head of Biosafety Course, Fundação Oswaldo Cruz (Fiocruz) – Oswaldo Cruz Foundation, Av. Brasil, 4365, Manguinhos, Rio de Janeiro, CEP 21045-900, RJ, Brazil. e-mail: valle@fiocruz.br*

J.M. Waquil, *Embrapa Maize and Sorghum, Rodovia MG 424 Km 65 S/N, Caixa Postal 151, Sete Lagoas, 35701-970 MG, Brazil. e-mail: waquil@cnpms.embrapa.br*

J.A. Johnston West, *Postdoctoral Associate, Plant Biology, University of Minnesota, 239 East Wilson Avenue, Salt Lake City, UT 84115, USA. e-mail:jillyjo13@yahoo.com.*

R.E. Wheatley, *Research leader, Soil Plant Dynamics Unit, Scottish Crop Research Institute (SCRI), Invergowrie, Dundee DD2 5DA, Scotland, UK. e-mail: R.Wheatley@scri.sari.ac.uk*

Series Foreword

The advent of genetically modified organisms (GMOs) offers new options for meeting food and agriculture needs in developing countries, but some GMOs used in agriculture can also affect biodiversity and natural ecosystems. These potential environmental risks and benefits need to be taken into account when making decisions about the use of GMOs. International trade and the unintentional transboundary spread of GMOs can also pose environmental risks depending on the national and regional contexts.

The complex interactions that can occur between GMOs and the environment heighten the need to strengthen worldwide scientific and technical capacity for assessing and managing environmental risks of GMOs.

The Scientific and Technical Advisory Panel (STAP) of the Global Environment Facility (GEF) provides strategic scientific and technical advice on GEF policies, operational strategies and programmes in a number of focal areas, including biodiversity. Its mandate covers *inter alia* providing a forum for integrating expertise on science and technology, and synthesizing, promoting and galvanizing state-of-the-art contributions from the scientific community. The GEF, established in 1991, helps developing countries fund projects and programmes that protect the global environment. GEF grants support projects related to biodiversity, climate change, international waters, land degradation, ozone layer and persistent organic pollutants.

Global environmental management of GMOs and the strengthening of scientific and technical capacity[1] for biosafety will require building policy and

[1] By scientific and technical capacity we mean 'the ability to generate, procure and apply science and technology to identify and solve a problem or problems' including 'the generation and use of new knowledge and information as well as techniques to solve problems' (Mugabe, 2000. Capacity Development Initiative, Scientific and Technical Capacity Development, Needs and Priorities. GEF-UNDP Strategic Partnership, October 2000).

legislative biosafety frameworks. The latter is especially urgent for developing countries, as the Cartagena Protocol on Biosafety of the Convention on Biological Diversity makes clear. The World Summit on Sustainable Development also identified the importance of improved knowledge transfer to developing countries on biotechnology. This point was stressed in recent international fora such as the Norway/UN Conference on Technology Transfer and Capacity Building, and in the capacity-building decisions of the first meeting of the parties to the Cartagena Protocol on Biosafety.

The STAP is collaborating with a number of international scientific networks to produce a series of books on scientific and technical aspects of environmental risk assessment of GMOs. This complements the projects being undertaken by the UN Environment Programme and the GEF to help developing countries design and implement national biosafety frameworks.

The purpose of this series is to provide scientifically peer-reviewed tools that can help developing countries strengthen their own scientific and technical capacity in biosafety of GMOs. Each book in the series will examine a different case study in developing countries. The workshops and writing teams used to produce each book are also capacity-building activities in themselves because they bring together scientists from the case-study country, other developing countries and developed countries to analyse and integrate the relevant science and technology into the book. The first book, a case study of Bt maize in Kenya, was published in 2004. This second book, on methodologies for assessing Bt cotton in Brazil, is written by 75 authors, including 48 scientists from Brazil as well as scientists and technical experts from Kenya, Vietnam, Australia, Europe and the USA. A third book, on methodologies for assessing transgenic fish, is in preparation. Each book provides methods and relevant scientific information for risk assessment, rather than drawing conclusions. The concerned organizations in each country will therefore need to conduct their own scientific risk assessments in order to inform their own biosafety decisions.

This book is the outcome of a scientific partnership between the STAP and the GMO Guidelines Project. This project was launched by scientists of the Global Working Group on Transgenic Organisms in Integrated Pest Management and Biological Control of the International Organization for Biological Control (IOBC). An international Advisory Board provided scientific and strategic advice, and included representatives from the STAP, the Secretariat of the Convention on Biological Diversity, and numerous agricultural, environmental, academic and governmental organizations, listed in the preface. The STAP then conducted an independent, international and anonymous scientific peer review.

We hope that this book will help governments, scientists, potential users of GMOs and civil society organizations in Brazil, in other parts of South America and in the rest of the world to strengthen their understanding of the scientific knowledge and methods that are available for conducting environmental risk assessments of GMOs. We encourage readers to draw their own insights to help them devise and conduct robust environmental risk assessments for their own countries.

Yolanda Kakabadze
Chair, Scientific and Technical Advisory Panel
Global Environment Facility
Quito, Ecuador

Anne R. Kapuscinski
Member, Scientific and Technical Advisory Panel
Global Environment Facility
St Paul, Minnesota, USA

Peter J. Schei
Member, Scientific and Technical Advisory Panel
Global Environment Facility
Trondheim, Norway

9 May 2005

Preface

The Cartagena Protocol on Biosafety (Biosafety Protocol) under the Convention on Biodiversity (CBD) identifies a need in both developing and developed countries for comprehensive, transparent, scientific methods for meaningful pre-release testing and post-release monitoring of transgenic plants to ensure their environmental safety and sustainable use. Most importantly, the needs of developing countries for capacity building and policy development must be addressed (UN DSD, 1999). Article 22 of the Biosafety Protocol requires that parties shall cooperate in the development and/or strengthening of human resources and institutional capacities in biosafety. It is also recognized that this capacity building will require significant investments, as many countries may not have the capability to make independent risk assessments or to evaluate independently submitted risk assessments on biosafety (CBD, 2000).

This Brazil case study is a product of the GMO Guidelines Project, 'Development of International Scientific Biosafety Testing Methodologies for Transgenic Plants'. This project was launched by scientists of the International Organisation for Biological Control (IOBC) Global Working Group on 'Transgenic Organisms in Integrated Pest Management and Biological Control'. It is funded by the Swiss Agency for Development and Cooperation (SDC) as a part of the Swiss government's commitment to the Biosafety Protocol. The project is advised by a 20-member advisory board representing a wide array of organizations from around the world. The board members function both as scientific advisers and as international mediators to the policy environment and relevant decision-making processes. The project is governed by a 15-member steering committee, which is responsible for all significant decisions.

The project addresses the environmental and agricultural effects of transgenic crops, and does not evaluate human health impacts or ethical implications. It has focused on available transgenic crop plants because there is more information available on this class of crop than any other, and it is possible to mobilize considerable expertise in this area.

One of the aims of the project is to improve the capacity of scientists in many developing countries to support environmental risk assessment of transgenic crop plants. To accomplish this, the project concentrates on scientist-to-scientist exchange, because these personal connections are likely to persist over time. To leverage these efforts, the project has focused on a few countries with reasonably developed scientific infrastructures, a desire to develop the scientific basis to support environmental risk assessment, and a need to do so. By strengthening the scientific capacities for risk assessment in these countries, expertise should be able to diffuse more readily to neighbouring countries. Brazil was the second focal country of the project and work conducted in Brazil forms the basis for this book. Among the countries of South America, Brazil has the largest agricultural science infrastructure, and a need to develop the scientific basis to support risk assessment of transgenic crops.

The other main aim of the project is to develop the scientific methodologies to support environmental risk assessment in developing countries. The Biosafety Protocol and many national and regional regulatory frameworks such as the European Union (EU) Directive 2001/18/EC specify that risk assessment should be conducted on a case-by-case basis. A case-by-case approach is necessary because sufficient experience is not available to allow aggregate analysis and assessment. Each transgenic plant and ecosystem must be looked at individually, because the relevant questions will differ on a case-by-case and country-by-country basis. Consequently, it would be difficult to propose generic risk-assessment guidelines, but it may be possible to develop general, robust scientific methodologies using case studies. Hence, the project has focused on developing scientifically sound, transparent case studies to instantiate the scientific principles supporting environmental risk assessment. This book is the final product from the case study of Bt cotton in Brazil. An earlier version of these results was published in Portuguese in 2004 (Capalbo and Fontes, 2004).

We would like to thank the Brazilian Agricultural Research Corporation (Embrapa Headquarters) and the Embrapa Center for Genetic Resources and Biotechnology (Cenargen) and their respective directors, for hosting the workshop on which this book is based, and the CTNBio (Brazilian National Biosafety Committee) for their support during our workshop in Brasília. We would also like to thank Embrapa, the National Council of Science and Technology (CNPq, Conselho Nacional de Ciência e Tecnologia) and the Funding Agency for Studies and Projects (FINEP, Financiadora de Estudos e Projetos) for their financial support of the workshop and project.

Dr Eliana Fontes (Embrapa Cenargen) and Dr Deise Capalbo (Embrapa Meio Ambiente) were instrumental in setting up the workshop in Brasília, DF, Brazil. Without their assistance we would not have been able to have the considerable Brazilian expertise at the workshop. We would like to extend our gratitude to Rodrigo Hermeto Correa Dolabella, Esq. (Camara dos Deputados, Chapter 3) and Dr Claudia Zwahlen (University of Minnesota, Chapter 10) for their substantial suggestions to the indicated chapters. We would also like to thank all of the members of the advisory board to the project for their support. These board members are Dr Ana Lucía Assad (Brazil Ministry of Science and Technology), Dr Joel Cohen (ISNAR), Dr Les E. Ehler (President, IOBC), Dr Les G. Firbank (Coordinator, UK Field Trials Programme), Dr Helmut Gaugitsch (Austrian Federal Environment Agency), Dr Hans R. Herren (Director General, ICIPE, Kenya), Mr Ryan Hill (Biosafety Protocol Secretariat, CBD), Dr Katharina Jenny (Swiss Agency for Development and Cooperation), Dr Anne Kapuscinski (GEF-STAP), Dr Peter Kenmore (Director, FAO Global IPM Facility), Dr Chris Ngichabe (Kenya Agricultural Research Institute), Dr William Padolina (International Rice Research Institute), Dr Francois Pythoud (Swiss Agency for Environment, Forest and Landscape), Dr Maria José Sampaio (Embrapa), Dr Julian Smith (CABI), Dr Wilson Songa (Kenya Plant Health Inspection Service), Dr Braulio de Souza Días (Brazil Environment Ministry), Dr Sutat Sriwatanapongse (Thailand Biotechnology Centre), Dr Hermann Waibel (University of Hannover) and Dr Jing Yuan Xia (Chinese Ministry of Agriculture). We also acknowledge the considerable efforts of the Steering Committee of the Project, without whom the project could not exist. In addition to us, these members are Drs Nick Birch, B.B. Bong, Deise Capalbo, Gary Fitt, K.L. Heong, Jill Johnston, Kristen C. Nelson, Ellie Osir, Allison Snow, David Somers, Josephine Songa and Fang-Hao Wan. Most importantly, we acknowledge Evelyn Underwood, without whose help and enthusiasm the Brazil workshop and this book would not have been possible.

We thank the Global IOBC and the IOBC Global Working Group on Transgenic Crops for their continued intellectual support of our work. In addition, some of the authors acknowledge the following sources of support: College of Natural Resources, University of Minnesota (KCN), SEERAD (NB and RW) and the European Commission (ECOGEN Project, NB). We also thank the six anonymous reviewers who reviewed parts or all of an early draft of this book on behalf of the STAP, and we especially thank Drs Anne Kapuscinski, Pierre Silvie and Ryan Hill for their extensive reviews of the book. Their comments were all extremely useful and greatly improved the contents. Finally, all of the authors thank partners and families for their support and understanding when meeting the many tight deadlines.

The overall task of the project is large and complex, and we invite the involvement of all public-sector scientists and encourage interested researchers to contact us via our website. At present, we are initiating efforts to develop teaching tools for broader educational activities, building bridges to other scientific projects to contribute to a consortium of coordinated efforts and involving both the private sector and the non-government organizations to further improve the quality of the project's products. The more people who become involved and engage in developing these products, the better they will become and the more widely they will be recognized. Interested public-sector scientists can enrol in the project at http://www.gmo-guidelines.info.

Angelika Hilbeck
Zürich, Switzerland
David A. Andow
St Paul, Minnesota, USA
Eliana M.G. Fontes
Brasília, DF, Brazil

7 March 2005

References

Capalbo, D.M.F. and Fontes, E.M.G. (2004) *GMO Guidelines Project, Algodão Bt.* Embrapa Document 38, Embrapa Meio Ambiente, Jaguariúna, SP, Brazil. Available at: http://www.cnpma.embrapa.br/download/documentos_38.pdf

CBD (Secretariat of the Convention on Biological Diversity). (2000) Cartagena Protocol on Biosafety to the Convention on Biological Diversity: Text and annexes. Montreal: Secretariat of the Convention on Biological Diversity. Available at: www.biodiv.org/doc/legal/cartagena-protocol-en-pdf

UN-DSD (Division for Sustainable Development). (1999) Agenda 21. Chapter 16: Environmentally Sound Management of Biotechnology. Available at: www.un.org/esa/sustdev/agenda21chapter16.htm

1   Improving the Scientific Basis for Environmental Risk Assessment through the Case Study of Bt Cotton in Brazil

D.A. ANDOW, P.A.V. BARROSO, E.M.G. FONTES, M.F. GROSSI-DE-SA, A. HILBECK AND G.P. FITT

Corresponding author: Professor D.A. Andow, Department of Entomology, University of Minnesota, 219 Hodson Hall, 1980 Folwell Avenue, St Paul, MN 55108, USA. Fax: +1 612 625 5299, e-mail: dandow@umn.edu

Introduction

This book provides scientifically sound and transparent methods for generating the scientific data necessary for a regulator to conduct an environmental risk assessment. The scientific basis for environmental risk assessment of transgenic crops has been controversial, because there is debate over what constitutes an environmental risk and what scientific methodologies effectively and efficiently provide the data needed for regulators to assess these risks (NRC, 2002). The purpose of this book is to present methodologies for conducting the scientific analysis that supports environmental risk assessment of transgenic crops that are in the process of commercial development, using *Bacillus thuringiensis* (Bt) cotton as a case study. This book offers scientists, regulators and educators a rich source of information and methodologies to help identify possible adverse environmental effects, formulate hypotheses to evaluate the likely occurrence of possible environmental risks, and plan appropriate scientific experiments to test these hypotheses. For regulatory personnel, it can be used as a tool to support decision making. The aim of the book is not to attempt to resolve the controversies regarding the scientific basis for risk assessment; not to conduct an actual environmental risk assessment of Bt cotton[1] in Brazil; and not to instruct regulators on how to

[1] Bt cotton varieties are transgenic cotton varieties expressing a gene coding for a crystalline protein that is toxic to certain insects. This gene comes from the bacterium, *Bacillus thuringiensis*.

©CAB International 2006. *Environmental Risk Assessment of Genetically Modified Organisms: Vol. 2. Methodologies for Assessing Bt Cotton in Brazil* (eds A. Hilbeck *et al.*)

conduct an environmental risk assessment. We believe that the methodologies outlined in this book have the benefits of prioritizing data needs, addressing risks that may actually occur in the environment of release (in this case Brazilian cotton agroecosystems and neighbouring ecosystems) and allowing flexibility to address the most critical concerns with greater effort. The book, however, does not attempt to compare and contrast the merits of alternative approaches.

Brazil is an ideal location for developing a case study on the environmental risks of transgenic organisms. The largest country in South America, it is also the largest agricultural exporter in South America and maintains the largest agricultural research infrastructure in the world after the USA. These natural and human endowments have enabled meaningful dialogue between Brazilian and international scientists to develop to address some of the issues of concern to Brazil (Capalbo *et al.*, 2003).

For the past several years, Brazil has been grappling with the many and diverse issues associated with transgenic crop plants. These issues have included important social questions about the ownership and distribution of the inputs to agricultural production; the values that motivate the development of and the cultural values that are affected by genetic engineering; and who benefits and how will genetic engineering affect the plight of the poor. In addition, there have been critical issues focused on who should decide that a transgenic crop[2] can be used in Brazil and on what basis should they make such a decision. These and other issues influence scientific discussions about transgenic crops in Brazil. As environmental scientists, we are all affected by and involved in many of these socio-cultural and ethical controversies, but, in our role as scientists, we have chosen to focus on the scientific issues underlying the risk assessment process in this book.

With the entry into force of the Cartagena Biosafety Protocol (CBD, 2000) in September 2003, it is essential that the scientific basis for risk assessment be developed so that Parties to the Protocol can develop the necessary infrastructure to evaluate transgenic organisms. In addition, as the International Plant Protection Convention (IPPC) begins to turn its eye to environmental risk assessment of transgenic organisms, it is crucial to develop a sound basis for risk assessment.

Brazil is a party to the Cartagena Biosafety Protocol and is interested in developing risk assessment methodology to implement the Protocol. A Biosafety Law was sanctioned in Brazil in January 1996 and the regulations began to be implemented in June 1996, when the National Technical Biosafety Commission was established (Fontes, 2003). Regulatory guidelines were developed later in 1996 (CTNBio, 1996) for any planned field release (not a contained experiment). These include various questions specific to the transgene characterization and the gene flow issues. Non-target issues and

---

[2] The terms genetically modified (GM) crop and transgenic crop will be used synonymously throughout this book. For environmental risk issues, these terms are roughly equivalent to living modified organism (LMO) as used in the Cartagena Protocol.

resistance are covered briefly with a focus on concerns about pollinators. The methodologies developed in this book are completely consistent with the CTNBio Guidelines.

The developments in biotechnology and biosafety throughout the years, and the specific conditions in Brazil, convinced many people that the current regulations need to be changed. A new biosafety law was approved in January, 2005 (for a discussion of this process, see Capalbo *et al.*, Chapter 3, this volume) and regulations are being developed. Both the existing regulations and the new law require a science-based risk assessment. Implementation of this requirement has created controversy in part because clear acceptable scientific methods for conducting a risk assessment have not been fully established (Fontes, 2003).

The efforts recorded in this book are initial steps towards addressing the scientific needs for environmental risk assessment. This book represents the efforts of public-sector scientists (mostly environmental scientists) to bring the collective wisdom of the scientific community together to help to develop an approach to scientific risk assessment that fits the conditions of the environment, the aspirations of the Protocol, the requirements of the IPPC and the regulatory requirements of the Brazilian regulatory structure as it exists and evolves. The emphasis throughout is on transparency and scientific rigour in the risk assessment methods.

The Bt cotton case study for Brazil

The question of which transgenic crop would be most suitable as a case study was approached through a variety of considerations: (i) the main biosafety concerns about each possible case study in the Brazilian environment; (ii) standing knowledge accumulated about the crop, the inserted gene(s) and the transgenic plant in Brazil and throughout the world; (iii) the potential benefits of changing from conventional varieties to transgenic varieties; and (iv) the need for science-based risk assessment of transgenic crops most likely to be commercialized in the near future.

Transgenic virus-resistant beans posed interesting challenges but were not chosen. Dry beans (*Phaseolus vulgaris* L.) are an important staple food, widely consumed throughout the country and are mainly cultivated by small holding farmers. Additionally, South America is one of the centres of origin of dry beans. However, we concluded that the expertise needed to analyse this case might be difficult to bring together to focus on risk assessment of virus-resistant crops.

Bt maize[3] has been tested in the field worldwide more extensively than most of the other transgenic varieties, but it was not chosen. Brazil is not a centre of origin of maize, and we wanted the case to include an evaluation of environmental effects in a centre of origin. Moreover, many of the transgenic

---

[3] Similar to Bt cotton, Bt maize varieties are transgenic maize varieties expressing a crystalline protein toxin gene from *Bacillus thuringiensis*.

varieties being tested were not developed specifically to protect maize against important maize pests in Brazil, such as fall armyworm *Spodoptera frugiperda*. In addition, Bt maize was used as the case study in the previous volume of this series featuring Kenya (Hilbeck and Andow, 2004) and we believed it would be more interesting and illuminating to focus on a different crop.

Herbicide-tolerant (HT) soybeans would be a particularly interesting case, given the controversies that surround it, especially in Brazil. One transformation event of HT soybean had been approved for commercialization by the National Competent Regulatory Authority on Biosafety, but a court injunction stopped the process and prohibited any planting of this variety in Brazil until very recently.[4] Despite this prohibition, there were reports of illegal importation of HT soybean seeds from Argentina and planting in the southern states of Brazil near Argentina. HT soybean had become a political issue. We decided against this case because the politics could interfere in the development of the independent, science-based methodologies that we have developed in this book.

Bt cotton was selected because it is a challenging case with a solid basis of scientific information, and may soon be commercialized in Brazil. In addition, Brazil feels the need to conduct a risk assessment on Bt cotton as soon as is feasible. The production of perennial cotton (*Gossypium hirsutum* var. *marie-galante* Hutch.) and upland cotton (*G. hirsutum* var. *latifolium* Hutch.) has historically been one of the productive and profitable agriculture systems in Brazil. After some decline during the late 1980s and early 1990s, cotton production has expanded throughout central Brazil and become one of the most profitable activities in Brazilian agriculture. Similar to HT soybean, it has recently been acknowledged (CTNBio, 2004) that transgenic cotton, including the Bt and HT traits, is being planted in Brazil prior to its legalization. In addition, transgene contamination in the cottonseed supply has been found. In response, CTNBio (2004) has recommended the establishment of a 1% seed-contamination limit on conventional cottonseed.

Cotton cropping presents some serious biosafety challenges. A large amount of pesticide is applied each year on cotton fields, because the crop is attacked by many pests. Pest-resistant cotton varieties could contribute to reduce the likelihood of environmental contamination and human poisonings, as long as other new risks that might be associated with the transgenic variety were minimal. A large number of arthropods, weeds and microbes are found in association with cotton fields. This raises the issue of potential adverse effects of Bt cotton on the large number of non-target organisms. Additionally, there is the critical issue of gene flow. Three species of the genus *Gossypium* occur in Brazil, all of them are allotetraploids and sexually compatible amongst themselves: *G. hirsutum* L., *G. barbadense* L. and *G. mustelinum* Mier ex Seem. The two crop species, *G. barbadense* and *G. hirsutum*, may occur in cultivated, feral, landrace or dooryard populations. Parts of the natural area of distribution of these species are being replaced by improved varieties of

---

[4] This ban was suspended in August 2004 by the court. The production and marketing of HT soybean is now authorized by the 2005 Biosafety Law.

G. hirsutum, and gene flow from new conventional or transgenic varieties of upland cotton may pose a threat to the long-term preservation of the species' genetic diversity. *G. mustelinum* is a rare species threatened with extinction. Gene flow effects on this species need to be evaluated (Johnston *et al.*, Chapter 1, this volume).

There are many specialists from different areas of research working with cotton in Brazil and around the world. Although the available Brazilian literature is still weak in many important issues, the expertise available in the country is strong and was certainly one of the most significant reasons to select Bt cotton for a case study.

Relation to the Kenya case study

In the first volume of this series, we addressed methods to conduct an environmental risk assessment for Bt maize in Kenya (Hilbeck and Andow, 2004). Unlike the white maize, which is a staple food for both rich and poor East Africans, cotton is a fibre crop from which only oils from the seed are used for human consumption. For maize, food-related issues constantly hover over discussions of environmental risk, but for cotton, the lesser importance of human food safety considerations mean the environmental issues are under more public focus. We recognize that there are concerns about the food safety of transgenic cottonseed oils for human consumption and feed safety of transgenic cotton meal for cattle, but we will not address these, or other food or feed safety issues, in this book, simply because these issues require a different expertise than has been assembled here to address the environmental concerns of transgenic crops.

Transgenic maize in East Africa is being developed largely by the public sector, led by the Insect Resistant Maize for Africa collaboration, with the aim of improving the livelihood of poor, economically marginal, small-scale farmers. Transgenic cotton in Brazil is being developed as a collaboration between the public and private sectors, and has multiple aims including helping poor cotton farmers and increasing Brazilian cotton exports. These contrasts frame the risk assessment differently, for example, in relation to the receiving environment (Capalbo *et al.*, Chapter 3, this volume).

Transgenesis and Bt Cotton

Worldwide use of Bt cotton

With pressing environmental and human health issues associated with high pesticide usage on cotton, there has been a strong imperative for more integrated approaches to pest management that minimize pesticide requirement. Transgenic cottons expressing insecticidal protein genes from Bt represent one new approach for integrated pest management (IPM) of cotton (Fitt, 2000; Wilson *et al.*, 2004).

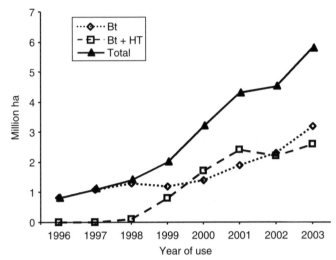

Fig. 1.1. Global area of Bt cotton (including Bt + herbicide tolerance (HT) traits) (after James, 2003).

Following a number of years of pre-release evaluation, Bt cotton varieties were first introduced in the USA, Mexico and Australia in 1996 (James and Krattiger, 1996). At that time, the global area of Bt cotton was approximately 800,000 ha. Figure 1.1 shows the worldwide pattern of adoption of Bt cotton since then. By 2003, the global area of GM cotton had reached 5.8 million ha, with GM cotton grown in nine countries (Table 1.1). This represents approximately 13% of the total world cotton area, but represents a significantly higher proportion of world production. Approximately 60% of the Bt cotton area is made up of varieties expressing only Bt, while the remaining 40% contains both Bt genes and a gene for herbicide tolerance.

The steady adoption of Bt technology reflects both an increase in the number of countries growing Bt cotton (Table 1.1) and steadily increasing areas in the major producing countries. In some cases (USA, China), Bt cotton has increased to over 50% of the cotton area, while in others (e.g. Australia)

Table 1.1. Adoption of Bt cotton by different countries.

Country	Year of commercial adoption
USA	1996
Australia	1996
Mexico	1996
China	1997
Argentina	1998
South Africa	1998
India	2002
Indonesia	2002
Colombia	2002

growth has been more controlled and within rigid management guidelines. Bt cotton was adopted by China, the world's largest cotton producer, in 1997. By 2003, Bt cotton accounted for 58% of China's 4.8 million ha cotton crop. In 1998, Argentina and South Africa commenced commercial production of Bt cotton, followed by India, Indonesia and Colombia in 2002 (Table 1.1).

The majority of Bt cotton, now under production, relies on a single Bt protein (Cry1Ac). In China, a combination of Cry1Ac and CPTi (cowpea trypsin inhibitor) has been grown for some years in one province, although accurate estimates of the area grown are scant. More recently, the first two-gene Bt varieties expressing both Cry1Ac and Cry2Ab have been approved for commercial use in Australia and the USA from 2003. With two independently acting insecticidal proteins, these two-gene varieties may add greater resilience against the risk of Bt resistance evolving in the target pests, while enhancing efficacy (Fitt *et al.*, Chapter 12, this volume).

Transgenesis in cotton

Advances in genetic transformation technology and useful economic characters have opened up tremendous possibilities for the improvement of commercial cotton varieties over the past two decades. Different methods have been applied for the genetic transformation of cotton using *Agrobacterium* and other direct DNA-transfer methods (Firoozabady *et al.*, 1987; Umbeck *et al.*, 1987; Finer and McMullen, 1990; Perlak *et al.*, 1990; McCabe and Martinell, 1993; Rajasekaran *et al.*, 2000; Sawahel, 2001). However, due to difficulties in cotton regeneration from undifferentiated callus tissue, most genetic transformations of cotton are restricted to Coker 312 and related varieties of G. *hirsutum*, none of which are widely cultivated in the world. Desirable traits, thus, are initially introduced into the Coker variety and transferred later to economically important cotton varieties. A newer genotype-independent, microparticle-mediated transformation method allows for direct transformation of differentiated tissue (typically hypocotyls) capable of growing into a plant of nearly any cotton genotype.

Currently, the two most widely used methods for transferring genes into plants are the *Agrobacterium*- and microparticle-mediated transformation methods. *Agrobacterium*-mediated genetic transformation of plants is based on the ability of the plant pathogen *Agrobacterium tumefaciens* (Smith and Townsend) to transfer a portion of a Ti (for tumour-inducing) plasmid into the plant's genome. This portion of the Ti plasmid is referred to as the T-DNA. *Agrobacterium* infection and gene transfers occur in nature at the site of a wound in the plant, and cause a characteristic growth referred to as a crown gall tumour, which produces a class of plant metabolites called opines on which the *Agrobacterium* has the unique ability to feed. By removing the genes responsible for tumour induction and opine production from the Ti plasmid, and by substituting in exogenous chimeric genes of interest, scientists have taken advantage of this naturally occurring DNA-transfer mechanism, and have designed DNA vectors capable of carrying desired genes into plant tissues.

Microparticle-mediated (biolistic) transformation refers to the delivery of DNA into target tissues by particle bombardment (Klein *et al.*, 1987; McCabe and Martinell, 1993), which is a physical delivery system whereby gold or tungsten beads coated with DNA are shot into cells that have the capacity to develop or differentiate into shoots or whole plants. This method can result in transgenic events with a higher copy number, complex integration patterns and fragmented inserts. *Agrobacterium*-mediated plant transformation can also result in transformed plants with multiple copies of inserts and complex integration patterns, but a simple pattern with non-rearranged T-DNA is obtained more frequently than in the microparticle method.

The transgenic BOLLGARD® cotton (transformation event 531) was developed via *A. tumefaciens*-mediated DNA transformation of a *cry1Ac* gene, which encodes for insecticidal activity towards certain Lepidopteran insect pests, into the cotton cultivar, Coker 312, using a binary plasmid vector (Perlak *et al.*, 1990). The BOLLGARD® cotton event also contains the bacterial antibiotic-resistance genes (*nptII* and *aad*) that were used in the genetic modification process to select cells containing the desired genes. The *nptII* gene confers resistance to kanamycin and neomycin, and the *aad* gene confers resistance to streptomycin and spectinomycin, which is linked to a bacterial promoter that does not function in plants, so the protein is not expressed in BOLLGARD® cotton plants (see Grossi *et al.*, Chapter 4, this volume for additional details).

BOLLGARD II® cotton contains two *Bt* genes that confer additional protection against Lepidopteran insect pests, particularly *Helicoverpa* species. BOLLGARD II® cotton was produced by inserting a second insecticidal gene (*cry2Ab*) and the marker *uid*A (GUS) gene (transformation event 15895) into the genome of BOLLGARD® cotton variety DP50B, which already expressed the Cry1Ac protein (event 531) by using the microparticle-acceleration plant transformation method (Doherdty *et al.*, 2000a,b).

The BOLLGARD® and BOLLGARD II® traits have been incorporated into commercial cotton varieties. Some of these varieties can be grown in Brazil, but many are susceptible to aphids and viruses (Fontes *et al.*, Chapter 2, this volume). Cooperative arrangements may be possible to enable the use of *cry1Ab*, *cry1F* and *vip3A* transgenes, which protect cotton from Lepidopteran attack, in Brazilian cotton varieties developed locally by several public institutions. Moreover, the Brazilian Agricultural Research Corporation (Embrapa) maintains an active cotton transformation research programme, which may provide a wider range of transgenes, including some that confer resistance to boll weevil, one of the key pests of cotton in Brazil.

Cotton breeding in Brazil

Cotton is cultivated in Brazil in a great diversity of environments with very different climates and soils (for additional details see Fontes *et al.*, Chapter 2, this volume). The level of technology and agricultural inputs such as herbicides, insecticides and chemical or organic fertilizers used is variable, from very low by small-scale farmers to very high by large-scale farmers, including modern machinery, irrigation and extensive use of agricultural inputs. Production varies from 2000 kg of fibre/ha in the highly industrialized large-scale farms to 400 kg of fibre/ha in most of the North-east states. The diversity of climatic conditions and technological profiles among regions necessitates at least three different breeding programmes to attend to the demands of cotton farmers (Fontes *et al.*, Chapter 2, this volume).

Cotton breeding of *G. hirsutum* in Brazil formally started at the beginning of the 1930s, and is conducted both by public and private institutions today. Traits evaluated by breeders can be divided into two categories: agricultural and industrial. The agricultural traits include resistance to pathogens, drought and arthropod pests, plant height, early maturation, boll size, retention of the lint in the boll and yield. The main industrial traits are lint percentage, fibre strength and fineness and fibre length.

Regional differentiation in cotton varieties
Three different kinds of varieties of *G. hirsutum* are used in the North-east region. The most traditional is the mocó cotton (*G. hirsutum* var. *marie-galante*) that was planted on more than 2 million ha during the 1970s. At that time, 5–6 years was needed to produce a new cotton variety. Today, mocó varieties are planted on only 9000 ha, restricted to the driest regions of Rio Grande do Norte and Paraiba States, and are cultivated as a semi-perennial crop with 3 years of production on average. The second class of varieties is formed by hybrids between mocó and upland cotton, and includes both white and coloured lint cottons. The third kind is upland cotton, which is planted under rain-fed conditions in areas with more than 600 mm and under irrigation. Coloured fibre varieties are also produced in these areas. Early maturation (130–150 days) and resistance to drought are the main features that distinguish the North-east varieties from the other upland Brazilian varieties.

About 70% of Brazilian cotton is planted in the Midwest region and this area is expanding rapidly. Cotton in this region is upland cotton and cultivated mainly under rain-fed conditions. Varieties from this region can yield more than 3000 kg/ha with lint percentage higher than 38%, and must be adapted to mechanical harvesting. Other important breeding goals for this region are resistance to various diseases.

Upland cotton is also planted in the Meridian region and cultivars are similar to those used in the Midwest. Fusarium wilt and nematodes are more widespread, and it is becoming obligatory to include genetic resistance in these cultivars. Larger bolls with loose lint are desirable because most of the harvest is hand-harvested. Recently, there has been an increase in large industrialized farms using mechanical harvesting.

Breeding transgenic cotton varieties
The transgenic varieties in Brazil probably will be produced by one of the three approaches. The first is based on the introduction of lines generated in

other countries and tested under Brazilian environmental and agricultural conditions. International cottonseed companies have used this method to produce
traditional (non-transgenic) varieties. Most of them have introduced transgenic
varieties into other countries and will use a similar breeding strategy to introduce Bt cotton to Brazil.

The second approach is backcrossing, which will be used to introduce a
transgene into the main advanced Brazilian varieties and lines. Backcrossing
is a breeding method used to transfer one or few genes into elite materials
and begins with the hybridization between an elite line (recurrent parent) and
another line with the transgene of interest (donor parent). The F_1 offspring
from this cross are themselves crossed with the recurrent parent to generate
the first backcross progeny (RC_1). The RC_1 plants are screened to find the
individuals that have the transgene, which are crossed again with the recurrent
parent. This procedure is repeated until the recovery of the recurrent parental
genotype with the transgene is achieved. Backcross transference probably will
be assisted by molecular markers to reduce the time necessary to accomplish
the transgene transfer. Two to three generations of backcrosses are expected
to be necessary for producing a new variety and at least four cycles to produce
sufficient seeds to plant commercial crops. Brazilian public and private breeding programmes will employ this method, if the owner companies license the
cotton transgenes used around the world to Brazil. These licensing arrangements have been accomplished in other cotton-producing countries, and
Brazilian breeders and farmers expect that similar arrangements will take
place in Brazil.

Finally, some transgenic lines will be used during the formation of
breeding populations. These transgenic lines will be crossed with non-transgenic lines in the same schemes used for breeding conventional varieties.
Various methodologies are used in cotton breeding in Brazil, such as single-
seed descent and mass selection, but the main scheme used is pedigree
breeding. The pedigree scheme used by Embrapa combines selection and
seed increase to make new cultivars available rapidly. Individual F_2 plants are
selected and the seeds from each plant are used to produce F_3 progenies.
Selections among and within progenies are made and the seeds of selected
plants and progenies are used to generate F_4 progenies. The step applied in
F_3 is repeated and the selected plants and progenies are used to form F_5
lines that are evaluated in experiments in one environment. At the same
time, the seed increase process begins. Selections from the F_5 are made and
the performances of F_6 lines are evaluated in at least three localities.
Regional evaluations (in the North-east and Midwest regions for Embrapa
cultivars) are achieved within the next 2 years and the best lines are released
as new cultivars.

The main difference among the conventional and transgenic cotton
breeding is the requirement of screening transgene expression in plants and
lines. Serological and biological assays can be used to verify the level of protein expression and resistance to key pests, respectively. For Bt transgenes,
the screening in F_2, F_3 and F_4 generations will be made by ELISA due to the
large number of individuals. Biological assays will be performed on some of

the individuals from the F_2 generation and on all F_5 lines to determine the resistance in the line to the target pests.

Scope of the Case Study

The case study published here is not a full risk assessment of Bt cotton in Brazil. Instead, it illustrates several methodologies for conducting an environmental risk assessment of Bt cotton. Every effort has been made to use the most recent publicly available scientific information so that the results from this case study will follow transparently from the information and be as accurate as possible without using private information. Because the private sector has significant sources of information that are not publicly available, some of the analyses in this book may partially reflect this limitation. Therefore, this case study of Bt cotton in Brazil is an insufficient support for a regulatory decision on Bt cotton in Brazil. A full risk assessment must update and complete the analyses outlined in this book. Most notably, we believe that experiments need to be conducted to provide the data needed to support the decision process.

The present study highlights several points that stem from the case study, our previous work on Bt maize in Kenya (Hilbeck and Andow, 2004), and previous research and applications of environmental risk assessment. Most significantly, we emphasize the importance of identifying possible adverse effects that are hypothetically feasible in the environment of concern. By this we mean that the effect should be hypothetically possible in association with a Brazilian cotton agroecosystem.

Specifically, we made concerted efforts to integrate the risk-assessment methodologies with the expected process of transgene development in what we refer to as 'staging' the risk-assessment process. Second, the gene-flow chapter (Johnston *et al.*, Chapter 11, this volume) addresses a significantly complicated problem; there are several wild, interfertile relatives of cotton in Brazil. The methods in this chapter may prove generalizable. Third, the resistance-risk chapter (Fitt *et al.*, Chapter 12, this volume) addresses the problem of low-dose resistance management directly, and provides a comprehensive approach to initiating the analysis of this risk. Fourth, the non-target methodology was further elaborated to develop and integrate exposure and adverse-effect scenarios to provide a route to evaluate specific non-target concerns. Some of these topics are introduced below; others are taken up in their respective chapters.

In addition, we incorporate one scientific approach for addressing scientific uncertainty so that decision makers can be clear about how science may relate to their right to take precautionary decisions. Although there are several sources of uncertainty in scientific advice, we focus on the uncertainty over the assumptions adopted in conducting our methodologies. We identified three sources of uncertainty for which we developed transparent, scientific approaches that can be incorporated into our methods to clarify how uncertainty may influence scientific conclusions. There is uncertainty about the relevant factors and relationships to be taken into account, such as the functional groups of non-target species considered (Hilbeck *et al.*, Chapter 5, this volume) and the dose of Bt toxin experienced by the target pests in relation to resistance risk (Fitt *et al.*, Chapter 12, this volume). Uncertainty arises from different value judgements about the existing scientific data, such as the expert ranking of species characteristics in the non-target methodologies (Sujii *et al.*, Chapter 6; Arpaia *et al.*, Chapter 7; Faria *et al.*, Chapter 8; Pallini *et al.*, Chapter 9; Mendonça *et al.*, Chapter 10, this volume). Finally, uncertainty pervades the assumptions adopted throughout the methods, such as the amount of gene-flow introgression to recipient populations (Johnston *et al.*, Chapter 11, this volume), and the completeness of the species lists for non-target species (Arpaia *et al.*, Chapter 7, this volume). The latter uncertainty arises both because not all of the species are known in all of the significant cotton-growing regions of Brazil, and because species of conservation concern are not completely described. The approach that we implemented to address uncertainty was to use the worst-case assumption in the absence of knowledge and continue with the methodology. In many cases, even under these worst-case assumptions, there was no effect on the outcome of the process, but in the other cases, this process allowed rapid identification of key knowledge gaps that could be addressed with additional research.

This case study does not address potential human or animal health risks associated with Bt cotton. Nutritional characterization of the products, such as oils or cotton meal, from Bt cotton is not considered here. There is considerable concern about potential health risks of transgenic crops in general, and while the methods described on transgene characterization (Grossi *et al.*, Chapter 4, this volume) should be relevant to such health assessments, it is not possible to consider these issues and the related issues associated with food labelling and informed consent of consumers in this book. Clearly, there is a need to consider these issues in depth.

This case study also does not conduct a sociological, philosophical or economic analysis of the use of Bt cotton in Brazil. The authors are nearly all environmental biologists and do not presume to have the necessary expertise to address these issues. It should go without saying that there is no attempt to conduct a cost-benefit analysis of Bt cotton for Brazil. This being acknowledged, many of the scientific issues embedded in these social science issues do connect to the environmental evaluations that are developed in this book.

Staging methodologies for environmental risk assessment

Methodologies for environmental risk assessment need to be integrated with the process of development of the transgenic crop. If such integration does not occur, the data needs may be addressed near the end of the product-development process when commercialization is imminent. At this stage of product development, experiments can be viewed as causing a delay or a loss of benefits, and there will be strong incentives to avoid these expenses of risk assessment, strong disincentives to reconsider previous decisions in developing the product, such as safe design of the transgenic crop (Grossi *et al.*, Chapter 4, this volume) and pressures to deliver a swift regulatory decision.

By integrating the methodologies for producing data for risk assessment with the product-development process, biosafety issues can actually inform and improve product development. In this case study, the methodologies are integrated with the stages of development of a transgenic plant. This allows efficient development of the transgenic crop and, simultaneously, timely development of information for the assessment and management of environmental risk. If certain information needed for risk assessment were collected too early in the development process, it could delay development without reducing possible risk. Conversely, if information needed for risk assessment were collected too late in the development process, it may be difficult to facilitate development of an effective, low-risk transgenic organism or to conduct an environmental risk assessment supported by sound scientific information.

We considered the development of a transgenic plant to follow a sequence of stages, each of which leads to a key decision or set of decisions for the organization or individual working to develop a commercializable transgenic plant (Table 1.2).

Stage 1: Design of plant transformation.
This stage includes consideration of factors such as plant species, target gene sequence, plasmid structure, transgene structure, use of markers, etc. Any of these could influence the way environmental risk assessment will be conducted. These ideas are developed more thoroughly in Grossi *et al.* (Chapter 4, this volume) on designing transgenes to minimize the requirements for risk assessment and Fitt *et al.* (Chapter 12, this volume) on transgene constructs that minimize the risk of resistance evolution in the target pests of Bt cotton.

Table 1.2. Summary of the key stages, subsequent decisions and consequent actions involved in the development of a transgenic organism for commercial use.

Stage	Leads to decision	Leads to action
1. Design of plant transformation	Decisions on how to conduct transformation	Transformation conducted
2. Focusing on a few transformation events based on laboratory assays	Decisions to focus on a few events	Discarding many events and continuing work on only a few
3. Characterizing focal events in laboratory	Whether to conduct small-scale field tests	Conducting field tests
4. Small-scale field testing	Whether to conduct large-scale field tests	Conducting large-scale trials
5. Large-scale field trials	Whether to allow commercial use	Commercial use
6. Commercial use		

Stage 2: Focusing on a small set of transformation events.
During transformation, perhaps hundreds of transformation events are created. An essential step in developing a transgenic product is to focus on a small
set of transformation events prior to conducting extensive testing on any one
of them. This will involve laboratory-based screening procedures, initial product testing, genetic characterization, etc., and each of these could affect the
subsequent environmental risk or risk assessment as discussed in Grossi *et al.*
(Chapter 4, this volume). It is also possible to anticipate some of the possible
concerns about adverse effects to non-target organisms. These are discussed
generally in Hilbeck *et al.* (Chapter 5, this volume) and in specific contexts in
Sujii *et al.* (Chapter 6, this volume), Arpaia *et al.* (Chapter 7, this volume),
Faria *et al.* (Chapter 8, this volume), Pallini *et al.* (Chapter 9, this volume) and
Mendonça *et al.* (Chapter 10, this volume).

Stage 3: Characterizing the small set of events in laboratory.
Additional laboratory-based and greenhouse-based testing procedures will be
needed to characterize the smaller set of events from stage 2 to determine
which of these are sufficiently promising to merit the additional expense of
field testing; these procedures could provide information that affects subsequent environmental risk or the need for risk assessment. In addition, product
testing in greenhouse and other contained environments will be important for
the initial steps of case-specific risk assessment. Many of the chapters of this
book address how laboratory and greenhouse experiments may be essential
for providing the proper data for risk assessment.

Stage 4: Small-scale field testing.
Field-based screening procedures, product testing, breeding protocols, etc. will
play a role in the selection of the best event(s) to concentrate further development. These evaluation procedures could affect how risk assessment needs to
be conducted. For example, the breeding process eliminates many poorly per-
forming plants. Some, but not all, environmental risks may be reduced during
this screening process (NRC, 2002). In addition, small-scale field trials coupled
with additional laboratory trials will be important for providing data for environmental risk assessment.

Stage 5: Large-scale field trials.
Variety testing and screening, etc. will further concentrate development on a
set of potentially commercial varieties containing the one or two most effective transgenes. Procedures for moving products to commercialization could
affect the magnitude of environmental risk. In addition, all work needed to
complete an environmental risk assessment must be completed during this
stage. Monitoring needs and effective monitoring tools should also be identified and developed during this stage.

Stage 6: Commercial use.
Implementing risk management must begin prior to this stage, but actual management practices must begin when the transgenic plant is commercialized. In

addition, the monitoring tools must be implemented and refined, and a process to adapt the management of risk to the information obtained during monitoring should be implemented.

Development of non-target biodiversity assessment

Non-target biodiversity environmental risk assessment for agricultural products tends to rely on the use of indicator species (Forbes and Forbes, 1994; CEC, 1996; EPA, 1998). As argued by many (Cairns, 1981; Cairns and Mount, 1990; Forbes and Forbes, 1994; Elmegaard and Jagers, 2000; Forbes and Calow, 2002), the use of indicator species has many serious scientific problems, including not being predictive of actual environmental risks in the target environment.

Assessing risks of transgenic crops to non-target biodiversity is challenging, both for risk assessors and scientists interested in providing data relevant to a risk assessment. Because total biodiversity can be examined only in field environments, the early stages of non-target assessment, prior to release in the environment, must rely on species-based procedures. However, there are so many species that one of the main challenges is determining which species should be evaluated. We have proposed a screening process to focus effort where it is most needed. Biodiversity should be divided into ecological functional groups so that species representing important ecosystem functions can be assessed (Andow and Hilbeck, 2004; Birch *et al.*, 2004). Within each functional group, the known species are listed and, using a series of ecological criteria, these species are ranked to identify those that would be most likely to be adversely affected.

In the present case study of Bt cotton in Brazil, we develop and test the scientific basis of this non-target biodiversity methodology. Specifically, we generate a rationale for selecting certain functional groups for further evaluation, formalize a process for prioritizing species, clarifying the critical ecological criteria used in the prioritization process, and develop a method for identifying and integrating exposure and adverse effects scenarios so that the scientific basis for the scenarios can be tested in the laboratory or greenhouse.

For this case study, Brazilian scientists selected five functional groups for evaluation. These selections were based on concrete concerns about how Bt cotton might adversely affect the environment. Non-target herbivores (Sujii *et al.*, Chapter 6, this volume), predators (Faria *et al.*, Chapter 8, this volume) and parasitoids (Pallini *et al.*, Chapter 9, this volume) were selected because of the concern that Bt cotton may exacerbate an existing pest problem or cause a secondary pest problem. Because boll weevil and cotton aphids are the key pests in cotton in Brazil and the Lepidopteran targets of Bt cotton are usually of secondary importance, it is important that Bt cotton does not exacerbate problems associated with boll weevil or cotton aphids. For the poor, small-scale farmers in the North-east and Meridian regions, it is important that Bt cotton does not cause a secondary pest problem, because this could threaten the livelihood of these farmers. Non-target pollinators and flower visitors

(Arpaia *et al.*, Chapter 7, this volume) were selected because of concerns that Bt cotton could adversely affect pollination of cotton or other crops. Indeed, present regulations (CTNBio, 1996) require evaluation of effects on pollinators. In addition, flower visitors were also considered because many natural enemies and species of potential conservation concern are flower visitors. Species of conservation concern can most readily be evaluated when they come in contact with Bt cotton. Soil ecosystem functions were selected for additional evaluation because of the importance of these functions for the sustained productivity of agriculture. A major concern was that Bt cotton does not create additional problems for the people involved in cotton production, especially the rural poor who might lose their livelihood should such problems occur. Implicit in each of these concerns is a desire to improve cotton production without harming the environment.

Developing methodologies to support risk assessment

Risk assessment typically combines an assessment of exposure with an assessment of hazard (NRC, 1983). These concepts take on different terms in different risk assessment frameworks (e.g. Annex 3; CBD, 2000); and rather than engage in discussing these terminologies, we focus on identifying possible adverse effects of Bt cotton on the environment, possible pathways by which the environment may be exposed to Bt cotton and/or its transgene products and possible ecological scenarios by which such exposure might result in an adverse effect. By constructing these hypothetical chains of possible exposure integrated with a consequent possible adverse-effect scenario, it is possible to experimentally test the possibilities frequently with simple experiments. In this book, we develop this framework in the context of Bt cotton in Brazil.

Chapter 2 reports on the agricultural and socio-economic context of cotton production in Brazil, which provides the needed background information for the rest of the chapters in the book. It describes the three main regional production areas, briefly describing productivity, farming systems and economic status of farmers. It also identifies the main biological and socio-economic constraints to cotton production and summarizes the main pest problems, contextualizing the target Lepidoptera within the broader pest complex. It also introduces IPM as a possible solution to pest problems, which will be taken up in Chapter 3.

Chapter 3 provides a framework for specifying the problem, evaluating the utility (including benefits) of the transgenic plant in specific Brazilian crop-production contexts and comparing it to other potential solutions to the problem. This problem formulation and options assessment (PFOA) helps frame the methodologies that follow in Chapters 5–12. By specifying the problem, it defines the target agroecosystems for which Bt cotton or an alternative solution is proposed, including the crop system, farming system and ecological and structural context, and the people who will be affected. It also addresses one of the most critical aspects for scientific experiments in support of risk assessment, the appropriate comparisons for the experiments. In general, risk assessment should compare alternative futures (NRC, 1983) rather than a possible future (a transgenic crop) with the past (conventional plant breeding) (NRC, 2002). The PFOA provides a means to identify and specify appropriate alternative futures through a deliberative process.

Chapter 4 identifies the genetic issues relevant for risk assessment and sketches methodological strategies to address these issues and thereby identify all of the possible transgene products and relevant phenotypic modifications of the transgenic plant. Characterization of the transgene is structured into four components: transgene design, genotypic characterization, phenotypic characterization and transmission between generations. Some of these characteristics should be measured in the field, and others can be measured in the laboratory. Particular attention is paid to when the various assessments should be conducted during the development of the transgene.

Chapters 5–10 provide and illustrate the steps of a scientific methodology for use during the pre-release stages of transgene development to support non-target risk assessment. The methodology screens the species and ecosystem processes associated with the full biodiversity, so that effort is focused on the most relevant parts. The first series of steps determines which non-target species, structural characteristics of the biota and/or ecosystem functions should be considered for additional evaluation. The result focuses on those species and functions that are more likely to be adversely affected based on the scientific information available prior to environmental release. It should be emphasized that this step does not demonstrate that an adverse effect will or will not occur. It prioritizes the species and functions so that additional effort is focused most efficiently before any experiments are conducted. The second series of steps specifies scientific procedures for testing the potential effect of Bt cotton on the selected species or functions. These procedures are primarily laboratory and greenhouse ones that can be used prior to field release. Some field procedures are also proposed, some of which would have to be conducted when the transgene is being tested in the field. Chapter 5 introduces the methodology and Chapters 6–10 illustrate the application of this methodology to non-target herbivores, pollinators, predatory natural enemies, parasitoids and soil organisms and ecosystem functions.

Chapter 11 addresses gene flow to wild relatives and its consequences. Brazil is one of the centres of origin for *Gossypium* species, and several sexually compatible species and varieties occur, especially in North-east Brazil. This chapter reviews the evolutionary history of the genus *Gossypium* in relation to gene flow from cotton. It establishes the likely recipient populations of *Gossypium* and evaluates the likelihood of gene flow from *G. hirsutum* var. *latifolium* to potential recipient populations, the possibility of subsequent geographical spread of transgenes and the potential ecological effects resulting from gene flow.

Chapter 12 assesses the risk of resistance evolution in the target Lepidopteran pests to Bt cotton. It lays out a simple procedure for identifying the target species most at risk of resistance evolution, and proposes effective and potentially workable plans to manage this resistance risk to delay the occurrence of resistance to Bt cotton. It also considers approaches for developing a practical monitoring and response system to detect resistance and to adapt management appropriately.

Finally, Chapter 13 summarizes the major achievements and key findings from this case study. It concludes with recommendations for completing a risk assessment of Bt cotton in Brazil and some challenges that remain for improving the scientific basis in support of risk assessment.

Acknowledgements

We thank Drs Eleusio Curvelo Freire (Embrapa Cotton), Luiz Paulo de Carvalho (Embrapa Cotton) and Ms Evelyn Underwood (Swiss Federal Institute of Technology) for their help.

References

Andow, D.A. and Hilbeck, A. (2004) Science-based risk assessment for non-target effects of transgenic crops. *BioScience* 54(7), 637–649.

Birch, A.N.E., Wheatley, R., Anyango, B., Arpaia, S., Capalbo, D., Getu Degaga, E., Fontes, D.E., Kalama, P., Lelmen, E., Løvei, G., Melo, I.S., Muyekho, F., Ngi-Song, A., Ochieno, D., Ogwang, J., Pitelli, R., Schuler, T., Sétamou, M., Srinivasan, S., Smith, J., Van Son, N., Songa, J., Sujii, E., Tan, T.Q., Wan, F.-H. and Hilbeck, A. (2004) Biodiversity and non-target impacts: a case study of Bt maize in Kenya. In: Hilbeck, A. and Andow, D.A. (eds) *Environmental Risk Assessment of Transgenic Organisms: A Case Study of Bt Maize in Kenya.* CAB International, Wallingford, UK, pp. 117–185.

Cairns, J. Jr (1981) Sequential versus simultaneous testing for evaluating the hazards of chemicals to aquatic life. *Marine Environment Research* 4, 165–166.

Cairns, J. Jr and Mount, D.I. (1990) Aquatic toxicology. Part 2 of a four-part series. *Environmental Science and Technology* 24, 154–161.

Capalbo, D.M.F., Hilbeck, A., Andow, D.A, Snow, A., Bong, B.B., Wan, F.-H., Fontes, E.M.G., Osir, E.O., Fitt, G.P., Johnston, J., Songa, J., Heong, K.L. and Birch, A.N.E. (2003) Brazil and the development of international scientific biosafety testing guidelines for transgenic plants. *Journal of Invertebrate Pathology* 83, 104–106.

CBD (2000) Cartagena Protocol on Biosafety to the Convention on Biological Diversity: Text and Annexes. Secretariat of the Convention on Biological Diversity, Montreal. Available at: www.biodiv.org/doc/legal/cartagena-protocol-en-pdf

CEC (1996) Technical guidance documents in support of the Commission Directive 93/67/EEC on risk assessment for new substances and the Commission Regulation No. 1488/94 on risk assessment for existing substances. Commission of the European Communities, Brussels, Belgium.

CTNBio (Comissão Técnica Nacional de Biossegurança) (1996) Instrução Normativa CTNBio No. 3 from 12.11.96. Publicada no D.O.U. de 13.11.96, Section I, p. 23.691.

CTNBio (Comissão Técnica Nacional de Biossegurança) (2004) Extrato de Parecer Técnico No. 480/2004.

Doherdty, S.C., Hamilton, K.A., Lirette, R.P. and Borovkova, I. (2000a) Amended report for molecular characterization of cotton event 15985. Monsanto Company Laboratory project ID study 99-01-36-04, MSL-16620.

Doherdty, S.C., Lirette, R.P. and Hamilton, K.A. (2000b) Molecular report of the stability of cotton event 15985. Monsanto Company Laboratory project ID study 00-01-36-09, MSL-16749.

Elmegaard, N. and Jagers op Akkerhuis, G.A.J.M. (2000) Safety Factors in Pesticide Risk Assessment. Differences in Species and Acute-Chronic Relations. NERI Technical Report 325. National Environmental Research Institute, Silkeborg, Denmark, 60 pp.

EPA (1998) Guidelines for Ecological Risk Assessment. EPA 630/R-95-002F. Environmental Protection Agency, Washington, DC.

Finer, J.J. and McMullen, M.D. (1990) Transformation of cotton (*Gossypium hirsutum* L.) via particle bombardment. *Plant Cell Reports* 8, 586–589.

Firoozabady, E., Deboer, D.L. and Merlo, D.J. (1987) Transformation of cotton (*Gossypium hirsutum* L.) by *Agrobactium tumefaciens* and regeneration of transgenic plants. *Plant Molecular Biology* 10, 105–116.

Fitt, G.P. (2000) An Australian approach to IPM in cotton: integrating new technologies to minimise insecticide dependence. *Crop Protection* 19, 793–800.

Fontes, E.M.G. (2003) Legal and regulatory concerns of transgenic plants in Brazil. *Journal of Invertebrate Pathology* 83, 100–103.

Forbes, V.E. and Calow, P. (2002) Extrapolation in ecological risk assessment: balancing pragmatism and precaution in chemical controls legislation. *BioScience* 52, 249–257.

Forbes, V.E. and Forbes, T.L. (1994) *Ecotoxicology in Theory and Practice.* Chapman & Hall, London.

Hilbeck, A. and Andow, D.A. (eds) (2004) *Environmental Risk Assessment of Genetically Modified Organisms: A Case Study of Bt Maize in Kenya.* CAB International, Wallingford, UK.

James, C. (2003) Preview: global status of commercialized transgenic crops: 2003. ISAAA Briefs No. 30. ISAAA, Ithaca, New York.

James, C. and Krattiger, A.F. (1996) Global review of the field testing and commercialization of transgenic plants, 1986 to 1995: the first decade of crop biotechnology. ISAAA Briefs No. 1. ISAAA, Ithaca, New York, 31 pp.

Klein, T.M., Wolf, E.D., Wu, R. and Sanford, J.C. (1987). High-velocity microprojectiles for delivering nucleic acids into living cells. *Nature* 327(6117), 70–73.

McCabe, D.E. and Martinell, B.J. (1993) Transformation of elite cotton cultivars via particle bombardment of meristems. *Biotechnology* 11, 596–598.

NRC (National Research Council) (1983) *Risk Assessment in the Federal Government: Managing the Process.* National Academies Press, Washington, DC.

NRC (National Research Council) (2002) *Environmental Effects of Transgenic Plants: The Scope and Adequacy of Regulation.* National Academies Press, Washington, DC.

Perlak, F.J., Deaton, R.W., Armstrong, T.A., Fuchs, R.L., Sims, S.R., Greenplate, J.T. and Fischhoff, D.A. (1990) Insect resistant cotton plants. *Biotechnology* 8, 939–943.

Rajasekaran, K., Hudspeth, R.L., Cary, J.W., Anderson, D.M. and Cleveland, T.E. (2000) High-frequency stable transformation of cotton (*Gossypium hirsutum* L.) by particle bombardment of embryogenic cell suspension cultures. *Plant Cell Reports* 19(6), 539–545.

Sawahel, W.A. (2001) Stable genetic transformation of cotton plants using plybrene-spermidine treatment. *Plant Molecular Biology Reporter* 19(4), 377a–377bf.

Umbeck, P., Johnson, G. and Barton, K. (1987) Genetically transformed cotton (*Gossypium hirsutum* L.) plants. *Biotechnology* 5, 235–236.

Wilson, L.J., Mensah, R.K. and Fitt, G.P. (2004) Implementing integrated pest management in Australian cotton. In: Horowitz, A.R. and Ishaaya, I. (eds) *Insect Pest Management: Field and Protected Crops*. Springer, Berlin, Heidelberg, New York, pp. 97–118.

2 The Cotton Agricultural Context in Brazil

E.M.G. Fontes, F. de Souza Ramalho, E. Underwood,
P.A.V. Barroso, M.F. Simon, E.R. Sujii, C.S.S. Pires,
N. Beltrão, W.A. Lucena and E.C. Freire

*Corresponding author: Eliana M.G. Fontes, Embrapa Cenargen
(Genetic Resources and Biotechnology), Caixa Postal 02372, Parque
Estação Biológica, Av. W3 Norte – Final, Brasília, DF-707 70-901,
Brazil. Fax: +55-61-34484673, e-mail: eliana@cenargen.embrapa.br*

Introduction

Brazil is one of the largest cotton producers and consumers in the world. In the agricultural year 2002/03, 739,200 ha of cotton were planted in the country producing more than 850,800 t fibre (CONAB, 2003). The best-quality fibre is exported. The remainder is supplemented by imported lower-quality fibre and supplies the internal industrial demand estimated at 950,000 t/year (IBGE, 2004). Both area and yield are expected to increase in the near future.

The crop is grown in a great diversity of environments with very different climates and soils. The level of technology and agricultural inputs used is variable, from very low by small-scale farmers in North-eastern Brazil to very high by large-scale farmers in the Brazilian Midwest. Production varies from 2000 kg fibre/ha in the highly industrialized farms to 400 kg fibre/ha in the majority of the small-scale farms in the North-east states. Most cotton production is rain fed.

Pest and disease damage is one of the main constraints to productivity. The group of pests attacking cotton is well known, comprising at least 30 species of insects and three species of mites. The most economically damaging species and the intensity of the damage they cause vary from region to region. Diseases caused by viruses, bacteria and fungi also seriously affect productivity in some regions, and some are transmitted by pest insects.

In this chapter, we divide the cotton-growing areas of Brazil according to common practice into three regions that differ significantly in their topographical, climatic, ecological and socio-economic characteristics. These production regions are organized according to the similarities of production systems used, and are different from the geographical regions of Brazil (Fig. 2.1). The largest production region is the Midwest, which comprises the states of Mato Grosso, Mato Grosso do Sul, Goiás, Distrito Federal, Minas Gerais and the western part of Bahia (76.5% of the cotton-producing area, 87% of cotton fibre production). Most of the states in this region were originally covered by natural vegetation of the biome Cerrado (Plate 1 – see colour frontispiece). The soil, climate and

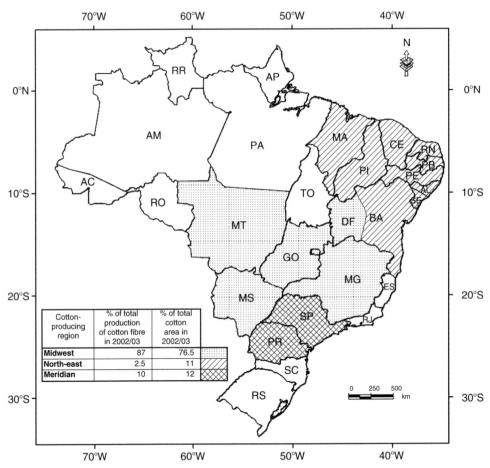

Cotton-producing region	% of total production of cotton fibre in 2002/03	% of total cotton area in 2002/03
Midwest	87	76.5
North-east	2.5	11
Meridian	10	12

Fig. 2.1. Cotton-producing regions in Brazil, with % of cotton area and cotton-fibre production in each region. (*Source*: adapted by the authors from CONAB, 2003.)
Key to Brazilian states: AC = Acre, AL = Alagoas, AM = Amazonía, AP = Amapá, BA = Bahía, CE = Ceará, DF = Distrito Federal, ES = Espírito Santo, GO = Goiás, MA = Maranhão, MG = Minas Gerais, MS = Mato Grosso do Sul, MT = Mato Grosso, PA = Pará, PB = Paraíba, PE = Pernambuco, PI = Piauí, PR = Paraná, RJ = Rio de Janeiro, RN = Rio Grande do Norte, RO = Rondônia, RR = Roraima, RS = Rio Grande do Sul, SC = Santa Catarina, SE = Sergipe, SP = São Paulo, TO = Tocantins.
Note: The remaining % that is not included in the table and figure represents the production and area in the state of Tocantins in the Amazon region. No data are available to the authors for the other cotton-producing states of the Amazon region (Amazonas, Rondônia, Pará and Acre).

savannah-like natural vegetation of this area are unique. Agricultural use has greatly expanded over the Cerrado in the last 30 years, mainly due to the favourable topography and to the development of crop varieties adapted to this region. Cotton production has also expanded north into the Amazon region (the state of Rondônia, and in the south of Amazonas, Pará and Tocantins).

The second largest production region is the Meridian (12% of the cotton-producing area, 10.2% of cotton fibre production), consisting of the states of Paraná and São Paulo (Fig. 2.1). The third largest production region is the North-east (11% of the cotton-producing area, 2.5% of cotton fibre production). Although the state of Bahia is geographically placed in the Brazilian North-east, most of it is included in the Midwest cotton-production region, as cotton farms in this state are being developed over natural areas of Cerrado vegetation and the level of technology applied is comparable to that of the Midwest region. Twenty years ago, the North-east region alone produced a significant part of the cotton grown in the world, mainly in small farms. This situation changed dramatically as a result of economic factors combined with the introduction of the boll weevil (*Anthonomus grandis* Boheman), which heavily damaged the susceptible local varieties.

This chapter gives an overview of the cotton agricultural context in Brazil to provide a broad perspective of the different geographical, social, ecological and economic circumstances under which cotton is cultivated. This information provides the background for the following chapters of this book.

Cotton Diversity in Brazil

Genetic diversity and geographical distribution

Three species of the genus *Gossypium* occur in the present cotton-growing regions of Brazil: *Gossypium hirsutum* L., *Gossypium barbadense* L. and *Gossypium mustelinum* Mier. ex Seem, all of them sexually compatible amongst themselves. *G. mustelinum* is a wild tetraploid species endemic to Brazil. A few natural populations remain at sites in Caicó (state of Rio Grande do Norte), and Macurerê and Jaguarari (state of Bahia) in the North-east (Plate 2 – see colour frontispiece). The domesticated cotton species (*G. hirsutum* and *G. barbadense*) are found as cultivated varieties, but also as volunteers and feral populations. *G. barbadense* varieties are also grown as 'dooryard' cottons (single plants found near habitations). This imposing array of morphological and ecological variation has resulted from the changing requirements and achievements of domestication and breeding (Fryxell, 1979).

G. barbadense is a perennial shrub endemic to South America. It has a wide geographical distribution in Brazil, from the Amazon region, the low lands of Maranhão and Piauí states, in mining cities of Mato Grosso State, in areas around the Pantanal and in the Atlantic rain forest from Rio Grande do Norte to Espírito Santo and Minas Gerais (Plate 2 – see colour frontispiece). Two varieties can be clearly differentiated by their seed form: *G. barbadense* var. *barbadense*, known as 'Quebradinho' or 'Maranhão' and *G. barbadense* var. *brasiliense*, known as 'Rim-de-Boi' (cattle kidney).

G. barbadense var. *barbadense* has unattached seeds. *G. barbadense* var. *brasiliense* is called kidney cotton, because the seeds of each locule are fused into a solid mass that is roughly kidney shaped. Its centre of diversity is in the Amazon region.

Two easily distinguished biotypes of *G. hirsutum* are cultivated in Brazil. The biotype *latifolium* (*G. hirsutum* var. *latifolium*), also known as herbaceous or upland cotton, is widely cultivated where rainfall is above 600 mm/year. The second biotype is *G. hirsutum* var. *marie-galante*, known as mocó or arbóreo cotton. During the 1970s, mocó cotton was grown on more than 2.5 million ha in the North-east, for its high fibre quality and drought resistance. Technical and economic constraints and the accidental introduction of the boll weevil in the region resulted in a dramatic reduction, and today only around 10,500 ha in the dry Seridó regions of Paraíba and Rio Grande do Norte are still cropped with this variety (Fig. 2.2). Immigrants from the North-east still plant mocó cotton in some counties of the Amazon region, mainly for medicine and domestic uses (Freire, 2000). The coloured lint cottons increasingly grown in Brazil are thought to be hybrids between mocó and upland cotton.

The genetic diversity, distribution and sexual compatibility of the Brazilian cotton species and varieties are described in more detail in Johnston *et al.* (Chapter 11, this volume).

Cottonseed use and cultivars

Research on genetic improvement of cotton (*G. hirsutum*) in Brazil began in 1924, with the creation of the Cotton Unit in the Instituto Agronômico de Campinas (IAC), in the state of São Paulo. The initial cotton-breeding programmes aimed to improve yield and fibre length (Carvalho, 1999). As the area of cotton production in São Paulo state increased, pest pressure on productivity increased in importance and pest resistance became a priority trait to be added to the high-yield varieties. During the same period in the North-east, the main objective of cotton breeding by the State Agriculture Secretaries was the adaptation of cultivars to the local environment, the improvement of local varieties to increase yield and fibre quality, followed by the introduction of traits conferring tolerance or resistance to fungal diseases such as fusarium wilt (*Fusarium oxysporum* f. sp. *vasinfectum* E.F. Sm & Swingle), and ramulose (*Colletotrichum gossypii* var. *cephalosporioides* A.S. Costa) and pests, mainly the root borer (*Eutinobothrus brasiliensis* (Hambleton)) (Carvalho, 1999). In 1976, Embrapa's National Centre for Cotton Research was created, and a new cotton-breeding phase began (Carvalho, 1999), with emphasis on the improvement of fibre production, higher lint weight, higher yield, and tolerance to diseases and to drought (Carvalho, 1999).

Current government cotton-breeding programmes in Brazil prioritize traits according to the problems and necessities of each cropping region, and provide a variety of materials to be cultivated in the wide range of environmental and socio-economic conditions under which cotton is grown in Brazil (Freire and Costa, 1999; Fuzatto, 1999).

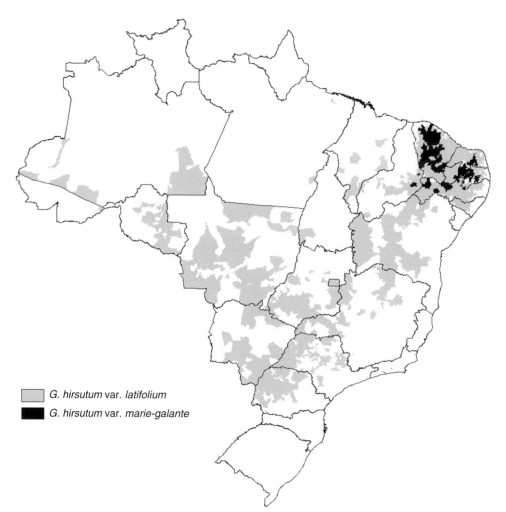

Fig. 2.2. Municipalities where there are records of cropping areas with *Gossypium hirsutum* var. *latifolium* (upland cotton) and *G. hirsutum* var. *marie-galante* (mocó cotton) in Brazil, from 2000 to 2002. (Source: Brazilian Institute for Geography and Statistics (IBGE), modified by the authors.)
Note: *G. hirsutum* var. *marie-galante* also occurs in the Midwest and the Amazon Basin as dooryard cotton, and in the North-east as feral populations (Johnston *et al.*, Chapter 11, this volume).

In the Midwest, breeding programmes select for cultivars that are adapted to the environmental conditions prevailing in the Cerrado, and that are resistant to diseases, including ramulose, bacterial blight (*Xanthomonas axonopodis* pv. *malvacearum* Starr and Garces), grey mildew (*Ramularia areola* G.F. Atk.) and fusarium wilt. The main problem that is being targeted is blue disease, caused by a luteovirus that is transmitted by aphids (Corrêa *et al.*, 2004). Several resistant varieties have been developed, both in the public and private domain, such as BRS Cedro and CD406. Resistance to the Lepidoptera pest complex (cotton leafworm, tobacco budworm and fall armyworm) is also being sought for the Midwest, but presents a difficult challenge. Varieties from this region can yield more than 3000 kg/ha, with lint percentage higher than 38%. They are adapted to mechanical harvesting and have little pilosity to avoid fibre contamination.

In the Meridian region, the main priority is the selection of varieties that present multiple resistances to fusarium wilt caused by the fusarium–nematode complex, and other fungal diseases. In the past, most cotton in this region was hand picked, so varieties with larger bolls with loose lint were preferred; recently, however, there has been an increase in large industrialized farms using mechanical harvesting.

In the North-east, early maturation (130–150 days) became a trait of primary importance to avoid the build-up of boll weevil populations during boll maturation. The current breeding programmes aim to introduce resistance to ramulose and to the boll weevil, aphids, whiteflies and three species of Lepidoptera (cotton leafworm, pink bollworm and tobacco budworm), into short-season varieties adapted to the local environmental conditions (Freire and Costa, 1999). In the Amazon region (including the state of Rondônia, and the south of the states Amazonas, Pará and Tocantins), the priority trait is adaptation to the environmental conditions, as well as resistance to diseases.

Before the mid 1990s, the Brazilian seed market was essentially supplied by the public sector. Freire and Costa (1999) described the two state-run models of seed production in Brazil. In the first, adopted by the majority of states, the cotton cultivars that come out of the public breeding programmes are maintained by various public institutions such as the Embrapa Cotton Research Centre, the Agronomic Institute of Paraná (IAPAR), the Agronomic Institute of Campinas (IAC) and the Agricultural Research Enterprise of Minas Gerais (EPAMIG). They multiply and transfer the seed to the Embrapa Basic Seed Production Service, which multiplies, certifies and commercializes the seeds, and also contracts farmers to multiply seed. The other model is operated in the state of São Paulo, where the whole process of cultivar development, production, maintenance, multiplication and selling of certified seed is controlled by the state's Secretary of Agriculture. The General Coordination of Cultivar Protection (Coordenação Geral de Proteção de Cultivares) of the Federal Ministry of Agriculture guarantees seed purity standards. The states (departments) may establish their own, more rigorous, seed-purity standards than those specified by the Federal Ministry.

With the Cultivar Protection Law that began to be implemented in 1997, the Brazilian Seed Service scene has changed, with a more dynamic and active participation in the seed market of groups and companies, public or private, national or international. Today, cotton breeding is carried out by private companies (Mato Grosso Foundation (FMT), Delta & Pine of Brazil, Aventis and Bayer), public institutions such as Embrapa, IAC and EPAMIG, cotton cooperatives such as the Central Agricultural Cooperative of Economic and Social Development (COODETEC), and by partnerships formed between cooperatives and publicly funded national and international institutions, for instance, the partnership between COODETEC and CIRAD (ABRASEM, 2004). In 2003, 23 (72%) of the registered (protected) varieties available in the market were developed by public institutions and 9 (28%) were developed by private companies (ABRASEM, 2004). However, the participation of public institutions in the cottonseed market in Brazil has decreased during the last few years (Wetzel, 2002). During the 2001/02 harvests, private companies had about 56% of the cottonseed production and market, with a private cultivar leading the market (Embrapa, 2003). A number of the varieties sold by the international seed companies have high and consistent productivity, but some are very susceptible to viral diseases (Santos *et al.*, 2004). The trend towards a larger seed-market share by private companies is being strengthened by the introduction of transgenic varieties, as public institutions are still in the early stages of developing transgenic cotton varieties. It is not clear if the private companies will enter into partnership with the public institutions, allowing the transfer of transgenic constructs owned by the companies to the locally adapted varieties developed by public institutions.

The percentage use of purchased seed varies among regions and states (Table 2.1), ranging from 20% to 90% of the cropped area in six states analysed (ABRASEM, 2004). A rough estimate based on data of these states suggests that saved seeds are planted on one-third of the cotton area in Brazil. There is an increasing trend to use of saved seeds instead of annual purchase (Wetzel, 2002). While farmers may be trying to minimize production costs by saving seeds, this attitude may result in planting lower-quality seeds with possible negative repercussion on crop quality and farmers' income.

Cotton Development, Morphology and Physiology

Cotton is a perennial with an indeterminate growth habit, so that vegetative and reproductive growth occurs at the same time; but four main growth stages can be distinguished: (i) germination and seedling establishment; (ii) leaf area and canopy development; (iii) flowering and boll development; and (iv) maturation. In Brazil, the cotton-growing season varies from 100 days to over 190 days according to environmental conditions (mainly climate and altitude), and according to the plant variety and cultivar (short-, intermediate- and long-season cultivars) (Beltrão, 2002). Flowering begins around 50 days after seedling emergence and continues until day 120 or longer (Fuzzato, 1999). In the following text, we describe the main morphological and physiological characteristics of the cotton plant growth stages, and the impacts, of pest damage (Cothren, 1999; Oosterhuis and Jernstedt, 1999).

Germination and seedling establishment

Under favourable conditions, the cotton radicle emerges within 2–3 days after the beginning of germination and forms the primary root or taproot of the cotton plant. The cotyledons emerge above the soil in 4–6 days. In the first month of growth, the cotton plant develops a substantial root system while growth of the stem and leaves above ground is relatively slow.

Table 2.1. Cottonseed production and rate of use of purchased seed in six Brazilian states. Based on the 2003 annual report produced by the Brazilian Association of Seed Producers (ABRASEM, 2004).

State	Seed production 2001/02 (tonnes)	Seed production 2000/01 (tonnes)	Area planted with cotton 2002/03 (10³ ha)	Seed demand[a] Potential (tonnes)	Seed demand[a] Effective (tonnes)	Rate of use of purchased seed (%)	Rate of use of saved seed (%)
Goiás	1423	300	95.4	1329	997	75	25
Mato Grosso	3911	4347	303.0	4242	2846	65	35
Mato Grosso do Sul	150	287	42.9	513	102	20	80
Minas Gerais	2962	2610	35.3	423	338	80	20
Paraná	946	—	30.6	920	828	90	10
São Paulo	137	180	59.9	599	240	40	60
Total	9529	7724	567.1	8026	5351	67	33

[a]Potential seed demand is calculated from the total cropping area. Effective seed demand is the actual use of certified seeds purchased for that growing season.

Leaf area and canopy development and response to pest damage

The main stem leaves produce most of the energy for vegetative growth, and are a primary oviposition site for certain insects. Below node 5, *G. hirsutum* typically produces one or two upright vegetative branches, which grow like the main stem and only bear leaves, though they also produce secondary fruiting branches. These fruiting branches may be important in compensation, if fruit is lost from the dominant main stem. Cotton leaves have glands in all areas on both sides of the leaf and petiole, and on the leaf underside a nectary on the midrib surface (some genotypes also have nectaries on the surface of the two major lateral veins). Different cotton cultivars show considerable variation in leaf shape, degree of waxiness of the leaf surface and number of trichomes and nectaries, and cotton-leaf morphology affects the behaviour of cotton insects and the predators and parasitoids that attack them. These structures have been widely investigated by breeders as source of pest-resistant traits (e.g. Jenkins and Wilson, 1996). For example, smooth leaves with few trichomes have less oviposition by *Heliothis* spp. that prefer pubescent cottons (Gillham, 1963; Fehr, 1987). Increased density of trichomes decreased the effectiveness of *Chrysopa rufilabris* larvae as predators of *Helicoverpa zea* (Boddie) eggs by inhibiting movement and increasing searching time (Treacy *et al.*, 1987). Another study found higher populations of predators on early maturing glabrous cotton than on pilose cultivars (Shepard *et al.*, 1972). Cultivars resistant to aphids were found to have much harder stem-tips, so that the aphids required a much greater force to insert the proboscis (Kadapa *et al.*, 1988). Early canopy closure is of advantage for weed control and control over water loss and soil erosion; however, the open canopy of the okra leaf-shape cotton provides some resistance to the boll weevil by increasing air movement and the amount of light that penetrates the canopy, which desiccates buds, and weevil eggs and larvae (Andries *et al.*, 1969; Reddy, 1974). A more open canopy also decreased whitefly attack (Sippell *et al.*, 1987), and resulted in slower build-up of mite populations (Wilson, 1994).

The glands on cotton roots, leaves and buds produce chemical compounds that play a role in defence against insects, including terpenoid compounds such as tannins, terpenoid aldehydes (leaves) and gossypol (flower buds and seed), phenolic compounds such as anthocyanins and isoflavonoids and alkaloids. Tannins act as feeding deterrents by binding to various chemical groups in proteins, inhibiting enzyme activity and digestion. The terpenoid aldehydes hemigossypolone and the heliocides H1, H2, H3 and H4 in cotton leaves are associated with resistance to *Heliothis* spp. (Stipanovic *et al.*, 1988). Gossypol also acts as a feeding deterrent for *Heliothis* spp. (Lukefahr and Houghtaling, 1969) and is reported to be highly toxic to aphids (Bottger *et al.*, 1964); however, *Alabama argillacea* (Hubner), a cotton specialist, was not affected by high gossypol levels in cotton cotyledons (Montandon *et al.*, 1986). The isoflavonoid quercetin suppresses the development of *Pectinophora gossypiella* (Saunders), *Heliothis virescens* and *H. zea* (Hedin *et al.*, 1975). Mechanical abrasion of cotton cotyledons and feeding by the two-spotted spider mite (*Tetranychus urticae* (Koch)) induce a response in the plant that leads to reductions in the later populations of mites feeding on cotton (Schuster and Maxwell, 1976; Karban and Carey, 1983, 1986, 1988). Cotton plants also release elevated levels of volatiles in response to insect attack from both the damaged and undamaged tissue, which attracts natural enemies of the pests to the plant (Paré and Tumlinson, 1997, 1998; Röse *et al.*, 1998). Genotypes of cotton may differ significantly in their production of specific volatiles (Elzen *et al.*, 1985, 1986).

Cotton-plant breeders have used many of these morphological and physiological traits (elevated defensive chemistry, leaf shape, bract shape and nectariless) to develop varieties resistant to specific pests (e.g. Gannaway, 1994; Lege *et al.*, 1995; Jenkins and Wilson, 1996), but few of these traits are widely adopted in cultivars (see section on cottonseed use and cultivars above). This is because many host plant insect-resistance traits have both advantages and drawbacks, for instance in relation to productivity (yield potential or fibre quality), or greater susceptibility to other diseases or pests. However, young cotton plants can withstand considerable damage from insect pests, often without loss of yield or delay in maturity. This includes loss of leaf area and loss of the apical meristem (Wilson *et al.*, 2003). There is also evidence of high variability in recovery capacity to vegetative-bud loss between different cotton cultivars, from complete recovery and no yield loss to high susceptibility (Sadras and Fitt, 1997).

Flowering and boll development and pest damage

In ancestral cotton species, flowering is initiated by short days, but in domesticated cotton cultivars, flowering is day-neutral. Floral buds (known as squares) start to form at the six- to seven-leaf stage. The square is composed of the developing flower and three large green bracts that enclose it completely. The bases of most or all of the bracts of *G. hirsutum* usually have an extrafloral nectary on the outer side of the bract and a floral nectary on the inside. The flower consists of a calyx of five fused sepals, five conspicuous petals fused at the base, the superior ovary and a staminal column of numerous filaments, each bearing a two-lobed anther, surrounding the style bearing the two- to five-lobed stigma which protrudes over the top of the staminal column. The first flowers to open are low on the plant, and then they open spirally upward and outward. *G. hirsutum* flower petals are creamy white on the day of flowering, and then become pink the following day and abscise at the base 1 or 2 days later. *G. barbadense* flowers are yellow at anthesis, but also turn pink.

Variations in flower morphology in different cotton cultivars influence the behaviour of flower-visiting cotton insects, and the predators and parasitoids that attack them. Flower bract size, shape and dentation are particularly variable in the *Gossypium* genus (Percy and Kohel, 1999). First instar *H. virescens* (Fabricius) (tobacco budworm) larvae begin feeding along the margin area of the calyx crown of the square and avoid consuming gossypol glands until they moult to the second instar at between 48 h and 72 h of age (Parrot, 1990). Boll weevil oviposition on the square is much lower on cotton cultivars with the 'frego-bract' trait (rolled-up floral bud bracts) than on cotton cultivars with normal bracts (Jenkins *et al.*, 1969). However, the frego-bract squares are much more susceptible to attack by plant bugs such as *Horciasoides nobilellus* Bergston (Jenkins and Parott, 1971). The effect of the trait on boll weevil parasitoids has been found to differ between species; for example *Catolaccus grandis* (Burks) oviposited less frequently in weevil larvae on frego-bracts fruits, but *Bracon mellitor* Say was able to parasitize 50% of the immature weevils on the frego-bract fruits compared to only 7% on normal bract fruit (Soares and Lara, 1993). Some cotton cultivars have been bred to be nectariless, which reduces bollworm and tobacco budworm egg laying (Lukefahr *et al.*, 1969), and damage to bolls by pink bollworm (Wilson and Wilson, 1976), plant bugs (McCarty *et al.*, 1983) and boll rots (Meredith *et al.*, 1973). However, extrafloral nectaries are essential for many predatory insects and parasitoids, to allow them to survive periods of host or prey scarcity, and as a supplement to a nutritionally balanced diet (Schuster and Calderon, 1986; Faria *et al.*, Chapter 8, this volume; Pallini *et al.*, Chapter 9, this volume). Such negative impacts on natural enemies may be masked in cultivar trials where pests are managed with broad-spectrum insecticides.

Cotton flowers open at or near dawn and remain open for only a single day, closing near sunset. The anthers open to release pollen soon after the petals open, though high humidity or cool temperatures can delay this by some hours. The cotton pollen is more or less spherical, very spiny and relatively large (81–143 microns) (Kaziev, 1964 quoted by McGregor, 1976). This makes wind transportation very unlikely. In the absence of insect pollinators, cotton flowers are self-pollinated, but when pollinators are present, cross-pollination can be significant (Arpaia *et al.*, Chapter 7, this volume). The stigma becomes receptive shortly after the flower opens. The pollen is viable for approximately 12–24 h (Cobley, 1956), and early on the second day only a small proportion of the pollen is still fertile. The pollen tube may fertilize a cotton ovule 12 h after pollination, but it may take up to 30 h, depending on the temperature.

The developing cotton fruit, or boll, consists of the capsule, seeds and cotton fibre. In modern cotton cultivars, the capsule dries and splits into three to five parts (the locules or ovaries) to reveal usually about 8 seeds per locule, attached to the central column of the boll. The main fibres start to grow on the surface of the ovules around the time of flowering, and the fuzz fibres, or linters, 5–10 days later. When the capsule splits and the boll opens, the fibres dry, flatten and twist.

Maturation

Cotton plants have indeterminate flowering and will continue producing flowers until changes in the weather cause the mature leaves to be shed. Most large-scale cotton producers will apply chemical defoliants when about 50–60% of the bolls are open. This removes the leaves so that they cannot cause green stains on the lint during machine harvesting and also hastens maturity of the remainder of the crop.

In general, the cotton plant has a high capacity to replace loss of reproductive structures (Brook *et al.*, 1992). A cotton plant usually sheds half to two-thirds of its squares and young bolls, even under unstressed crop-growing conditions, as it produces more fruit than it can retain. Some shedding can stimulate vegetative and root growth and increase flower production (Dale, 1959); however a large amount of shedding from insect damage in high-yield cotton systems can change the plant morphology due to increased secondary branching and vegetative growth, delaying ripening and possibly reducing boll weight and so yield (Brook *et al.*, 1992). The ripening of a larger proportion of the fruit on the secondary branches may affect mechanical harvesting. Damage to the bolls more than 14 days after fruit set but before the bolls open can cause yield loss because the bolls are not shed and continue to absorb and assimilate nutrients, yet they are too damaged to harvest (Matthews, 1994a).

Insect damage to buds, flowers or bolls also allows pathogens to enter, causing boll rot. Cotton stainers (*Dysdercus* spp.) and some other sucking Hemiptera introduce fungi that kill the seed and some of the developing lint hair, 'staining' the boll. The honeydew produced by whitefly (*Bemisia tabaci*), cotton aphids (*Aphis gossypii*) and other sucking insects such as scale insects, causes 'sticky cotton', which creates significant problems for cotton-spinning machinery, and provides nutrients for fungal infections of the boll (see section on diseases). Insect damage can also affect boll opening, slowing its drying and attracting secondary fungal infection. Damaged bolls will also attract scavenger insect species.

Species Diversity Associated with Cotton in Brazil

Cotton in Brazil is grown in regions with a wide variety of natural flora and fauna. The Midwest production area is on soils originally covered with vegetation of the Cerrado biome (Plate 1 – see colour frontispiece), a Brazilian savanna that contains a high biodiversity of trees and shrubs, almost half of which may be endemic to the biome (Rizzini, 1963; Heringer *et al.*, 1977), and a very rich herbaceous flora as well as fauna (Dias, 1992; Ratter *et al.*, 1997). Since the 1970s, the Cerrado has been extensively cleared for large-scale crop cultivation and improved pasture. It is estimated that in 1997 a bit less than half the original vegetation area has been cleared (Ratter *et al.*, 1997). As cotton fields in the Midwest are still replacing the Cerrado, it is common to find patches of original Cerrado vegetation in the middle or around large cotton plots. Cotton production has also expanded into the Amazon biome region.

It has been reported that the cotton agro-ecosystem community worldwide has more than 1000 insect species (Luttrell *et al.*, 1994; Matthews, 1994b). A study conducted in Paraguay, adjacent to Brazil, reports more than 300 species of acari and herbivorous insects (Michel and Prudent, 1985; Michel, 1989). A survey of the fauna associated with cotton in Brazil was made many years ago by Silva *et al.* (1968), who listed 259 insect species associated with cotton in the country. More recent reports are mainly focused on economically important pest species, and do not report on groups such as flower-visiting and pollinating species, and species with no clear ecological function, which are relevant for the risk assessment of transgenic crops. Presently, additional studies are under way. A 1-year survey of a cotton field surrounded by Cerrado vegetation resulted in more than 500 morpho-species[1] of arthropods living on the plant or on the soil, including more than 80 morpho-species of spiders and a significant number of flower-visiting insects that have never been reported as associated with cotton in Brazil (Pires *et al.*, 2004). Other fauna from the same area includes annelid worms, amphibians, reptiles, birds and mammals (Schmidt *et al.*, unpublished results). It is likely that additional studies will reveal even more species.

Pests and diseases

Arthropod pests

The main pest species of cotton in Brazil are listed in Sujii *et al.* (Chapter 6, this volume). The pest status of a species in each region may be influenced by the crop variety, the cultural practices used, particularly the pest-control methods, and the environmental conditions. Two species are reported as key primary pests in all three regions where cotton is produced – the cotton aphid (*A. gossypii* Glover) and the boll weevil (*A. grandis*). The Lepidoptera species complex on cotton may also cause serious damage in some years and regions. Some species are predominant in the Midwest (*Spodoptera* spp. and *A. argillacea*); other species (*H. virescens, P. gossypiella*) are potential pests in the Meridian region (Silvie *et al.*, 2001a). The cotton leafworm *A. argillacea* is also a problem in the North-east (Almeida and Silva, 1999). For more information on the Lepidopteran pests see Fitt *et al.* (Chapter 12, this volume).

Aphis gossypii *Glover [Hemiptera: Aphididae] (pulgão – cotton aphid)*

This aphid feeds on the phloem of young shoots and buds and on the undersurfaces of leaves by means of sucking mouthparts. It may appear in varied colours, from light yellow to dark green (Santos, 2001) and black (M. Frizzas, Brasília, 2004, personal communication). Neither a sexual form nor eggs of this species has been described in Brazil, and reproduction is carried on exclusively by viviparous parthenogenetic females. This species has a very fast life cycle and under normal cotton-growing conditions can complete a generation in 5–10 days, allowing populations to build very quickly if natural enemies are eliminated. They also have a wide host range, allowing them to persist when cotton is out of season. *A. gossypii* feeds only upon the phloem of young growing tissues throughout the growing period of cotton, reducing vegetative growth during the early stages of the crop, and later directly feeding on squares, flowers

---

[1] Species classification made by para-taxonomist based on overall appearance. The confirmation of the species status by a specialist taxonomist is under way.

and developing bolls. Its feeding causes the edges of young leaves to curl down-wards and the sugars it excretes as honeydew cover the foliage, interfering with leaf respiration. Honeydew also provides a substrate for a sooty mould (*Capnodium* spp.), which blackens the leaves, retarding photosynthesis, and which can discolour the cotton lint in open bolls, and causes 'sticky cotton' that creates problems for cotton spinners. Most significantly, aphids are vectors of virus and mycoplasma diseases. These diseases are a major threat to production when they are transmitted early in vegetative growth. Calcagnolo and Sauer (1954) reported that the yield losses caused by cotton aphid range from 24% to 44%. Another analysis on the damage caused by this pest, which accommo-dates recent changes on cotton-cultivation systems, is needed. The chemical control threshold for aphids on blue disease-susceptible cultivars in Brazil is between 5% and 10% of plants with aphid attack (Degrande, 1998a,b; Santos, 2001; Silvie *et al.*, 2001a) due to the need to prevent aphids from spreading the disease. Such a low threshold increases the risk of selecting for resistance to insecticides. Cotton aphids persist on many other host plants, particularly in the Malvaceae, Fabaceae and Cucurbitaceae.

Anthomonus grandis Boheman [Coleoptera: Curculionidae] (bicudo – cotton boll weevil)

The adults puncture the squares and bolls by chewing into them with their long slender rostrum to feed on the tissues inside and to lay their eggs in the holes. Each female may lay from 100 to 300 eggs. The eggs hatch after 3–4 days, and the larva feeds inside the square or boll, consuming the developing flower or seeds and fibres. The damaged squares and bolls either drop off the plant or hang withered and dry. Most larval development occurs while the square or boll is on the ground. The larva undergoes two or three moults over about 8–9 days. Then it transforms into the pupal stage inside the hollowed-out cavity created by its feeding. The pupal stage lasts about 5 days, after which the adult emerges and eats its way out of the square or boll. The aver-age time for a generation, from egg to egg, is about 25 days (Ramalho and Santos, 1994). Because of the indeterminant flowering of cotton, the boll weevil can complete between four and six generations on a single crop (Santos, 2001). The yield losses caused by boll weevil in the North-east range from 58% to 84% (Ramalho, 1994). In the Midwest, when control methods are not properly applied, yield losses may reach 70% (Degrande, 1998a). The weevils feed and reproduce almost exclusively on the cotton plant, and between cotton crops the adults migrate from the field and diapause in sur-face litter in field edges and around gins, and feed on the pollen of a wide vari-ety of plants (Jones and Coppedge, 1996). At the start of the next cotton season, they migrate back into the crop. A review of host plants of the boll weevil in South America was written by Ramalho and Wanderley (1996).

It is difficult to quantify the significant social and economic losses caused by the boll weevil in North-eastern and south-eastern Brazil due to the role of other concurrent factors such as depreciated land values, closure of cotton gins and oil mills, unemployment, emigration of rural workers to urban areas and other indi-rect results of the weevil introduction (Ramalho and Santos, 1994).

Alabama argillacea *(Hübner) [Lepidoptera: Noctuidae] (curuquerê do algo-doeiro – cotton leafworm)*
The cotton leafworm has slender, greenish, looping larvae with black-and-white stripes and a number of small black spots scattered over the body. This pest attacks very early, when the crop has only a few leaves. The larvae strip the leaves of cotton plants, and late in the season, when leaves are scarce, larvae may eat the squares or bolls. The larva feeds only on cotton or wild cotton. The adult female deposits an average of 500 eggs on the underside of the larger cotton leaves (Santos, 2001), laying eggs singly. The larva passes through six instars over 2–3 weeks and pupates in a flimsy, silken cocoon inside a folded leaf or fastened by the tip of the abdomen to the foliage. A generation is completed in about 4 weeks. In the Meridian and Midwest regions, this species may occur from November to May (Santos, 2001), allowing it to complete up to seven generations in a single cotton crop. In the North-east, Ramalho *et al.* (1990) observed three population peaks in 1 year. At the onset of the rainy season, many adults migrate north in the Americas towards warmer temperatures (Medeiros *et al.*, 2003). Marchini (1976) estimated the yield losses caused by cotton leafworm at 21–35%. The percentage of damage varies according to the crop variety and production region. Santos (2001) recommends the use of shake cloths or direct counts of larvae as sampling methods for population monitoring. Insecticides should be applied when more than five small (1 cm or less) or two medium-sized (1–2 cm) caterpillars are found on one-third of the plants.

Spodoptera *spp. [Lepidoptera: Noctuidae] (lagarta militar – armyworms)*
A complex of *Spodoptera* species may feed on cotton in Brazil, including *Spodoptera frugiperda* (J.E. Smith), *Spodoptera ornithogalli* (Guenée), *Spodoptera exigua* (Hübner), *Spodoptera cosmioides* (Walker) and *Spodoptera sunia* (Guenée) [*Spodoptera albula* (Walker)]. All these species damage the leaves, but *S. frugiperda* prefers the bolls (Silvie *et al.*, 2001a). The larvae of these species frequently crawl in large aggregations, and may cause considerable damage to field and vegetable crops by consuming the foliage and stems. The larvae feed on cotton leaves and also on squares, flower and bolls. In Brazil, the most significant species is *S. frugiperda*, the fall armyworm. About 1000 eggs are laid, usually on green plants, by each female fall armyworm, in masses ranging from 50 to 100 eggs, which are covered with hairs and scales from the female's body (Santos, 2001). The number of fall armyworm generations on a cotton crop in Brazil has not yet been reported. This species has only recently been observed attacking cotton, particularly in the Cerrado area. It is becoming an increasingly important pest. Plants of the grass family are probably the preferred food (Pashley *et al.*, 1985; Pashley, 1986), and the insect is often called the 'grassworm'; but it also attacks all grain crops such as maize and sorghum, and lucerne, beans, groundnuts, potato, sweet potato, turnip, spinach, tomato, cabbage, cucumber, tobacco, clover and cowpeas. There is no reliable information on the yield loss caused by these armyworms on cotton in Brazil (see Fitt *et al.*, Chapter 12, this volume, for further discussion).

Heliothis virescens *(Fabricius) [Lepidoptera: Noctuidae] (lagarta das maçãs – tobacco or cotton budworm)*
The budworm is a widely distributed and destructive cotton pest. In addition to cotton, it feeds on many cultivated and wild plants in the Solanaceae, including tobacco and ground cherry, and also geranium and ageratum. The eggs of this pest, about 600 per female (Santos, 2001), are laid singly on the leaves and outside of the squares. Larvae feed on developing bolls, but do not remain in a single boll, boring into bolls, feeding, then moving to damage other bolls. Crawling among bolls and squares, a single larva can destroy, on average, six squares and two bolls per plant (Santos, 1977). The pupal stage is passed in the soil at a depth of 2.5–10 cm. There are two or three generations of this pest on a cotton crop. Estimated yield losses caused by *H. virescens* range from 18% to 32% in the Meridian region (Santos, 1977) to 20% in the North-east (Degrande, 1998a).

Pectinophora gossypiella *(Saunders) [Lepidoptera: Gelichiidae] (lagarta rosada – pink bollworm)*
The female moth deposits about 200–400 white, oval eggs, usually singly or in groups of five to ten all over the cotton plant; but they tend to be concentrated on the terminals early in the season, and on the developing bolls later in the season. The eggs hatch in 3–5 days, and the larvae bore into squares and eat developing flowers, or into bolls, where they consume both lint and seeds. The larva pupates in the upper 5 cm of soil, in surface trash, or in the seed or boll. It emerges as an adult after about 9 days, and under optimum conditions, the entire life cycle requires about 31 days from egg to egg, allowing it to complete up to five generations on a single cotton crop. It affects cotton yield in several ways (Santos, 2001). In severe infestations, damaged squares and small bolls may be shed, leaving no visible evidence of injury on the plant itself. The preferred food of the larva is the kernel of the seed. Usually the tiny larva, upon entering a boll, travels a short distance just under the inner surface of the boll covering, making a characteristic path that is commonly referred to as a mine. It soon leaves the lining of the boll and cuts through the immature lint to a seed. It devours the inside of the seed; then proceeds to the next seed, ruining the lint as it passes through. Many larvae are heavy feeders and eat all the seed of a locule or cell of the boll before they reach maturity. They lower the grade and staple of the lint because of the stained and cut fibres. There is also the added loss of oil content of the seed from heavily infested bolls. In extremely heavy infestations, the greatest loss is the reduced yield of pickable cotton. Davidson and Seara (1966) reported that yield losses caused by pink bollworm in Brazil range from 5% to 45%. There are no recent evaluations of the damage caused by this insect in the different cotton-cultivation regions. The plants attacked by pink bollworm are cotton and rarely okra, hollyhock, hibiscus and other malvaceous plants.

Helicoverpa zea *(Boddie) [Lepidoptera: Noctuidae] (lagarta da espiga – maize earworm)*
Although this species can attack cotton in Brazil (Santos, 1999; Silvie *et al.*, 2001a), it is insignificant and not reported as a primary pest in the main

cotton-growing regions (Ramalho, 1994; Degrande, 1998a; Almeida and Silva, 1999; Santos, 2001; Gallo *et al.*, 2002; P. Silvie, Brasília, 2004, personal communication).

Diseases

Many diseases affect cotton cultivation in Brazil (Cia and Fuzzato, 1999; Lima *et al.*, 1999). Some of the pathogens are cosmopolitan, but others are found only in certain regions or specific circumstances that favour their development.

The main viral disease is 'blue disease' transmitted by *A. gossypii*. It causes stunting due to shortening of the internodes, leaf rolling, an intensive green colour of the foliage and vein yellowing. These symptoms are caused by the cotton leaf roll dwarf virus (CLRDV), a luteovirus (Luteoviridae) of the genus *Polerovirus* (Corrêa *et al.*, 2004a,b, 2005). The relationship between the aphid and the virus is persistent, i.e. the vector remains infectious for a long period of time and can infect numerous plants (Costa and Carvalho, 1965; Costa, 1966). The 'blue disease' is one of the main disease problems in Brazil, but does not occur in the North-east region. *B. tabaci* Gennadius (whitefly) transmits the Abutilon mosaic virus, a geminivirus, but this is not transmitted in a persistent manner and is not therefore considered a serious disease of cotton in Brazil.

Three important fungal diseases attack cotton roots (Bell, 1999). Fusarium wilt is a soil- and seed-borne disease caused by *F. oxysporum* f. sp. *vasinfectum* E.F. Sm & Swingle that can enter the root via damage caused by root knot nematode (*Meloidogyne incognita* (Kofoid and White) Chitwood) infestation. The severity of this disease is therefore usually related to the severity of nematode infestation, which is highest in sandy soils. In general, *G. hirsutum* cottons are susceptible to *M. incognita* infestation whereas *G. barbadense* cottons are resistant (Ponte *et al.*, 1998). Verticillium wilt is caused by *Verticillium dahliae* Kleb., and is also soil- and seed-borne. *G. barbadense* is also more resistant to this disease than *G. hirsutum*. The disease is most common in neutral to alkaline silt and clay soils of arid regions, particularly in irrigated cotton. Most other fungal diseases are transmitted to the growing cotton plant from infected plant residues via wind, water or insects. *Rhizoctonia solani* (Kühn) and other seedling pathogens can attack cotton seedlings or leaves. Leaves are also damaged by the alternaria spot diseases caused by *Alternaria macrospora* (Sacc.), grey or areolate mildew caused by *R. areola* G.F. Atk., leaf blight caused by *Stemphylium solani* G.F. Weber (Mehta, 2001) and ramulose or witches broom caused by *C. gossypii* var. *cephalosporioides* A.S. Costa. Several pathogenic and saprophytic fungi and bacteria, such as *Ashbya gossypii* (S.F. Ashby & W. Nowell) Guillierm., *Diplodia gossypina* Cooke and *Fusarium moniliforme* J. Sheld, cause cotton boll rot. Most of the boll rot diseases require insect punctures to enter bolls (Lima *et al.*, 1999). More recently, a new fungal disease caused by *Myrothecium rodium* Tode was found infecting cotton in the Midwest. On some cotton farms, this disease resulted in about 50% crop loss due to leaf shedding and boll rot (Chitara and Meyer, 2005).

Bacterial blight, caused by the bacterium *X. axonopodis* pv. *malvacearum* Starr and Garces, is restricted to cotton. It can attack the cotton seedling and stem causing symptoms called the 'blackarm' phase; the cotton plant foliage with symptoms called the angular leaf-spot phase; and the cotton bolls causing the boll-disease phase (Bell, 1999). It is carried over in infested seed and in crop residues on the soil surface.

Weeds

Weeds can seriously constrain cotton plant growth, particularly in the first month, as they compete for limiting production factors such as water and nutrients. They can also reduce fibre quality and interfere with harvesting. According to Deuber (1999), productivity in the Meridian region can be reduced from 68% to 95% if weeds are not properly managed during the whole crop cycle. The composition of the weed community found in cotton fields is complex, as it includes annual and perennial species, monocots and dicots (Table 2.2) (Azevedo *et al.*, 1999; Deuber, 1999). Control methods include cultural control (soil preparation, crop rotation, intercropping and management of plant density), physical control (hand pulling and hoeing, mostly used by small farmers, cultivators such as animal traction and tractors, complemented with hoeing) and chemical control (pre- and post-emergence herbicides) (Azevedo *et al.*, 1999).

Cotton-Producing Regions and Cropping Systems

Cotton production in Brazil has over 100 years of history and its spatial distribution has been very dynamic. During the 1980s, the main centres of production were located in the Meridian and North-east regions. More than 2.5 million ha of cotton were planted in the North-east region of Brazil alone, mainly in small farms averaging 3 ha. This situation started to change early in the 1980s, when the textile industry in Brazil began to look for and import fibre from other countries, motivated by the cheaper prices caused by a Brazilian government policy that reduced importation taxes. As a result, the imported fibre became cheaper than the mocó fibre produced in the North-east region. At the same time, in 1983, the boll weevil (*A. grandis*) was accidentally introduced into the region, a disaster for the small-holding cotton farmers (Ramalho and Santos, 1994). The cotton production area in the North-east decreased from over 2 million ha of mostly mocó cotton to only 101,068 ha in 2001 (IBGE, 2002).

The main cotton production in the country then moved to the states of Paraná, São Paulo, Goiás and Minas Gerais. By the beginning of the 1990s, the state of Paraná was the main cotton producer in Brazil, with 44% of national production, and, together with the other three cited states, produced 76% of Brazilian cotton (Maronezzi *et al.*, 2001). However, since 1998 the situation has changed again, as large areas in central and western Brazil have been cleared for agriculture with the help of government subsidies and tax breaks. By 2002, the Midwest region (Mato Grosso, Mato Grosso do Sul,

Table 2.2. Principal weed species in cotton in Brazil[a]

Weed group	Taxonomic assemblage	Common name (Brazilian)	Common name (English)	Type of damage[b]	Host plant for pest or disease	Limiting factor in cotton fields
Broadleaf annual	Bidens pilosa L. (Asteraceae)	Picão preto	Hairy beggar-ticks	C, FQ	Aphis gossypii	Aphids
Sedge	Cyperus rotundus (Cyperaceae)	Tiririca	Purple nutsedge	C		
Grass annual	Cenchrus echinatus L. (Poaceae)	Capim carrapicho	Southern sandbur	C, FQ		
Broadleaf annual	Acanthospermum hispidum DC. (Asteraceae)	Carrapicho de carneiro	Bristly starbur	C, FQ		Aphids
Broadleaf annual	Sida rhombifolia L. (Malvaceae)	Guaxumas	Southern sida		Cotton anthocyanosis and Aphis gossypii, boll weevil, Dysdercus spp., Pectinophora gossypiella	Viruses
Broadleaf annual	Sida glaziovii K. Schum. (Malvaceae)	Guaxumas	Southern sida	C	Cotton anthocyanosis and Aphis gossypii, boll weevil, Dysdercus spp., Pectinophora gossypiella	Viruses
Broadleaf annual	Sida acuta Burm. f. (Malvaceae)	Guaxumas	Southern sida		Cotton anthocyanosis and Aphis gossypii, boll weevil, Dysdercus spp., Pectinophora gossypiella	
Broadleaf annual	Acanthospermum australe (Loefl.) Kuntze (Asteraceae)	Carrapicho de carneiro	Bristly starbur	C, FQ		Aphids

Continued

Table 2.2. Principal weed species in cotton in Brazil[a] – *cont'd*.

Weed group	Taxonomic assemblage	Common name (Brazilian)	Common name (English)	Type of damage[b]	Host plant for pest or disease	Limiting factor in cotton fields
Broadleaf annual	*Commelina benghalensis* L. (Commelinaceae)	Comelina	Tropical spiderweed, spreading dayflower	C		Soil fertility (increasing in no-tillage systems)
Broadleaf annual	*Amaranthus* spp. (Amaranthaceae) – mainly *A. hybridus* L.	Caruru de espinho	Smooth amaranthus/ pigweed	C		Soil fertility
Broadleaf annual	*Senna* (*Cassia*) *obtusifolia* (L.) Irwin & Barneby (Fabaceae)	Fedegoso	Sicklepod	C		
Broadleaf annual	*Tridax procumbens* L. (Asteraceae)	Erva de touro				
Broadleaf annual	*Rhynchelitrum roseum* (Nees) Stapf & C.E. Hubb. Ex Bews (Poaceae)	Capim favorito	Ruby grass			
Grass perennial	*Digitaria* spp. (Poaceae) – mainly *D. horizontalis* Willd	Capim colchão	Jamaican crabgrass	C		
Grass annual	*Eleusine indica*	Capim pé de galinha	Goose grass	C		
Broadleaf annual	*Ageratum conyzoides* L. (Asteraceae)	Mentrasto	Goat weed	C, FQ		Aphids
Broadleaf perennial climber	*Ipomaea* spp. (Convolvulaceae)	Corda-de-viola	Morning glory	C		Chrysomelidae

Growth habit	Species (family)	Common name (Portuguese)	Common name (English)	Code	Pests and diseases	Diseases
Broadleaf annual	*Euphorbia heterophylla* L. (Euphorbiaceae)	Leiteiro	Wild poinsettia	C		
Grass annual	*Cynodon dactylon* (L.) Pers. (Poaceae)	Capim da cidade	Bermuda grass	C		
Broadleaf annual	*Sida cordifolia* L. (Malvaceae)	Guaxumas	Southern sida	C	Cotton anthocyanosis and *Aphis gossypii*, boll weevil, *Dysdercus* spp., *Pectinophora gossypiella*	Viruses
Broadleaf annual	*Sida santaremensis* Monteiro (Malvaceae)	Guaxumas	Southern sida	C	Cotton anthocyanosis and *Aphis gossypii*, boll weevil, *Dysdercus* spp., *Pectinophora gossypiella*	Viruses
Broadleaf annual	*Sida spinosa* L. (Malvaceae)	Guaxumas	Southern sida	C	Cotton anthocyanosis and *Aphis gossypii*, boll weevil, *Dysdercus* spp., *Pectinophora gossypiella*	Viruses
Broadleaf annual	*Sida linifolia* Cav. (Malvaceae)	Guaxumas	Southern sida		Cotton anthocyanosis and *Aphis gossypii*, boll weevil, *Dysdercus* spp., *Pectinophora gossypiella*	
Broadleaf annual	*Malva* spp. (Malvaceae)	Guaxumas	Southern sida		Cotton anthocyanosis and *Aphis gossypii*, boll weevil, *Dysdercus* spp., *Pectinophora gossypiella*	Viruses
Broadleaf annual	*Wissadula subpeltata* (Kuntze) R.E.Fr. (Malvaceae)	Malva estrela		C		
Broadleaf annual	*Blainvillea rhomboidea* Cass. (Asteraceae)	Picão grande, Erva palha		C, FQ		Lepidoptera

Continued

Table 2.2. Principal weed species in cotton in Brazil[a] – *cont'd.*

Weed group	Taxonomic assemblage	Common name (Brazilian)	Common name (English)	Type of damage[b]	Host plant for pest or disease	Limiting factor in cotton fields
Broadleaf annual	*Desmodium tortuosum* (Swartz) DC. (Fabaceae)	Carrapicho-beiço-de-boi	Florida beggarweed			Aphids
Grass perennial	*Brachiaria decumbens* Stapf. (Poaceae)	Capim brachiaria	Surinam grass, signal grass, sheep grass			Cicads

[a]This table is a contribution by R. Pitelli, F. Muyekho and R. Carmona, Brasília, 2003.
[b]C = Competition; FQ = Fibre quality.

Minas Gerais, Goiás, Distrito Federal and west Bahia) was responsible for 87% of the total cotton production in Brazil. The state of Mato Grosso has now become the principal cotton producer, representing 53% of all cotton-fibre production in the country (IBGE, 2004).

The North-east region

This area comprises all the cotton-producing states of the semiarid region in the North-eastern part of Brazil (Fig. 2.1). Excluding the western part of the state of Bahia (which is included in the Midwest cotton-growing regions due to environmental and cultivation similarities), in the North-east in 2002/03 upland cotton was cultivated on 67,100 ha (11% of the total cotton area in Brazil), and production was 19,200 t fibre (CONAB, 2003). The average yield varies greatly in the different states of the North-east, from 460 kg/ha to 3450 kg/ha (CONAB, 2003). The majority of farmers are small holders on predominantly family holdings averaging 3 ha in size. Most of these farmers obtain improved seeds from the Federal government (P. Barroso, Campina Grande, 2004, personal communication). During 1995/96 there were 54,050 farmers in the North-eastern region (IBGE, 1996), although this number has declined substantially to the present.

The predominant climate is xerophytic with high temperatures (BSwh, according to the Koeppen classification). Cotton is cultivated at altitudes of 30–400 m above sea level, where annual precipitation varies from 600 mm to 1000 mm (Plate 3 – see colour frontispiece), and is irregularly distributed with a high probability of dry periods of variable intensity resulting in a high risk of drought. This periodically causes variable losses in yield and contributes to the relatively low average productivity. Solar radiation is high at more than 3000 h of sunlight a year, particularly in the dry Sertão region in the Caatinga biome in the state of Rio Grande do Norte and part of Paraíba (Plates 1 and 3 – see colour frontispiece), where 500 cal/cm^2 reach the soil surface on sunny days (Amorim Neto *et al.*, 2003). Cotton is mainly planted on soils of medium to high fertility, with a good balance in chemical properties and neutral pH (Medeiros *et al.*, 1999; Amorim Neto *et al.*, 2003). Mocó cotton (*G. hirsutum* var. *marie-galante*) is adapted to the arid and shallow soils of the semiarid regions of the North-east.

In the 1970s, most farmers in the North-east grew mocó cotton varieties; but most of the area is now cultivated with upland cotton, mostly short-season cultivars (110–130 days from sowing to maturity) that fit in the rainy season. The cultivars used are resistant to viruses and other diseases. Mocó cotton is cropped on about 10,500 ha in the Seridó region (Fig. 2.2). This cotton is grown in a perennial, usually 3-year cycle; as the next rainy season approaches, each plant is pruned to 20 cm above the soil in order to stimulate regrowth (Beltrão *et al.*, 2003). Cotton varieties with naturally coloured fibres have been preferably chosen for cultivation in the last 3 years due to their higher price, which can be at least 50% over the white-fibre price (Carvalho, 2003).

Small-scale farmers in the North-east have little access to information and financial resources that would enable them to acquire and support modern

machinery. Management is generally done manually, except for soil preparation, which is done mechanically using animal traction or tractors. Herbicides
and chemical fertilizers are not used (Carvalho and Chiavegato, 1999). About
30% of the cotton production is intercropped with beans and maize. Where
perennial cotton (mocó) – upland hybrids are cultivated, the plant residues are
fed to cattle after harvest, and their manure is returned to the soil. Most of the
cotton-production area is sown to cotton every year, and few farmers practice
crop rotation (E.C. Freire and N. Beltrão, Campina Grande, 2004, personal
communication). Large-scale farmers do their own ginning on the farm.
Small-scale farmers usually sell their cotton to others for ginning, but there are
a few ginning cooperatives for small-scale farmers (N. Beltrão, Campina
Grande, 2004, personal communication).

Major limiting factors for cotton production in the North-east

The main limiting factor is water, usually scarce during critical periods of crop
growth such as flowering. Modern soil-conservation practices such as no-
tillage are practically non-existent, but soil degradation is probably minimal
under the predominant crop system in the region. Diseases such as ramulose,
areolate mildew and bacterial blight occur at low levels, and fungicides are not
used (Carvalho and Chiavegato, 1999).

In the North-east, the main pests are the cotton aphid (*A. gossypii*), the
cotton leafworm (*A. argillacea*), the boll weevil (*A. grandis*) and the pink bollworm (*P. gossypiella*) (Ramalho, 1994). The North-east is characterized by
having fewer cotton pests and more abundant natural enemy populations than
other regions. In the perennial mocó cotton-cropping system, the heat and
dryness during the dry season are effective limiting factors for many insects,
particularly the cotton stem borer and boll weevil, in the semiarid Caatinga
biome (Plate 1 – see colour frontispiece) (Ramalho, 1994).

The low level of education among the farmers makes technology adoption very difficult compared to other cotton-production regions of the country.
Lack of access to financial capital is another important limiting factor, despite
the fact that the Federal government has released more funds in the 2003/04
crop season than in the past years, through a special programme for small
farmers via development banks such as Banco do Nordeste and Banco do
Brasil. Other actions are needed to facilitate technology adoption and to
transform the cotton farming by small holders into an attractive and profitable
activity by family and small farmers in the region, including community ginning and processing to aggregate value before selling (Sobrinho *et al.*, 2003).
Some initiatives to this end are being taken by local NGOs in partnership with
the Brazilian government and international organizations such as the
International Fund for Agricultural Development (FIDA).

The Meridian region

The Meridian cotton-production region consists of the states of São Paulo and
Paraná, where 89,200 ha of cotton (12% of the total area in Brazil) produced

87,000 t fibre/year (10.2% of the total Brazilian production) in the agricultural year 2002/03 (CONAB, 2003). In this region, cotton is cropped at altitudes of 400–700 m above sea level, with a predominant climate of hot and humid summers and cold and dry winters (Cfa, according to the Koeppen classification). Precipitation ranges from 900 mm to 1000 mm during the cropping season, with an average of 1500 mm/year (Plate 3 – see colour frontispiece). Small- and medium-scale farmers, organized in cooperatives, are predominant in this region. The productivity level is intermediate between those found in the Midwest and in the semiarid areas of the North-east. The size of the majority of cotton farms varies from 10 ha to 15 ha, but some farms may have more than 500 ha cultivated with cotton, mainly in the flatter areas. According to IBGE (1996), during 1995/96, 26,481 farmers grew cotton in the Meridian region. Since 2001 in the state of São Paulo, the system of many small farms is being replaced by a system of a few large farms that adopt higher technology with higher financial returns, and the number of cotton farmers has decreased.

Cotton is sown from October to November, and harvested from March to April. The crop cycle is of intermediate length, from 140 to 150 days. Use of purchased seed is variable. In the state of Paraná, most farmers bought and planted certified seed in the 2002/03 season (Table 2.1), but in São Paulo state 60% of farmers planted saved seed. The most commonly used varieties are those resistant to the most common diseases in the region, especially viruses (blue disease), bacterial blight and the fusarium wilt–nematode complex (Cia and Fuzatto, 1999).

The soils have been exploited for agriculture for more than 70 years. Chemical fertilizers are widely used. Farmers commonly grow cotton in the same area for decades without any crop rotation. No irrigation is used. Weed control is similar to that described below for the Midwest region. In the Meridian region, other crops such as maize and soybeans are preferably cultivated, but the cotton area increased by 29% in the state of São Paulo in 2003/04. Among the annual crops, cotton is one of the most profitable, thanks to the application of new technological advances both in the production system and in ginning and fibre processing (Estado de São Paulo, 2004). A few farmers in Paraná now use organic production methods (P. Silvie, Brasília, 2005, personal communication).

The level of mechanization is intermediate to the other two main regions, although it has been increasing recently. Mechanical traction is used, but harvesting is carried out manually. The quality of the cotton fibre is inferior, as it stains due to the soil properties, and contamination levels are higher than in cotton produced in the Midwest region. Ginning is carried out mainly in cooperatives and the cottonseed is used for cattle feed or oil production.

Major limiting factors for cotton production in the meridian region

An important limiting factor is the high incidence of diseases and pests. The boll weevil (*A. grandis*), cotton leafworm (*A. argillacea*), tobacco budworm (*H. virescens*) and pink bollworm (*P. gossypiella*) are significant pests (Santos, 1999). The fusarium wilt–nematode complex is widespread and farmers are increasingly using resistant varieties. Blue disease transmitted by aphids is a problem. Leaf blight (*S. solani*) has also caused severe yield losses in recent years, especially in the state of Paraná (Mehta, 2001). Pest control is carried out in an integrated pest management (IPM) framework based on pest-population monitoring, preventive cultural control practices and six to twelve insecticide sprays, including systemic insecticides, without the use of fungicide (see chemical control below).

The Midwest region

The Midwest region now has the largest cotton-producing area in Brazil. In 2002/03, the region produced 740,200 t fibre (87% of the total production) on 565,900 ha (76.5% of the total cotton area) (CONAB, 2003). Cotton is grown in the states of Mato Grosso, Mato Grosso do Sul, Goiás, Distrito Federal, Minas Gerais and in the west of Bahia (see Fig. 2.1). The average cotton farm is larger than 500 ha, but farm sizes range from 5 ha to over 10,000 ha. In the state of Mato Grosso, where more than 50% of the cotton fibre is produced, most of the farms are larger than 1000 ha (Freire et al., 1999). In 1995/96 there were 11,195 farmers in the Midwest region (IBGE, 1996), but this has probably increased significantly in the last 5 years. The level of technology used is high and yields can reach up to 2000 kg fibre/ha.

Cotton and other agricultural crops have replaced the Cerrado biome, which is located on a plateau at altitudes from 400 m to 1000 m above sea level, covering most of the central and western parts of Brazil (Plate 1 – see colour frontispiece). The climate is tropical stationary, with humid summers and dry winters (Aw, according to the Koeppen classification). The average temperature range is from 20°C to 26°C, the duration of solar radiation is between 7 h/day and 10 h/day, and averages 450 cal/cm^2/day. Precipitation ranges from 1300 mm/year to 1500 mm/year, with an average of 1100 mm during the wet season, which generally starts in October and ends in April (Plate 3 – see colour frontispiece). The high rainfall enables farmers to attain high cotton yields without irrigation. Cotton is sown in November and December and harvested from May to August. At mid-altitudes (400–600 m above sea level), the cotton growth cycle takes 150 days, whereas at higher altitudes (700–1100 m above sea level) it takes 180 days. From May to September, low temperatures and lack of rain prevent the cultivation of cotton.

In the Midwest, cotton production has increasingly expanded onto the acidic, strongly weathered Latosols and Argisols under the Cerrado vegetation (Plate 4 – see colour frontispiece)[2] (Medeiros et al., 1999). After clearing the Cerrado vegetation, cotton crops can be introduced only after the cultivation of four to six cycles of other crops such as rice, soybean and maize, during which time the soil is treated to correct the excessive acidity and bind aluminium, which is present at toxic levels. Fertilizer use is necessary.

---

[2] In the US Soil Taxonomy, these soil types are described as Oxisols and Ultisols, respectively (Soil Survey Staff, 1975, quoted by Camargo et al., 1987).

Soil erosion is a problem due to the soil type and the heavy machinery used. Pre- and post-emergence herbicides are frequently used. Very few farmers use partial-tillage systems consisting of intercropping soybean and cotton. Most farmers use the minimal-tillage system that consists of planting millet in the rainy season before cotton, and then desiccating this crop with herbicides prior to planting cotton. Before planting the millet, the soil is protected by the cotton-crop residues and weeds. Crop rotation, which is normally an integral part of the no-tillage system, is not commonly used. The minimal-tillage system of cultivation improves the biological quality of soils, protecting against erosion and providing good physical properties favourable to cotton-cropping systems (Maronezzi *et al.*, 2001).

Modern, high-yield cultivars of *G. hirsutum* var. *latifolium* adapted to the region and to the mechanical harvesting are most frequently used, which allow plant densities of 80,000 plants/ha to 130,000 plants/ha. These cultivars produce a high percentage of fibre adapted to mechanical harvesting and have multiple resistances to diseases. Growth regulators based on inhibitors of the gibberellin biosynthetic pathway are applied to regulate and channel plant development to facilitate harvesting and improve yield. In these systems, the mean productivity can reach more than 2000 kg fibre/ha. About two-thirds of the farmers in Mato Grosso, Minas Gerais and Goiás states use purchased certified seed, whereas the majority of farmers in Mato Grosso do Sul and west Bahia use farm-saved seed (Table 2.1).

Modern machinery is used for all operations, including ginning (seed removal), on the farm. In general, the Midwest cotton farmers have their own system of ginning cotton, and to transform seed cotton into fibre cotton. The seeds are processed to produce cattle feed and oil. In farms where modern technologies are adopted, manual labour is reduced. In spite of this, cotton cropping is more labour intensive than other cropping systems such as soybean and maize. It is a source of jobs for 10 months in a year, and helps to employ populations in rural areas.

The very few small-scale farmers in the region are joined in Farmers Associations or Settlements. Besides the standard white-fibre cotton, these farmers have also started cultivating brown coloured cotton since 2000. The naturally coloured cultivars provide a fibre of excellent intrinsic quality, with high weavability, which is targeted at specific niche markets for natural fibres and environmentally sound technologies.

Major limiting factors for cotton production in the Midwest

Limiting factors to cotton production in the Midwest are rainfall, fungal and viral diseases and insect pests. Around 8000 ha of Midwest cotton are irrigated by central pivot irrigators (1% of the total cotton area in Brazil), but the rest of the cotton area is cultivated without irrigation, and crop productivity is affected yearly by the pattern of rainfall. In years with high rainfall, there is an increased incidence of the fungal diseases areolate mildew (*R. aureola*) and witches' broom (*C. gossypii*), which are transmitted via crop residues in the soil. Depending on the resistance level of the cotton cultivar planted, up to seven fungicide sprays are necessary to control the disease. Other important fungal diseases include fusarium wilt (*F. oxysporum* f. sp. *vasinfectum*), alternaria spot (*A. macrospora*) and bacterial blight (*X. axonopodis*). Recently, there have been severe epidemics of a cotton-attacking genotype of leaf blight, *S. solani* (Mehta, 2001). Farmers usually apply two to four preventive applications of fungicide against leaf diseases.

The predominant Lepidopteran pests are fall armyworm species (*Spodoptera* spp.) and cotton leafworm (*A. argillacea*) (Silvie *et al.*, 2001a), followed by tobacco budworm (*H. virescens*) and pink bollworm (*P. gossypiella*) (Degrande, 1998a; Santos, 2001). Others include aphids, thrips, boll weevil, cotton stem borer, cotton stainers and mites. Few farmers use scouting and threshold methods, and insecticide use is high (see section on pest-control systems). Among the most significant factors limiting production are viral diseases, in particular blue disease transmitted by aphids (*A. gossypii*). Many Midwest farmers prefer Embrapa's high-yielding variety CNPA Ita 90 even though it is susceptible to blue disease (E. Brunetta, Brasília, 2004, personal communication; Santos *et al.*, 2004), because of familiarity and good yield and fibre qualities. They try to combat the disease by spraying six to eight insecticide applications against aphids.

The Amazon region

Cotton production has expanded into the Amazon region, including the state of Rondônia, the west and south of the state of Amazonas, and the south of the states of Pará and Tocantins (see Fig. 2.2). In 1994, the area planted to cotton in the Amazon region reached 65,500 ha, but it decreased to around 18,500 ha in 2003 (IBGE, 2004). In Tocantins, the 2002/03 production area was recorded as 2400 ha, producing 2000 t fibre (see Fig. 2.1) (CONAB, 2003). It is possible that there will be a further expansion in cotton production towards the Amazon basin. Little information is currently available on cotton production in this region; it may be important to examine environmental effects of cotton here in the future.

Insect Pest-Control Systems

IPM of cotton in Brazil went through periods of stagnation and development until 1979, when the First Brazilian Meeting on Cotton Pest-Control in Goiânia initiated a significant new phase of development (Ramalho, 1994). Cotton IPM requires the planned use of a group of crop-management strategies, aiming at high productivity and early harvesting. Factors to consider include: correct soil preparation and nutrition, the choice of appropriate varieties, space between plants, control of plant density and height, time of sowing, weed and disease control, use of trap plants and insect-pest refuges, crop rotation, destruction of crop residues and other mechanical, physical and biological methods, harvesting procedures and direct protection against the attack of insects and mites through chemical insecticides. The choice of appropriate strategies will depend mainly on the region where the crop is grown and on the size of the farm. Pest-management strategies reported by Ramalho (1994) were mainly developed for upland cotton and are discussed below, complemented by recent developments in the Midwest and Meridian regions, reported by Degrande (1998a,b), Santos (1999, 2001) and Gallo *et al.* (2002).

Selection of cultivar

Cotton researchers have long perceived rapid fruiting and early maturity as traits that would help growers cope with arthropod pests, particularly the boll weevil. The importance of these characteristics became evident in Brazil when the boll weevil invaded the North-eastern region. A primary advantage of these rapid-fruiting cultivars is that they escape the late-season build-up of boll weevil, tobacco budworm and other arthropod pests that often decimate the crop, which allows the important bolls in the middle and bottom of the plant to develop unharmed. Low-input short-season cotton management systems coupled with appropriate pest management are now recognized as promising pest-control methods that avoid high chemical use, particularly in the rainy conditions of the North-east, where the growth cycle can be shortened (Ramalho, 1994).

Many cotton varieties available in the Brazilian market are resistant or tolerant to diseases such as ramulose, fusarium wilt and viruses (ABRASEM, 2003). The aphid-transmitted blue disease is another very relevant pest problem on cotton (Cia and Fuzatto, 1999). It occurs in the Midwest and Meridian regions and is currently one of the most serious pest problems in the Midwest. The presence of aphids on the crop is problematic throughout the cropping season, but aphid-population control is crucial when the disease appears early during crop establishment (E. Brunetta, Brasília, 2004, personal communication). Genetic resistance is the most appropriate method to suppress the disease. The use of cotton varieties with multiple resistance to diseases, including tolerance or resistance to blue disease, allows for a reduction in the number of pesticide applications. As pesticide application is delayed, natural enemy populations are boosted. More than 50% of the cotton varieties recommended for the Midwest and Meridian regions are resistant to the blue disease including Deltaopal, IAC 23, CD406, Brs Cedro and Brs Jatobá. The susceptible varieties that are mostly planted are CNPA Ita 90, Makina, Fibermax 966, Fibermax 977 and Sure Grow 821 (ABRASEM, 2003). Of these, only CNPA Ita 90 was developed in Brazil; the others were developed in other countries and introduced in to Brazil by international companies. The primary factors that influence the use of susceptible cultivars by farmers are productivity and familiarity (E. Brunetta, Brasília, 2004, personal communication). Some of the susceptible varieties have high productivity that is realized consistently from year to year, provided disease vectors can be controlled.

Cultural control methods
Modifications in production practices can change the attractiveness and suitability of the plant and its environment to the pest, aggravate or ameliorate

certain pest problems or even create new problems. Likewise, changes in the timing, application or quantities of irrigation water or fertilizers, plant and row spacing, cultivation practices or the time of planting and crop maturity may all have an impact on pest or beneficial arthropod populations. Early harvest and stalk destruction remove the food and breeding sites of the cotton stem borer, boll weevil and pink bollworm. Cultural practices that could be used to reduce pest problems in upland cottons in Brazil include: areawide uniform planting dates; cotton-free periods; destruction of infested squares, bolls and alternate hosts; early and uniform stalk destruction; trap crop and trap crop residues; and crop rotation (Ramalho, 1994; Ramalho and Wanderley, 1996; Santos, 1999; Degrande, 2000).

In most of the Meridian and Midwest regions, the destruction of herbaceous cotton-crop residues after harvest is mandatory to prevent pest multiplication during cotton-free periods, particularly the pink bollworm, aphids and the boll weevil, and to avoid pest migration to neighbouring areas (Vieira et al., 1999). In direct sowing systems, this is mainly done by cutting down the stalks and controlling regrowth with broad-spectrum herbicides. The stalks are usually left lying in the field and incorporated into the soil during the next ploughing. If cotton is rotated with soybean, the residues are removed because volunteer cotton will negatively affect the soybean crop.

Minimal-tillage and no-tillage systems could cause progressive increases in fungal diseases that are transmitted via crop residues on the soil surface, and farmers use resistant varieties and systemic fungicides to deal with this.

Sunlight and heat play an important role in regulating cotton insect activity and abundance in the perennial cotton-cropping system in the North-east. Desiccation of boll weevil larvae, pupae and pre-emergent adults is the most important natural mortality factor in the North-east (Ramalho et al., 1989). Lower planting densities (increased plant and row spacing) will increase mortality by giving a more open canopy. This natural mortality factor in conjunction with natural biological control, cultivar manipulation and cultural practices has reduced the boll weevil in the 'Seridó' region in the Caatinga biome in the state of Rio Grande do Norte (Plate 1 – see colour frontispiece) to the status of a minor pest, and insecticides are seldom required for its control (Ramalho, 1994).

Biological control

Many entomologists have discussed the possible ecological and economic importance of using parasitoids and predators as a strategy for the integrated control of cotton pests in Brazil (Gravena et al, 1991; Ramalho, 1994; Almeida and Silva, 1999; Santos, 1999; Silvie et al., 2001b). Augmentation and conservation of native natural enemies is particularly promising in Brazil because most cotton agroecosystems, particularly those of the North-eastern region, have a rich complex of naturally occurring entomophagous arthropods and entomopathogenic microorganisms (reviewed by Gravena et al., 1991; Ramalho, 1994; Almeida and Silva, 1999; Silvie et al., 2001b; Gallo et al., 2002; Pieroto et al., 2002). The predator and parasitoid species most frequently sampled in cotton fields are listed in Faria et al., Chapter 8, this volume and Pallini et al.,

Chapter 9, this volume. Several species of birds, mammals, reptiles and amphibians also prey on cotton pests, but their effectiveness has not been quantified. This rich, complex fauna presents an interesting potential for biological control of arthropod pests of cotton. However, information on the main food webs and data on the ecological importance of individual species and their role on pest-population regulation is still weak, and insufficient to propose good agricultural practices that might conserve and implement natural biological control.

Some observations on the significant role of a few natural enemy species have been reported in Brazil. Some authors report observations on the significant role of natural enemies in regulating aphid populations (Degrande, 1998a; Santos, 1999). Important parasitoids of the cotton boll weevil in Brazil include Hymenopteran wasps from the Pteromalidae (*Catolaccus* spp.), Braconidae (*Bracon* spp. and *Urosigalphus rubicorpus* Gibon), Eurytomidae (*Eurytoma* spp.) and Eupelmidae (*Eupelmus* spp.). For further information, see Pallini *et al.*, Chapter 9, this volume. *Trichogramma* species parasitize eggs of Lepidopteran pests such as *A. argillacea, H. virescens, P. gossypiella* and *S. frugiperda* (Almeida *et al.*, 1998). Santos (1999) reports that *Trichogramma* parasitization rates on *H. virescens* eggs can reach up to 86% in the state of Paraná (Meridian region), and recommends that it is not necessary to apply chemical pesticides when monitoring procedures detect rates of parasitism over 60%. Similarly, Almeida (2000) reported high rates of *Trichogramma* parasitism on eggs of *A. argillacea* and *H. virescens* feeding on perennial cotton on the North-east. A system of large-scale *Trichogramma* production and a programme for inundative releases has been developed to control *A. argillacea*, particularly targeting small farms of the North-east, but still needs to be implemented (Almeida *et al.*, 1998; Almeida, 2001). Also, the Fundação Centro Oeste has a laboratory colony of *Trichogramma* spp. for research on cotton pests.

The North-east is characterized by having more abundant natural enemy populations than other regions. In the Midwest, natural enemy populations tend to be less abundant, probably due to the very high number of broad-spectrum insecticide applications, particularly early in the cotton season. This is a constraint to biological control in the region. The use of varieties resistant to the blue disease (CVMV) and to the cotton leafworm may minimize the need of insecticide applications and favour the colonization and build-up of natural enemy populations.

Pathogens are important regulatory agents of some cotton pests, both as natural controls and as microbial insecticides. However, more information is needed on the naturally occurring entomopathogens in Brazilian agroecosystems. Almeida and Silva (1999) report that *Bacillus thuringiensis* Bellinger is used as a microbial insecticide to control *A. argillacea* and *H. zea*, but no details are given about the percentage of use and the type of farm systems in which this technology is adopted.

Pheromones

A synthetic sexual pheromone (gossyplure) is available in the market and can be used to suppress pink bollworm populations or for monitoring this pest population (Santos, 1999; Gallo *et al.*, 2002). It functions by 'confusing' the males by attracting them to the traps, which interferes with their mating with the females.

Pheromone traps treated with insecticide are used to manage boll weevil infestations (Silva, 2003). Traps are used at the edge of the crop fields, starting about 10 days before sowing, to trap adult boll weevils that are immigrating from the surrounding areas. In Goiás (Midwest region), pheromone traps are used to suppress boll weevils (Ferreira and Echer, 2003). Boll weevil presence and movements are monitored using pheromone traps during the period when cotton is not grown, from September to November. Farmers' associations and private foundations distribute the traps free of charge. The number of insects in the traps is counted every week and this information is used to determine the timing and number of preventive-insecticide applications when the first flower buds appear on the plants. The number of applications varies from none to three (Ferreira and Echer, 2003).

Chemical control

Most cotton farmers, especially in the Midwest region, use scouting and economic thresholds for making control decisions; but they do not follow the recommendations on numbers and mixtures of insecticide applications and use is very high. See Table 2.3 for a list of the wide range of insecticides and acaricides that are presently used on cotton in Brazil. Farmers in the Midwest currently use an average of around 12 insecticide applications per cropping season, varying from 12 to 20. Two to four of these applications are against Lepidoptera (*A. argillacea*, *H. virescens* and *S. frugiperda*), two to four are against the boll weevil (*A. grandis*) and one to two are against cotton stinkbugs or other species of stinkbug that migrate from soybean crops. Virus-susceptible varieties receive six to eight insecticide applications against sucking insects (aphids, cicadellids and whitefly) in order to prevent viral disease, whereas, on virus-resistant varieties, only two insecticide sprays are applied against sucking insects (E. Brunetta, Brasília, 2004, personal communication). In the North-east, insecticides are mainly applied against cotton leafworm and boll weevil.

There are only a few studies on pest resistance to pesticides in Brazil (see Fitt *et al.*, Chapter 12, this volume). With the establishment of the Brazilian Insecticide Resistance Action Committee (IRAC-BR) in 1997, insecticide resistance-monitoring programmes on cotton have been conducted on cotton leafworm (*A. argillacea*) and fall armyworm (*S. frugiperda*). Preliminary data have shown that the frequencies of resistance to some organophosphate and pyrethroid insecticides have been increasing in both pests (Santos, 1999; Diez-Rodriguez and Omoto, 2001).

Integrated pest management systems

IPM using cultivar selection, natural biological control, cultural control, climatic control and chemical control by selective pesticides if pest populations reach an economic threshold, represents the optimum strategy for producing a profitable crop of cotton in Brazil. Many specialists agree that a modern IPM

Table 2.3. Insecticides and acaricides that can be used on cotton in Brazil according to current regulations.

Active ingredient	Chemical group	Commercial name	Target pests
I. Acetyl cholinesterase inhibitors			
Aldicarb	Carbamate	Temik	Aphids, thrips
Carbofuran	Carbamate	Furadan	Cotton stem borer, aphids
Cartap	Carbamate	Thiobel; Cartap	Cotton stem borer, cotton leafworm
Furathiocarb	Carbamate	Promet	Thrips, aphids
Methomyl	Carbamate	Lannate	Aphids, cotton leafworm, cotton budworm, thrips
Phorate	Carbamate	Granutox	Aphids, thrips, pink bollworm
Thiodicarb	Carbamate	Larvin	Cotton leafworm
Acephate	Organophosphate	Orthene	Aphids, cotton leafworm, cotton budworm, thrips
Carbosulfan	Organophosphate	Marshal	Aphids, cotton leafworm, thrips
Clorpyrifos	Organophosphate	Lorsban	Cotton stem borer, aphids, cotton leafworm
Dimethoate	Organophosphate	Perfekthion; Dimetoato	Cotton stainer bug, cotton plant bug, aphids, whitefly, cotton leafworm, thrips
Fenthion	Organophosphate	Lebaycid	Aphids, cotton leafworm, two-spotted spider mite
Malathion	Organophosphate	Malatol	Aphids, cotton leafworm, thrips
Metamidophos	Organophosphate	Hamidop; Tamaron	Aphids, cotton leafworm, cotton budworm, thrips
Methidathion	Organophosphate	Supracid	Boll weevil, aphids, cotton leafworm, cotton plant bug
Monocrotophos	Organophosphate	Azodrin; Nuvacron	Aphids, cotton leafworm, cotton budworm, pink bollworm
Parathion methyl	Organophosphate	Folidol	Aphids, cotton leafworm, cotton budworm, boll weevil, pink bollworm, spider mite, two-spotted spider mite, thrips, looper, stainer bug, cotton plant bug, stem borer, root borer, maize rootworm
Profenofos	Organophosphate	Curacron	Cotton leafworm
Pyridaphention	Organophosphate	Ofunack	Broad mite
Triazophos	Organophosphate	Hostathion	Brazilian cotton borer, aphids, cotton leafworm, cotton budworm
Triclorfon	Organophosphate	Dipterex	Cotton leafworm
II. GABA-gated chloride channel antagonists			
Endosulfan	Cyclodiene	Thiodan	Boll weevil, cotton stainer bug, cotton plant bug, aphids, whitefly, cotton leafworm

Continued

Table 2.3. Insecticides and acaricides that can be used on cotton in Brazil according to current regulations – cont'd.

Active ingredient	Chemical group	Commercial name	Target pests
III. Sodium channel modulators			
Alfacypermethrin	Pyrethroid	Fastac	Cotton leafworm, cotton budworm, boll weevil, pink bollworm
Betacyfluthrin	Pyrethroid	Bulldock	Boll weevil, cotton plant bug, aphids, cotton leafworm, cotton budworm, pink bollworm
Bifenthrin	Pyrethroid	Talstar	Cotton leafworm
Cyfluthrin	Pyrethroid	Baytroid	Aphids, cotton leafworm, cotton budworm, boll weevil, pink bollworm
Cypermethrin	Pyrethroid	Arrivo; Cymbush	Boll weevil, aphids, cotton leafworm, cotton budworm, pink bollworm, thrips
Deltamethrin	Pyrethroid	Decis	Boll weevil, cotton plant bug, aphids, cotton leafworm, cotton budworm, pink bollworm
Esfenvalerate	Pyrethroid	Sumidan	Boll weevil, cotton stainer bug, aphids, cotton leafworm, cotton budworm
Fenpropathrin	Pyrethroid	Danimen; Meothrin	Aphids, cotton leafworm, cotton budworm, boll weevil, pink bollworm, spider mite, two-spotted spider mite, thrips, looper
Fluvalinate	Pyrethroid	Mavrik	Cotton leafworm, cotton budworm, pink bollworm
Lambda-cyhalothrin	Pyrethroid	Karate	Boll weevil, cotton plant bug, cotton leafworm, cotton budworm, pink bollworm
Permethrin	Pyrethroid	Pounce	Aphids, cotton leafworm, cotton budworm, pink bollworm
Zeta-cypermethrin	Pyrethroid	Fury	Boll weevil, cotton leafworm, cotton budworm
IV. Acetylcholine-receptor agonists/antagonists			
Acetamiprid	Neonicotinoid	Mospilan	Whitefly
Imidacloprid	Neonicotinoid	Gaucho	Whitefly, aphids, thrips
Thiametoxam	Neonicotinoid	Cruiser	Brazilian cotton borer, aphids, thrips
Thiacloprid	Neonicotinoid	Calypso	Aphids, thrips
V. Acetylcholine-receptor modulators			
Spinosad	Naturalyte	Tracer	Cotton leafworm, cotton budworm
VI. Chloride-channel activators			
Abamectin	Lactone	Vertimec	Cotton leafworm
Clofentezine		Acaristop	Two-spotted spider mite

VIII. Inhibition of chitin biosynthesis

Novaluron	IGR	Gallaxy; Rimon	Cotton leafworm
Buprofezin	IGR	Applaud	Whitefly
Chlorfuazuron	IGR	Atabron	Cotton leafworm
Diflubenzuron	IGR	Dimilin	Cotton leafworm
Lufenuron	IGR	Match	Cotton leafworm, cotton budworm
Teflubenzuron	IGR	Nomolt	Cotton leafworm
Triflumuron	IGR	Alsystin	Cotton leafworm

IX. Inhibition of chitin biosynthesis type 1-Homopteran

Buprofezin	IGR	Applaud	Whitefly

X. Ecdysone agonist/disruptor

Methoxifenozide	IGR	Intrepid	Cotton leafworm, cotton budworm
Tebufenozide	IGR	Mimic	Cotton leafworm

XI. Microbial disrupters of insect midgut membranes

Bacillus thuringiensis	Microbial insecticide	Dipel; Thuricide	Cotton leafworm, cotton budworm, pink bollworm

XII. Inhibition of oxidative phosphorylation, disrupters of ATP formation

Diafenthiuron	Thiourea	Polo	Aphids, cotton leafworm

XIII. Uncoupler of oxidative phosphorylation via disruption of H proton gradient

Chlorfenapyr	Pyroles	Pirate	Two-spotted spider mite, cotton budworm

XIV. Inhibition of magnesium-stimulated ATPase

Propargite	Propargite	Omite	Two-spotted spider mite, broad mite

XV. Others

Sulphur	Sulphur	Enxofre; Rapid; Thiovit	
Tetradifon	Tetradifon	Tedion	Spider mite, two-spotted spider mite

system should emphasize three key components (Ramalho, 1994; Ramalho and Wanderley, 1996; Degrande, 1998a; Soares and Almeida, 1998; Almeida and Silva, 1999; Santos, 2001; Gallo *et al.*, 2002). First, it should rely on the development and use of cotton varieties with resistance to aphid-vectored viruses. Second, there should be a strong emphasis on cultural control of boll weevil. This may include planting windows, crop-residue destruction and available control tactics that conserve natural enemies. Finally, there should be appropriate emphasis on monitoring and thresholds for deciding when to make pesticide application and on conservation of natural enemies. Each of these components has been developed and recommended in farmer communication bulletins. Sometimes a single component is used, sometimes a combination of a few components. Most of the proposed IPM strategies prioritize control of the boll weevil, due to the severity of its attack and its negative impact on cotton production (Gutierrez *et al.*, 1991). More active initiatives need to be taken to integrate these components in a way that is easily communicated and disseminated among farmers through training courses and extension programmes.

Although there is considerable entomological expertise available on cotton in Brazil, the published literature is not easily accessible and is weak in some important areas. More information is needed on the ecology and population dynamics of the various pests and on crop/pest/natural enemy interactions in the major cotton-growing regions. Most significantly, crop loss for some important pest species has not been quantified at all (e.g. *S. frugiperda*); and for other important pests, the studies are old, associated with more susceptible older varieties. The information on crop losses from individual pests add up to more than 100% damage (Ramalho, 1994), which suggests that there are interactions between the pest species and that up-to-date, comprehensive studies of yield loss are needed.

The costs of pest control are very high, and the farmer's annual gains or losses depend, to a large degree, on the efficacy of pest-control tactics applied. This creates a risk-averse attitude that leads to a heavy reliance on preventative applications of chemical pesticides, disrupting natural biological control that could reduce or prevent pest-population outbreaks.

In the Midwest region, where large cotton farms predominate, some IPM recommendations are enforced, both because the law requires it (in the case of boll weevil) and because the farmers are well informed and hire agronomists for the monitoring and decision-making procedures based on threshold population levels. However, there is a need to review some of the threshold levels indicated in the available literature, considering regional differences on pest attack and the new developments in the cotton agricultural system during the last 10 years, and based on a wider vision of the ecological functions and processes of cotton agroecosystems and on the urgent need to preserve biodiversity and prevent environmental contamination by pesticides. These comments are also valid for the Meridian region, where a few large farms are replacing the many small cotton farms, particularly in the state of São Paulo. In the North-east and the small-farm cotton systems still present in the other two regions, the main constraint to the implementation of IPM strategies is the lack of extension services and the need for a continuous effort to develop pest-control strategies adapted to local environmental and socio-economic conditions. To be able to select appropriate cultivars and to successfully apply cultural control methods that will reduce pest populations, farmers need farmer community training programmes and continuous assistance. An extension-system programme to carry this out could be developed in cooperation between the government and the private sector.

Summary and Discussion

In the last 20 years, cotton production in Brazil has moved from a tropical xerophytic climate with high temperatures, passing to regions with a subtropical climate, and now reaching the warm and humid tropics with high volumes of rainfall (1800–2000 mm distributed throughout 7–8 months in a year). Production has moved from basic soils with high agricultural potential to acidic soils with lower agricultural potential. The transfer of the main cotton-production area from the North-east and Meridian regions to the Midwest was a move from subtropical regions with good potential for crop growth that was undermined by continuous cotton monoculture which led to severe soil degradation, to a warm and humid tropical region with less fertile soils that have been better managed by minimum-tillage cropping. It was also a change from a type of production system based on many small farms to a production system based on a smaller number of very large farms.

Pests and pest-control methods

Cotton is a crop with severe phytosanitary problems that require constant monitoring and prompt farmer intervention. The main problems are insects and diseases. Pest control is one of the most significant items in total cotton-production costs and, in most years, if appropriate measures are not taken to minimize damage caused by even one of the main pest problems, the farmers can lose all their profits. Current control strategies rely heavily on chemical pesticides, sometimes associated with the use of resistant cultivars and cultural control methods. The amount of pesticide used on this crop is still excessively large and the undesirable environmental effects of this practice are not being quantified or monitored. Very few farmers in Brazil grow cotton using organic methods. More research information on pest ecology and management is needed, but the available knowledge is sufficient to indicate that better pest-control practices can be implemented that would significantly reduce the amount of pesticide applied. The use of *Bacillus thuringiensis* (Bt) cotton varieties can be useful to reduce the damage from Lepidopteran pests, but many other pest problems will remain. There is clearly the need for a concentrated effort to analyse the overall pest problems on cotton and to propose a programme to address simultaneously the different facets of this complex situation in a concerted way, which must include a farmer's education plan and better extension services.

Predicted impact of Bt cotton on pesticide use

Lepidopteran-active Bt crops will still require insecticide applications to control aphids, mites, whiteflies, boll weevil, bugs and Coleopteran borers. The use of acaricides, fungicides and herbicides will also not change because of the use of these crops. Taking as a basis the number of insecticide applications described in the section on chemical control above, we estimate that, with the introduction of Lepidopteran-active Bt cotton containing the stacked proteins Cry1Ac and Cry2Ab in the Midwest, the number of insecticide sprays might be reduced to an average of ten applications on virus-susceptible varieties (a reduction of 10–20%), and to five applications on virus-resistant varieties (a reduction of 20–40%). This is an average reduction of two insecticide applications. We predict that in the North-east, the use of transgenic Bt cultivars might reduce pesticide application to two to three sprays per season, to half of the currently applied pesticide. The introduction of Bt cotton in the IPM programmes of the Meridian region might eliminate only two insecticide applications due to frequent high infestations of the boll weevil. While this projected reduction in insecticide use can be beneficial to farmers and the environment, especially in the North-east, it is clear that other significant insect pest problems remain to be addressed.

Acknowledgements

We express our thanks to Drs Lewis Wilson (CSIRO), Pierre Silvie (CIRAD-CA) and Marc Giband (CIRAD) for their careful reviews and helpful comments on earlier versions of the chapter.

References

ABRASEM (2003) *Anuário Abrasem 2002*. Associação Brasileira dos Produtores de Sementes, Brasília, Brazil.

ABRASEM (2004) *Anuário Abrasem 2003*. Associação Brasileira dos Produtores de Sementes, Brasília, Brazil.

Almeida, R.P. de (2000) Distribution of parasitism by *Trichogramma pretiosum* on the cotton leafworm. *Proceedings of the Section Experimental and Applied Entomology of the Netherlands Entomological Society* 11, 27–31.

Almeida, R.P. de (2001) Cotton insect pest control on a small farm: an approach of successful biological control using *Trichogramma*. *Proceedings of the Section Experimental and Applied Entomology of the Netherlands Entomological Society* 12, 81–84.

Almeida, R.P. de and Silva, C.A.D. (1999) Manejo integrado de pragas do algodoeiro. In: Beltrão, N.E. de M. (ed.) *O Agronegócio do Algodão no Brasil*, Vol. 2. Comunicação para Transferência de Tecnologia Embrapa, Brasília, Brazil, pp. 753–820.

Almeida, R.P. de, Silva, C.A.D. and Medeiros, M.B. (1998) Biotecnologia de produção massal e manejo de *Trichogramma* para o controle biológico de pragas. Documentos 60, Embrapa-CNPA, Campina Grande, Brazil, 61 pp.

Amorim Neto, M. da S., Medeiros, J. da C., Beltrão, N.E., Freire, E.C., Novaes Filho, M.B. and Gomes, D.C. (2003) Zoneamento para cultura do algodão herbáceo no Nordeste. Available at: http://algodão.cnpa.embrapa.br/algodão/zoneamento/zoneherba.html (accessed 24 May 2004).

Andries, J.A., Jones, J.E., Sloane, L.W. and Marshal, J.G. (1969) Effects of okra leaf shape on boll rot, yield and other important characters of Upland cotton, Gossypium hirsutum L. Crop Science 9, 705–710.

Azevedo, D.M.P. de, Beltrão, N.E. de M., Vieira, D.J. and Nóbrega, L.B. da (1999) Plantas daninhas. In: Beltrão, N.E. de M. (ed.) O Agronegócio do Algodão no Brasil, Vol. 2. Comunicação para Transferência de Tecnologia Embrapa, Brasília, Brazil, pp. 553–586.

Bell, A.A. (1999) Diseases of cotton. In: Smith, C.W. and Cothren, J.T. (eds) Cotton: Origin, History, Technology, and Production. John Wiley & Sons, New York, pp. 553–594.

Beltrão, N.E.M. (2002) Que planta é esta? Cultivar (Agosto de 2002), 17–18.

Beltrão, N.E.M., Sobrino, C., Fereira, F., Araújo, J.M., Cardoso, G.D. and Severino, L.S. (2003) Variabilidade nas fibras intrínsicas de algodão de cor BRS 200 marrom em função das tonalidades do marrom e da fibra branca. In: IV Congresso Brasileiro de Algodão, 4, 2003, Goiânia, Brazil. CD-ROM.

Bottger, G.T., Sheehan, E.T. and Lukefahr, M.J. (1964) Relation of gossypol content of cotton plants to insect resistance. Journal of Economic Entomology 57, 283–285.

Brook, K.D., Hearn, A.B. and Kelly, C.F. (1992) Responses of cotton, Gossypium hirsutum L., to damage by insect pests in Australia: compensation for early season fruit damage. Journal of Economic Entomology 85, 1378–1386.

Calcagnolo, G. and Sauer, H.F.G. (1954) Influência do ataque dos pulgões na produção do algodão (Aphis gossypii Glover, 1876, Hom. Aphididae). Arquivos do Instituto Biológico 21, 85–99.

Carmago, M.N., Klamt, E. and Kauffman, J.H. (1987) Sistema brasileiro de classificação de solos. Boletim da Sociedade Brasileira de Ciência do Solo, Campinas, 12(1), 11–33. Titulo correcto: Classificação de solos usada em levantamentos pedológicos no Brasil, Sociedade Brasileira de Ciência do Solo, Campinas, Brazil.

Carvalho, L.H. (1999) Contribuição do melhoramento do cultivo do algodão no Brasil. In: Beltrão, N.E.M. (ed.) O Agronegócio do Algodão no Brasil Embrapa, Brasília, Brazil, pp. 253–269.

Carvalho, L.H. (2003) Cor e valor. Cultivar 46, 28–29.

Carvalho, L.H. and Chiavegato, E.J. (1999) A cultura do algodão no Brasil: fatores que afetam a produtividade. In: Cia, E., Freire, E.C. and Santos, W.J. dos (eds) Cultura do Algodoeiro. Piracicaba, Potafos, Brazil, pp. 1–8.

Chitara, L.G. and Meyer, M.C. (2005) Novo e sem controle. Cultivar 69, 44–45.

Cia, E. and Fuzzato, M.G. (1999) Manejo de doenças na cultura do algodão. In: Cia, E., Freire, E.C. and Santos, W.J. dos (eds) Cultura do Algodoeiro. Piracicaba, Potafos, Brazil, pp. 121–132.

Cobley, L.S. (1956) An Introduction to the Botany of Tropical Crops. Longmans, London.

CONAB (2003) Algodão em pluma informativo especial Julho 2003. Available at: http://www.conab.gov.br/download/cas/especiais/CONJ%20ESPECIAL%20JULHO%2003.pdf (accessed September 2004).

Corrêa, R.L., Franca, T.S., Barroso, P.A.V., Vidal, M.S. and Vaslin, M.F.S. (2004a) Development of molecular diagnosis assays for the cotton blue disease. In: XV Encontro Nacional de Virología, 2004, São Pedro, p. 244.

Corrêa, R.L., Franca, T.S., Bonis, M. de, Barroso, P.A.V., Vidal, M.S. and Vaslin, M.F.S. (2004b) Identification of a polerovirus (Luteoviridae) associated with cotton blue disease. In: *XV Encontro Nacional de Virología*, 2004, São Pedro, pp. 243–244.

Corrêa, R.L., Silva, T.F., Araújo, J.L.S., Barroso, P.A.V., Vidal, M.S. and Vaslin, M.F.S. (2005) Molecular characterization of a virus from the family Luteoviridae associated with cotton blue disease. *Archives of Virology* 150(7), 1357–1367.

Costa, A.S. (1966) Moléstias de vírus do algodoeiro. In: *Divulgação Agronômica*. Rio de Janeiro, No. 21, pp. 27–29.

Costa, A.S. and Carvalho, A.M.B. (1962) Moléstias de Vírus do Algodoeiro. *Bragantia, Campinas*, 21(2), 50–51.

Cothren, J.T. (1999) Physiology of the cotton plant. In: Smith, C.W. and Cothren, J.T. (eds) *Cotton: Origin, History, Technology, and Production*. John Wiley & Sons, New York, pp. 207–268.

Dale, J.E. (1959) Some effects of the continuous removal of floral buds on the growth of the cotton plant. *Annals of Botany* 23, 636–649.

Davidson, A. and Seara, H.S. (1966) The incidence and losses caused by pink bollworm and other pests on cotton yield in northeast Brazil. *FAO Plant Protection Bulletin* 14, 80–85.

Degrande, P.E. (1998a) *Guia prático de controle das pragas do algodoeiro*. UFMS, Dourados, Brazil.

Degrande, P.E. (1998b) Manejo integrado de pragas do algodoeiro. In: *Algodão Informações Técnicas*. Embrapa, Circular Técnica No. 7, Dourados, Brazil.

Degrande, P.E. (2000) Manejo de pragas: realidade e desafios. In: Anais do V Seminário estadual da cultura do algodão, 'Negócios e tecnologias para melhorar a vida' 31/08-2/09/2000, Cuiabá, Mato Grosso, Brazil, pp. 229–244.

Deuber, R. (1999) Manejo integrado de plantas infestantes na cultura do algodoeiro. In: Cia, E., Freire, E.C. and Santos, W.J. dos (eds) *Cultura do Algodeiro*. Piracicaba, Potafos, Brazil, pp. 101–120.

Dias, B.F. de S. (1992) Cerrados: Uma Caracterização. In: Dias, B.F. de S. (ed.) *Alternativas de Desenvolvimento dos Cerrados: Manejo e Conservação dos Recursos Naturais Renováveis*. FUNATURA, Brasília, Brazil.

Diez-Rodriguez, G.I. and Omoto, C. (2001) Herança da Resistência de *Spodoptera frugiperda* (J.E. Smith) (Lepidoptera: Noctuidae) à Lambda-Cialotrina. *Neotropical Entomology* 30, 311–316.

Elzen, G.W., Williams, H.J., Bell, A.A., Stipanovic, R.D. and Vinson, S.B. (1985) Quantification of volatile terpenes of glanded and glandless *Gossypium hirsutum* L. cultivars and lines by gas chromatography. *Journal of Agriculture and Food Chemistry* 33, 1079–1082.

Elzen, G.W., Williams, H.J. and Vinson, S.B. (1986) Wind tunnel flight responses by hymenopterous parasitoid *Campoletis sonorensis* to cotton cultivars and lines. *Entomologia Experimentalis et Applicata* 42, 285–289.

Embrapa (2003) *A produção de sementes no Brasil. Relatório da safra: 2001/2002*. Convênio MAPA/Embrapa Transferência de Tecnologia/Abrasem, Brasília, Brazil.

Estado de São Paulo (2004) Available at: http://txt.estado.com.br/editorias/2004/05/09/eco023.html (accessed 10/05/2004).

Fehr, W.R. (1987) *Principles of Cultivar Development*. Macmillan, New York.

Ferreira, N.S. and Echer, C.R. (2003) Goiás mobilizado contra o bicudo. *Cultivar Especial* (May 2003), 3–10.

Freire, E.C. (2000) *Distribuição, coleta, uso e preservação das espécies silvestres de algodão no Brasil*. Embrapa Algodão, Campina Grande, Brazil.

Freire, E.C. and Costa, J.N. (1999) Objetivos e métodos utilizados nos programas de melhoramento do algodão no Brasil. In: Beltrão, N.E.M. (ed.) *O Agronegócio do Algodão no Brasil*. Embrapa, Brasília, Brazil, pp. 271–293.

Freire, E.C., Farias, J.C. de E. and Aguiar, P.H. (1999) Algodão de alta tecnologia no cerrado. In: Cia, E., Freire, E.C. and Santos, W.J. dos (eds) *Cultura do Algodeiro*. Piracicaba, Potafos, Brazil, pp. 181–199.

Fryxell, P.A. (1979) *The Natural History of the Cotton Tribe: (Malvaceae, tribe Gossypieae)*. Texas A & M University Press, College Station, Texas.

Fuzzato, M.G. (1999) Melhoramento genético do algodoeiro. In: Cia, E., Freire, E.C. and Santos, W.J. dos (eds) *Cultura do Algodeiro*. Piracicaba, Potafos, Brazil, pp. 15–34.

Gallo, D., Nakano, O., Carvalho, R.P.L., Baptista, G.C. de, Filho, E. Berti, Parra, J.R.P., Zucchi, R.A., Alves, S.B., Vendramin, J.D., Marchini, L.C., Lopes, J.R.S. and Omoto, C. (2002) *Entomologia Agrícola*. Fealq, Piracicaba, Brazil.

Gannaway, J.R. (1994) Breeding for insect resistance. In: Matthews, G.A. and Tunstall, J.P. (eds) *Insect Pests of Cotton*. CAB International, Wallingford, UK, pp. 431–454.

Gillham, F.E.M. (1963) A study in the response of bollworm, *Helioverpa zea* (Boddie) to different genotypes of upland cotton. In: *Proceedings of the Beltwide Cotton Production Research Conference*. Memphis, Tennessee, pp. 80–88.

Gravena, S. and Cunha, H.F. da, 1991. Artrópodos predadores da cultura algodoeira: atividade sobre *Alabama argillacea* (Hub.) com breves referências a *Heliothis* sp. (Lepidóptera: Noctuidae). FUNEP, Jaboticabal, 41 pp. (Boletim Técnico, 1).

Guttierez, A.P., Santos, W.J. dos, Pizamiglio, M.A., Villacorta, A.M., Ellis, C.K., Fernandes, C.A.P. and Tutida, I. (1991) Modelling the interaction of cotton and the cotton boll weevil: II. boll weevil (*Anthonomus grandis*) in Brazil. *Journal of Applied Ecology* 28(2), 398–418.

Hedin, P.A., Thompson, A.C. and Gueldner, R.C. (1975) Cotton plant and insect constituents that control boll weevil behavior and development. In: Wallace, J.W. and Mansell, R.L. (eds) *Biochemical Interaction Between Plants and Insects*, Vol. 10. Proceedings. Plenum Press, New York.

Heringer, E.P., Barroso, G.M., Rizzo, J.A. and Rizzini, C.T. (1977) A Flora do Cerrado. In: Ferri, M.G. (ed.) *Simpósio Sobre o Cerrado*. Editora Universidade de São Paolo, São Paolo, Brazil, pp. 211–232.

IBGE (1996) *Levantamento Sistemático da Produção Agrícola*. Instituto Brasileiro de Geografia e Estatística, Campo Grande, Brazil.

IBGE (2002) *Levantamento Sistemático da Produção Agrícola*. Instituto Brasileiro de Geografia e Estatística, Campo Grande, Brazil.

IBGE (2004) *Levantamento Sistemático da Produção Agrícola*. Instituto Brasileiro de Geografia e Estatística, Campo Grande, Brazil.

Jenkins, J.N. and Parrott, W.I. (1971) Effectiveness of frego bract as a boll weevil resistance character in cotton. *Crop Science* 11, 736–743.

Jenkins, J.N. and Wilson, F.D. (1996) Host plant resistance. In: King, E.G., Phillips, J.R. and Coleman, R. (eds) *Cotton Insects and Mites: Characterization and Management*. The Cotton Foundation, Memphis, Tennessee, pp. 563–597.

Jenkins, J.N., Maxwell, F.G., Parrott, W.L. and Buford, W.T. (1969) Resistance to the boll weevil (*Anthonomus grandis* Boh.) oviposition in cotton. *Crop Science* 9, 369–372.

Jones, G.D. and Coppedge, J.R. (1996) Pollen feeding by overwintering boll weevils. *Proceedings of the Beltwide Cotton Conference*, Vol. 2. National Cotton Council, Memphis, Tennessee, pp. 976–977.

Kadapa, S.M., Vizia, N.C. and Patil, N.B. (1988) A note on stem-tip stiffness in aphid tolerant cottons (*Gossypium hirsutum* L.). *Current Science, India* 57, 265–266.

Karban, R. (1986) Induced resistance against spider mites in cotton: field verification. *Entomologia Experimentalis et Applicata* 42, 239–242.

Karban, R. (1988) Resistance to beet armyworms (*Spodoptera exigua*) induced by exposure to spider mites (*Tetranychus turkestani*) in cotton. *American Midland Naturalist* 119, 77–82.

Karban, R. and Carey, J.R. (1983) Induced resistance of cotton seedlings to mites [*Tetranychus urticae, Tetranychus turkestani, Tetranychus pacificus*]. *Science* 225(4657), 53–54.

Lege, K.E., Cothren, J.T. and Smith, C.W. (1995) Phenolic acid and condensed tannin concentrations of six cotton genotypes. *Environmental and Experimental Botany* 35(2), 241–249.

Lima, E.F., Batista, F.A.S. and Vieira, R.M. (1999) Principais doenças do algodoeiro e seu controle. In: Beltrão, N.E.M. (ed.) *O Agronegócio do Algodão no Brasil.* V2. Embrapa Comunicação para Transferência de Tecnologia, Brasília, Brazil, 715–752.

Lukefahr, M.J. and Houghtaling, J.E. (1969) Resistance of cotton strains with high gossypol content to *Heliothis* spp. *Journal of Economic Entomology* 62, 588–591.

Lukefahr, M.J., Shaver, T.N. and Parrott, W.L. (1969) Sources and nature of resistance in *Gossypium hirsutum* to bollworm and tobacco budworms. In: *Proceedings of the Beltwide Cotton Production and Resistance Conference 1969*, pp. 81–82.

Luttrell, R.G., Fitt, G.P., Ramalho, F.S. and Sugonyaev, E.S. (1994) Cotton pest management. Part 1. A worldwide perspective. *Annual Review of Entomology* 39, 517–526.

Marchini, L.C. (1976) Avaliação de dano do curuquerê do algodoeiro (*Alabama argillacea*) (Hubner, 1818) (Lepidoptera: Noctuidae) em condições simuladas e redução de sua população através de iscas tóxicas. MSc. thesis. Universidade de São Paulo, Piracicaba, Brazil.

Maronezzi, A.C., Belot, J.-L., Martin, J., Seguy, L. and Bouzinac, S. (2001) *A safrinha de algodão: opcão de cultura arriscada ou alternativa lucrativa dos sistemas de plantio direto nos tropicos?* COODETEC, Boletim Técnico No. 37. Cascavel, Paraná, Brazil.

Matthews, G.A. (1994a) The effects of insect attack on the yield of cotton. In: Matthews, G.A. and Tunstall, J.P. (eds) *Insect Pests of Cotton.* CAB International, Wallingford, UK, pp. 427–430.

Matthews, G.A. (1994b) Insect and mite pests: general introduction. In: Matthews, G.A. and Tunstall, J.P. (eds). *Insect Pests of Cotton.* CAB International, Wallingford, UK, pp. 29–37.

McCarty, J.C. Jr, Meredith, W.R., Jenkins, J.N., Parrott, W.L. and Bailey, J.C. (1983) Genotypes × environment interaction of cottons varying in insect resistance. *Crop Science* 23, 970–973.

McGregor, S.E. (1976) *Insect Pollination of Cultivated Crop Plants.* Agriculture Handbook No. 496, U.S. Government Printing Office, Washington, DC.

Medeiros, J. da C., Silva, O.R.R.F. da and Carvalho, O.S. (1999) Edafologia. In: Beltrão, N.E. de M. (ed.) *O Agronegócio do Algodão no Brasil*, Vol. 1. Comunicação para Transferência de Tecnologia Embrapa, Brasília, Brazil, pp. 117–171.

Medeiros, R.S., Ramalho, F.S., Zanúncio, J.C. and Serrão, J.E. (2003) Estimate of *Alabama argillacea* (Hübner) (Lepidoptera: Noctuidae) development with nonlinear models. *Brazilian Journal of Biology* 63, 589–598.

Mehta, Y.R. (2001) Genetic diversity among isolates of *Stemphylium solani* from cotton. *Fitopatologia Brasileira* 26(4), 703–709.

Meredith, W.R. Jr, Ranney, C.D., Laster, M.L. and Bridge, R.R. (1973) Agronomic potential of nectariless cotton. *Journal of Environmental Quality* 2, 141–144.

Michel, B. (1989) Nouvelle contribution à la connaissance des insectes et arachnides rencontrés en culture cotonnière au Paraguay. *Coton et Fibres Tropicales* 44, 51–54.

Michel, B. and Prudent, P. (1985) Acariens et insectes déprédateurs du cotonnier (*Gossypium hirsutum* L.) au Paraguay. *Coton et Fibres Tropicales* 40, 219–224.

Montandon, R., Williams, H.J., Sterling, W.L., Stipanovic, R.D. and Vinson, S.B. (1986) Comparison of the development of *Alabama argillacea* (Hubner) and *Heliothis virescens* (F.) (Lepidoptera: Noctuidae) fed glanded and glandless cotton leaves. *Environmental Entomology* 15, 128–131.

Nyakatawa, E.Z., Reddy, K.C. and Lemunyon, J.L. (2001) Predicting soil erosion in conservation tillage cotton production systems using the revised universal soil loss equation (RUSLE). *Soil and Tillage Research* 57, 213–224.

Oosterhuis, D.M. and Jernstedt, J. (1999) Morphology and anatomy of the cotton plant. In: Smith, C.W. and Cothren, J.T. (eds) *Cotton: Origin, History, Technology, and Production.* John Wiley & Sons, New York, pp. 175–206.

Paré, P.W. and Tumlinson, J.H. (1997) De novo biosynthesis of volatiles induced by insect herbivory in cotton plants. *Plant Physiology* 114(4), 1161–1167.

Paré, P.W. and Tumlinson, J.H. (1998) Cotton volatiles synthesized and released distal to the site of insect damage. *Phytochemistry* 47(4), 521–526.

Parrot, W.L. (1990) Plant resistance to insects in cotton. *Florida Entomologist* 73(3), 392–396.

Pashley, D.P. (1986) Host-associated genetic differentiation in fall armyworm (Lepidoptera: Noctuidae): a sibling species complex? *Annals of the Entomological Society of America* 79, 898–904.

Pashley, D.P., Johnson, S.J. and Sparks, A.N. (1985) Genetic population structure of migratory moths: the fall armyworm (Lepidoptera: Noctuidae). *Annals of the Entomological Society of America* 78, 756–762.

Percy, R.G. and Kohel, R.J. (1999) Qualitative genetics. In: Smith, C.W. and Cothren, J.T. (eds) *Cotton: Origin, History, Technology, and Production.* John Wiley & Sons, New York, pp. 319–360.

Pieroto, N.W., Lara, R.I.R., Santos, J.C.C. dos and Selegatto, A. (2002) Himenopteros parasitoids (Insecta, Hymenoptera) coletados em cultura de algodao (*Gossypium hirsutum* L.) (Malvaceae), no municipio de Ribeirao Preto, SP, Brazil. *Revista Brasileira de Entomologia* 46(2), 165–168.

Pires, C., Sujii, E., Borges, M., Schmidt, F., Faria, M., Silveira, F., Moraes, M.C.B., Laumann, R.A., Barroso, P., Pereira, E., Portilho, T., Pinheiro, E., Ribeiro, P., Frizzas, M. and Fontes, E. (2004) Risk assessment of Bt cotton in Brazil-challenges, opportunities and preliminary results. *Proceedings of the 8th International Symposium on the Biosafety of Genetically Modified Organisms.* Montpellier, France, pp. 105–110.

Ponte, J.J. da, Silveira, F.J., Lordello, R.R.A. and Lordello, A.I.L. (1998) Synopsis of the Brazilian literature of the *Meloidogyne*-cotton association. *Summa Phytopathologia* 24(2), 101–104.

Ramalho, F.S. (1994) Cotton pest management. Part 4. A Brazilian perspective. *Annual Review of Entomology* 34, 563–578.

Ramalho, F.S. and Santos, R.F. (1994) Impact of the introduction of the cotton boll weevil in Brazil. In: Constable, G.A. and Forrester, N.W. (eds) *Challenging the Future – Proceedings of the World Cotton Research Conference 1*. Brisbane, Australia, pp. 466–474.

Ramalho, F.S. and Wanderley, P.A. (1996) Ecology and management of the boll weevil in South American cotton. *American Entomologist* 42, 41–47.

Ramalho, F.S., Jesus, F.M.M. and Bleicher, E. (1989) Manejo integrado de pragas e viabilidade do algodoeiro herbáceo no Nordeste. In: Sociedade entomológica do Brasil (eds) *Seminário sobre controle de insetos*. Fundação Cargill, Campinas, Brazil, pp. 112–123.

Ramalho, F.S., Jesus, F.M.M. and Gonzaga, J.V. (1990) Táticas de manejo integrado de pragas em áreas infestadas pelo bicudo-do-algodoeiro. *Pesquisa Agropecuária Brasileira* 25, 677–690.

Ratter, J.A., Ribeiro, J.F. and Bridgewater, S. (1997) The Brazilian Cerrado vegetation and threats to its biodiversity. *Annals of Botany* 80, 223–230.

Reddy, P.S.C. (1974) Effects of three leaf shape genotype of *Gossypium hirsutum* L. and row types on microclimate, boll weevil survival, boll rot and important agronomic characters. PhD dissertation. Louisiana State University, Baton Rouge, Louisiana.

Rizzini, C.T. (1963) Flora do Cerrado. Análise Florística dos Savanas Centrais. In: Ferri, M.G. (ed.) *Simpósio sobre o Cerrado*, Editora Universidade de São Paolo, São Paolo, Brazil, pp. 127–177.

Röse, U.S.R., Lewis, W.J. and Tumlinson, J.H. (1998) Specificity of systemically released cotton volatiles as attractants for specialist and generalist parasitic wasps. *Journal of Chemical Ecology* 24(2), 303–319.

Sadras, V.O. and Fitt, G.P. (1997) Resistance to insect herbivory of cotton lines: quantification of recovery capacity after damage. *Field Crops Research* 52(1–2), 127–134.

Santos, W.J. (1977) Efeitos da simulação dos danos da lagarta das maçãs *Heliothis virescens* (Fabr. 1781) (Lepidoptera: Noctuidae) na produção do algodoeiro. MSc thesis. Universidade de São Paulo, Piracicaba, São Paulo, Brazil.

Santos, W.J. (1999) Monitoramento e controle das pragas do algodoeiro. In: Cia, E., Freire, E.C. and Santos, W.J. dos (eds) *Cultura do Algodeiro*. Piracicaba, Potafos, Brazil, pp. 133–174.

Santos, W.J. dos (2001) Identificação, biologia, amostragem e controle das pragas do algodoeiro. In: Embrapa Agropecuária Oeste (Dourados) (ed.) *Algodão: Tecnología de Produção*. Embrapa Agropecuária Oeste/Embrapa Algodão, Dourados, Mato Grosso do Sul, Brazil, pp. 181–226.

Santos, K.B. dos, Neves, P.M.J. and Santos, W.J. dos (2004) Resistência de cultivares de algodeiro ao Virus do Mosaico das Nervuras transmitido pelo pulgão *Aphis gossypii* (Glover) (Hemiptera: Aphididae). *Neotropical Entomology* 33(4), 481–486.

Schuster, D.J. and Calderon, M. (1986) Interactions of host plant resistant genotypes and beneficial insects in cotton ecosystems. In: Boethel, D.J. and Eikenbarry, R.D. (eds) *Interactions of Plant Resistance and Parasitoids and Predators of Insects*. Ellis Horwood, Chichester, UK, pp. 84–97.

Schuster, M.F. and Maxwell, F.G. (1976) Resistance to two-spotted spider mite in cotton. *Mississippi Agricultural and Forestry Experiment Station Bulletin* 821, 1–13.

Shepard, M. and Sterling, W. (1972) Incidence of parasitism of *Heliothis* spp. (Lepidoptera: Noctuidae) on some cotton fields of Texas. *Annals of the Entomological Society of America* 65, 759–760.

Silva, A.G.A., Gonçalves, C.R., Galvão, D.M., Gonçalves, A.J.L., Gomes, J., Silva, M. do N. and Simoni, L. de (1968) *Quarto catálogo dos insetos que vivem nas plantas do Brasil-Seus parasitos e predadores*, Vol. 1. Ministério da Agricultura, Rio de Janeiro, Brazil, p. 2.

Silva, C.A.D. (2003) Eficiência de diferentes tipos de armadilha de feromônios na captura de adultos do bicudo do algodoeiro. *Congresso Brasileiro de Entomologia* 5–9 September 2003, Goiânia, GO, Brazil. Electronically published on CD-ROM.

Silva, N.M. da, Carvalho, L.H. de, Cia, E., Fuzzato, M.G., Chiavegato, E.J. and Alleoni, L.R.F. (1995) *Seja o doutor do seu algodoeiro*. Arquivo do Agrônomo 26. Potafos, Piracicaba, Brazil.

Silvie, P., Leroy, T., Belot, J.-L. and Michel, B. (2001a) *Manual de identificação das pragas e seus danos no algodoeiro*. Boletim Técnico No. 34, COODETEC/CIRAD-CA, Cascavel, Brazil.

Silvie, P., Leroy, T., Michel, B. and Bournier, J. (2001b) *Manual de identificação dos inimigos naturais no cultivo de algodão*. Boletim Técnico No. 35, COODETEC/CIRAD-CA, Cascavel, Brazil.

Sippel, D.W., Biindra, O.S. and Khalifa, H. (1987) Resistance to whitefly (*Bemisia tabaci*) in cotton (*Gossypium hirsutum* L.) in the Sudan. *Crop Protection* 6, 171–178.

Soares, J.J. and Almeida, R.P. de (1998) *Manejo integrado de pragas do algodoeiro, com ênfase aos efeitos colaterais dos pesticidas e o uso de controle biológico*. Embrapa – CNPA. Documentos, 62, Campina Grande, Brazil.

Soares, J.J. and Lara, F.M. (1993) Influence of cotton genotypes on levels of boll weevil, *Anthonomus grandis* Boh., parasitism by *Bracon mellitor* Say. *Anais da Sociedade Entomologica do Brasil* 22(3), 541–545.

Sobrinho, F.P.C., Araujo, J.M. and Silva, M.B. (2003) Avaliação do sistema de cultivo do algodoeiro herbáceo integrado a indústria de beneficiamento ano 2002. *Congresso Brasileiro de Entomologia* 5–9 September 2003, Goiânia, GO, Brazil. Electronically published on CD-ROM.

Stipanovic, R.D., Altman, D.W., Begin, D.L., Greenblatt, G.A. and Benedict, J.H. (1988) Terpenoid aldehydes in upland cottons: analysis by analine and HPLC methods. *Journal of Agricultural and Food Chemistry* 36, 509–515.

Thomson, N.J., Reid, P.E. and Williams, E.R. (1987) Effects of the okra leaf, nectariless, frego bract and glabrous conditions on yield and quality of cotton lines. *Euphytica* 36, 545–553.

Treacy, M.F., Benedict, J.H., Lopez, J.D. and Morrison, R.K. (1987) Functional response of a predator (Neuroptera: Chrysopidae) to bollworm (Lepidoptera: Noctuidae) eggs on smoothleaf, hirsute and pilose cottons. *Journal of Economic Entomology* 80, 376–379.

Vieira, D.J., Nóbrega, L.B., Azevedo, D.M.P., Beltrão, N.E.M. and Silva, O.R.R.F. (1999) Destruição dos restos culturais. In: Beltrão, N.E.M. (ed.) *O agronegócio do Algodão no Brasil*. Embrapa, Brasília, Brazil, pp. 603–615.

Wetzel, C.T. (2002) Sementes de algodão no Centro-Sul (1988–2001). Unpublished report. Embrapa Transferência de Tecnologia, Brasília, Brazil.

Wilson, L.J. (1994) Resistance of okra-leaf cotton genotypes to two-spotted spider mites (Acari: Tetranychidae). *Journal of Economic Entomology* 87, 1726–1735.

Wilson, L.J., Sadras, V.O., Heimoana, S.C. and Gibb, D. (2003) How to succeed by doing nothing: cotton compensation after simulated early season pest damage. *Crop Science* 43, 2125–2134.

Wilson, R.L. and Wilson, F.D. (1976) Effects of nectariless and glabrous cottons on pink bollworm in Arizona. In: *Proceedings of the Beltwide Cotton Production and Resistance Conference 1976*, pp. 91.

3 Consideration of Problem Formulation and Option Assessment for Bt cotton in Brazil

D.M.F. CAPALBO, M.F. SIMON, R.O. NODARI, S. VALLE, R.F. DOS SANTOS, L. CORADIN, J. DE O. DUARTE, J.E. MIRANDA, E.P.F. DIAS, LE QUANG QUYEN, E. UNDERWOOD AND K.C. NELSON[1]

Corresponding author: Dr Kristen C. Nelson, Forest Resources/Fisheries, Wildlife and Conservation, University of Minnesota, 115 Green Hall, 1530 Cleveland Avenue North, St Paul, MN 55108, USA. Fax: +1 612 6255212, e-mail: kcn@umn.edu

Introduction

During the past decade, Brazil has been a hotbed of discussion over the benefits and risks of genetically modified organisms (GMOs). While critics question the contribution and safety of these organisms and proponents argue for their benefits, some scientists struggle to develop good evaluations of their potential beneficial and adverse effects. To contribute to the societal deliberation over GMOs, Brazil needs to develop an environmental risk assessment based on scientific information and analysis. In this chapter, we consider a model for problem formulation and option assessment (PFOA) as a cornerstone of environmental risk assessment. In a 3-day workshop session the authors were introduced to the idea of a PFOA in environmental risk assessment, considered the types of questions and characteristics of the process,

---

[1] The authors and the participants of the PFOA discussion group at the workshop acknowledge Dr Maria José A. Sampaio of the Embrapa Secretariat for Intellectual Property Rights and Dr Ana Lúcía Assad of the Ministry of Science and Technology at the time of the workshop, for their contribution of ideas during the workshop.

and made consensus recommendations about the value of a PFOA. Of particular interest in the Brazilian case study, the authors support findings from previous publications (Nelson *et al.*, 2004), discuss how the PFOA might fit within existing Brazilian laws and regulations, and evaluate the staging of the PFOA as it blends with other components of a GMO's environmental risk assessment. The authors of this chapter have focused on science and environmental issues but recognize that others will need to be involved to understand how legal, ethical, social and economic issues can be included in risk assessment. The next step for a Brazilian team is to design the process for the PFOA based on a broader period of consultation and implement the PFOA within environmental risk assessment.

Since the global debate about the possible benefits and risks of GMOs has largely been framed by stakeholders from the developed world, governments in developing countries have found themselves in a dilemma. In some countries, a general apathy emerged and slowed adoption of GMO technology in agricultural systems. In many countries, political arenas provided the only avenue for debate, making it increasingly difficult to design effective and responsible GMO regulation that strikes a balance between safety and competitiveness, taking into account each country's unique context of society and ecosystems.

Certainly, transgenesis is a complex topic that embodies difficult technical, social and economic issues played out against a backdrop of human hunger, economic marginalization and environmental degradation (Traynor *et al.*, 2002). In Brazil, transgenesis has not enjoyed the same acceptance by consumers as it has in the USA or Argentina, where it is viewed by some sectors of society as a major tool to aid in the development of agribusiness. Scientific, legal, ethical, environmental, social and economic issues are hotly debated. Concerns are also associated with complicated issues such as labelling, market barriers and global commerce. Media information sources are full of contradicting opinions from scientists, government authorities and citizen groups. Within this context, it is very difficult for Brazilian consumers to understand or trust any given comment or decision (Nutti *et al.*, 2004, among others).

As a rule, the development of any novel technology emphasizes market issues. The current Brazilian situation indicates that the importance of safety issues associated with transgenesis will influence the path of development of transgenic crops; this path will follow a different route than technologies in other industrial sectors. Development of agricultural transgenic products will require consideration of other aspects, such as precise information about this new technology available to the consumer, with trustworthy scientific analysis as a reference point in the process.

Although experience has been acquired with products developed and commercialized in other countries, safety protocols should be developed and/or adapted to local conditions. In some countries there are limited monitoring and postharvest results (Department of Agriculture, Western Australia, 2004), but in most countries we lack this information. To ensure environmental safety, it will be necessary for each country to establish its own protocols, since results may differ from those already obtained in other regions of the world. Food and Agriculture Organization (FAO) officials emphasize this point in the following statement:

> FAO supports a science-based evaluation system that would objectively determine the benefits and risks of each individual GMO. This calls for a cautious case-by-case approach to address legitimate concerns for the biosafety of each product or process prior to its release. The possible effects on biodiversity, the environment and food safety need to be evaluated, and the extent to which the benefits of the product or process outweigh its risks assessed. The evaluation process should also take into consideration experience gained by national regulatory authorities in clearing such products. Careful monitoring of the postre-lease effects of these products and processes is also essential to ensure their continued safety to human beings, animals and the environment.[2]
> (2004, p. 225)

The report goes on to recognize that there are few ecological studies in tropi-cal areas and that more research will be necessary to address ecological risks in these areas. Brazil should review existing studies, using the findings when appropriate and establish its own robust research programme when necessary.

In 2003, a governmental decision unified the positions of the agencies regarding the necessity of health and environmental risk assessment analysis prior to large-scale GMO use. The prospect of increasing profits in major agri-cultural-commodity markets has prompted multinational companies to invest heavily in the acquisition of seed companies located in large Latin American countries such as Brazil. Consequently, major genes of economic importance that were used during the first wave of transgenic products in the USA and Europe are being transferred to Brazilian tropical varieties and are being experimentally tested in the country in accordance with the biosafety law passed in 1995 (Fontes, 2003). In June 2003, only one herbicide-tolerant (HT) soybean event had been approved for two planting seasons (2002/03 and 2003/04); approval was granted by the Brazilian National Congress because the decision-making authority of the National Technical Biosafety Commission had been challenged judicially. This Congressional approval was a temporary emergency measure to address the problems originating from illegal plantings of HT soybeans while the judicial challenge was pending.[3]

The focus of this chapter is motivated by two important concerns: (i) concern that premature release of GMO varieties will preclude critical steps

---

[2] There is an active debate about whether anything can be or should be considered 'objective'. Stakeholder values are specifically not 'objective' and yet they are impor-tant components of the societal evaluation of risk. The authors have a worldview that science has something valuable to contribute to societal discussions, and to risk assessment in particular. The PFOA process is designed to include the broader soci-etal discussion but enhance the way science informs this discussion.

[3] The production and marketing of HT soybean has been authorized in the new biosafety law, as well as the planting of saved seed in the season 2004–2005. See Background on Brazil Biosafety Legislation for GM crops (this chapter).

in environmental risk assessment; and (ii) the desire of Brazilian scientists and regulators for a carefully designed 'best science model' for the wise management of transgenic products. The challenge for chapter authors was to recognize and understand the conflicted nature of the GMO discussion in Brazilian society, while finding a common ground from which it is possible to move forward. In the PFOA model, multi-stakeholder participants have to formulate the problem that may be addressed by the GMO, as well as understand its scale and priority for society. Once this is accomplished the group has to compare possible options for solving the problem and evaluate the relative merit of the GMOs.

In Brazil, the science to assess GMOs is just beginning to be established as a foundation for review. As an example of one argument in the debate, Contini *et al.* (2003) used a simulation model to assess the potential economic benefits to the Brazilian economy resulting from unrestricted commercial approval of HT soybeans and other key crops. Model results indicated that economic benefits of legally adopting glyphosate-tolerant soybeans may be very high, and may also be significant for other transgenic crops such as maize and cotton. Contini *et al.* (2003) concluded that even though the country is fully capable of developing the necessary food-, feed- and environmental-safety tests, political and ideological reasons have prevented any legal commercialization of transgenic crops. Conversely, other scholars (Pelaez *et al.*, 2004) concluded that transgenic HT soybeans are not more profitable than conventional soybeans. Overall, there are still insufficient scientific data to reach a common conclusion.

Some stakeholders believe government agencies in Brazil should have taken the lead to clarify legal requirements for research and commercial release of GMOs. In media reports, journalists highlight the concern that it was difficult for Brazilian researchers involved with biotechnology to accept the state of chaos and confusion created by the disputes between pro- and anti-GMO campaigners, while the illegal use of GM seeds emerged (Bonalume Neto, 2003). Clarifying regulations is the main objective of the new biosafety law.

While the societal discussion continues, farmers are making decisions about their production systems, and often pursue technologies that give them the best agricultural and economic returns. However, some stakeholders argue that a balance between environmental and human health concerns and competitive agricultural practices, including use of transgenic crops with high economic return, must soon be reached to avoid the growth of illegal planting areas (not only as occurred for glyphosate-tolerant soybeans). Another alternative small farmers are increasingly adopting is the agroecological approach,[4] in which transgenic varieties are not part of the system. Compared with the traditional, conventional agricultural system or with systems that use transgenic varieties, the agroecological one can be more envi-

---

[4] In this case, the agroecological approach refers to a philosophy and system of farming that uses ecological principles, such as natural enemies, to guide decision making in production for a more sustainable system.

ronmentally safe, because it strives to reduce the use of pesticides, and other negative environmental impacts such as soil erosion and nutrient runoff (Altieri, 2000). Recently, the 7th Conference of the Parties of Convention on Biological Diversity, held in Kuala Lumpur, 2004, approved a programme of work on the ecosystem approach to be adopted by parties across the world (CBD, 2004).

Despite the urgency of the scientific assessment to support safe agricultural technology, scientists want to carefully design appropriate research protocols and regulatory processes that will be received with confidence by decision makers and the general public. In a field with tremendous uncertainty, testing regimes can serve as a framework for improved understanding and analysis for public discussion. But how can each country come to its own agreement on an acceptable risk-assessment strategy? Critics argue that there has already been too much done with risk assessment. Brody (2002) argues that the money invested around the world in compliance and enforcement of biosafety rules, regulations and laws might be better spent in ways that would save many more lives. Others think we have only begun to understand what will be necessary in risk assessment. Some stakeholders argue for international standards in environmental protocols that would, at a minimum, harmonize requirements for similar cases and lead to better understanding of risk assessment, communication and management of GMOs. Such standards and assessment would differ dramatically from the more commonly known environmental-impact analysis done before construction of roads, dams or buildings.

The GMO Guidelines Project entered the Brazilian discussion with a workshop in June 2003, designed to encourage discussion among international and Brazilian scientists and regulators.[5] The purpose of this discussion was to develop recommendations for environmental risk assessment protocols and processes that would serve as the basis for public deliberation and sound national decision making. This chapter represents consensus of the authors based on the workshop and subsequent discussions.

PFOA Method[6]

PFOA is one cornerstone of GMO's environmental risk assessment. Countries must create a responsive system to facilitate socially acceptable choices (Stern and Fineberg, 1996). At its core, the discussion focuses on the critical societal need[7] that will be addressed by the GMOs, i.e. which needs will be satisfied and at what risk? Societal needs and risks require societal reflection. Countries

---

[5] For more on the scope, purpose and sponsorship of the GMO Guidelines Project, see the preface or visit the project website: http://www.gmo-guidelines.info
[6] The PFOA method section has been modified from the original methods text in Nelson *et al.* (2004).
[7] Societal benefits are distinct from societal needs in that they could be any positive contribution to the broader society, such as increased income or more choice, and based on different values.

accept the possibility that something negative might happen with most major policy decisions but they never do so lightly. A deliberative process with multi-stakeholder participation allows members of society to participate in the evaluation of critical needs and risks. A cross section of society – farmers, consumer groups, industry, environmental representatives, policy makers, etc. – must have a vehicle to express their concerns and evaluate the future alternatives for addressing basic needs. Finally, this deliberative process[8] will become increasingly important for resource-scarce nations if public investment is involved, because a comparative reflection by a cross section of society may be beneficial in prioritizing and targeting resources.

Given present uncertainties surrounding GMOs and regular reports of new discoveries about GMOs, the system to conduct a deliberative discussion should be flexible and able to respond to a society's core values, concerns and needs. At the same time, the discussion is best served if it is driven by sound, scientifically guided assessment and review (Gibbons, 1999; CBD, 2000; NRC, 2002). A robust environmental risk assessment clearly delineates when scientific knowledge, information and analysis can effectively respond to key questions and when it cannot.

In most natural-resource arenas, practitioners and scholars are implementing and evaluating diverse options for societal discussion about critical issues (O'Brien, 2000; Wondolleck and Yaffee, 2000). Specific contributions within the biosafety arena have detailed potential approaches to multi-stakeholder dialogues and participation (Skorupinski, 2001; McLean *et al.*, 2002; Glover *et al.*, 2003, Kapuscinski *et al.*, 2003) and the evaluation of their key attributes (Irwin, 2001) at international and national scales. Societal discussion is advocated in setting priorities and strategies for agricultural research and development, in formulation of national biosafety frameworks and in environmental risk assessment of specific biotechnologies. A PFOA is applicable to all of these contexts, and can be employed in an iterative fashion to incorporate feedback from changing societal values and the scientific state of the art. In the Brazilian *Bacillus thuringiensis* (Bt) cotton case study presented in this chapter, PFOA creates the context for societal dialogue concerning the potential use of a proposed technology based on a transgenic organism such as Bt cotton. It is public deliberation about the transgenic organism that provides a rational, science-driven planning process by which multiple stakeholders can assess their needs, evaluate risks related to multiple future options, and make recommendations to decision makers about policies to reduce societal risks and enhance the benefits provided by various options.

Certainly the requirements set forth to accomplish a PFOA is a complex process – but this should not be considered as a possible argument against its use. Most important societal decisions are complex, messy and controversial, and because they are complex, messy and controversial we should work to improve them with transparent, systematic and scientifically based discussion. By doing so, the decision-making process gains social legitimacy and society

---

[8] In this case a deliberative process refers to one based on an 'organized or careful discussion and debate'.

gains greater confidence in the decisions taken. A PFOA can be used to provide such a discussion, and may play a significant role in environmental risk assessment.

Relation of PFOA to environmental risk assessment

Practitioners and scholars have tested numerous techniques that serve as a methodological foundation for the PFOA in environmental risk assessment (Grimble and Wellard, 1997; Kessler and Van Dorp, 1998; Schmoldt and Peterson, 1998; Biggs and Matsaert, 1999; Loevinsohn *et al.*, 2002, to name a few). Two crucial steps in risk assessment are addressed by many of these techniques, and this PFOA model is designed specifically to address them. The first critical step in risk assessment is problem identification (NRC, 1983, 1996). What is the scope of the problem, how is it defined? Problem identification frames the entire risk assessment. A second critical step is the identification of potential alternative solutions to the problem (NRC, 1983, 1996). The proposed action – in this case, the use of Bt cotton in Brazil – is never the only possible way to address the problem. Risk assessment depends entirely on an appropriate specification of alternatives (including taking no action and doing nothing), so that comparative risk can be assessed, and appropriate controls for risk assessment science can be defined and used (see other chapters in this volume).

This PFOA model is comprised of specific brainstorming discussion and analytical components (Table 3.1). First, formulating the problem serves as the core foundation. The problem is defined as an unmet need that requires change (Goldstein, 1993). Basic human needs are most commonly identified as food, shelter and safety. Other human interests are stakeholder-specific such as enhanced economic opportunity, positive social interactions or cultural richness. For example, as a minimum foundation for well-being, individuals have a basic need for a certain amount of calories per day or the security that their children will continue to live healthy lives. Once the needs for food, shelter and safety are met, individuals can expand their interests to include numerous options for well-being. These interests will differ from one individual to another and from one group to another.

After a problem is identified, the PFOA model requires a comparative approach to risk assessment. The participants clarify the relative importance of this problem as compared to other problems or issues. Once the group agrees the problem is sufficiently important to merit an analysis, the range of future alternatives for solving the problem are compared in relation to their attributes, potential ability to address the problem, changes required to implement the option and potential adverse effects. The PFOA is planning for alternative futures, not for the current conditions against one option; rather, the PFOA makes a comparative assessment of alternative futures. After a complete analysis by a multi-stakeholder group, a recommendation is made to decision makers to continue research and development (in some cases risk-assessment research) with the technology or to halt the development of the technology.

Table 3.1. Problem formulation and options assessment process.

Initiating proposal
A. *Proposal to use GMOs*
Any PFOA for transgenic organisms will be initiated by the request or suggestion that a particular GMO would be a beneficial alternative to the way things are currently being done in a particular cropping system.
B. *Decision by regulatory body*
Is there merit to moving forward to evaluate the GMO as a possible option or is the initiating proposal premature? Yes/No

PFOA process: questions to be answered by all representatives and shared in the deliberative process

Step 1: Problem formulation

Formulation of problem	Basic human needs	Interests
An unmet need that requires change	Food, shelter, safety	A stakeholder group's values, goals and perspectives

A. *Whose problem is it? Whose problem should it be?*

1. What needs of the people are not being met by the present situation?
2. What aspects of the present situation must be changed to meet the needs?

Step 2: Prioritization and scale

A. *Is this problem a core problem for the people identified?*

1. Do the people recognize the problem as important to their lives?
2. What are the potentially competing needs of these people?
3. How do the identified needs rank in importance to these other competing needs?

B. *How extensive is the problem?*

1. How many people are affected?
2. In what part of the country are these people located?
3. How large an area is affected by the problem?
4. How severe is the problem (local intensity)?

Step 3: Problem statement

Shared understanding of the unmet need and its relative importance for a particular group of people.

Step 4: Recommendation by regulatory body

Do we move forward to identify options and conduct and options assessment?

Table 3.1. Problem formulation and options assessment process – *cont'd.*

Option identification and assessment

Step 5 Options	Step 6 Characteristics	Step 7 Changes	Step 8 Effect on the system
Future alternatives	For problem solving	Required/Anticipated	Internal External (Social, environmental, economic)
Option A Option B Option C Etc.			

Step 5: Option identification

Brainstorm possible future alternatives to solve the identified problem; transgenic organisms would be one option. This step can be completed by the multiple stakeholder group for the initial identification of options. The multi-stakeholder group can do steps 6–8 or a technical committee can develop a report that covers steps 6–8 and the multi-stakeholder group can use the document to begin their evaluation of options and modify the assessment.

Step 6: Assessment of the options in relation to the problem

Assessing capability of potential solution to solve problem.

1. What are the characteristics of the 'technology' option i.e. transgene, intercropping system?
2. What is the range of crop-production systems and what is the geographical region the option is likely to be used in or have an effect on?
3. What is the efficacy of the 'technology' on the target?
4. What are the costs of the technology within the crop-production system?
5. What barriers to use exist? For instance, is the seed-distribution system in place; can the potential solution be integrated into present production; can the farmers afford the potential solution?
6. How might the use of the option change cropping practices, such as tillage systems or pesticide use (including impacts on non-target pests)? What useful practices are reinforced by the potential?
7. What information is needed to show that the changes are likely to occur? Baseline data associated with the diversity of present practices should be used if it is available.
8. How will anticipated changes in agricultural practices affect the needs identified in steps 1 and 2?

Step 7: Changes required and anticipated for a specific option

1. What changes in *farm management* practices might contribute to the solution?
2. What changes in the *local community* might contribute to the solution?
3. What changes in *government support* for farmers might contribute to the solution?
4. What changes in the *structure of agricultural production* might contribute to the solution?
5. What other changes would likely be *needed to facilitate widespread use* of this option?

Continued

Table 3.1. Problem formulation and options assessment process – *cont'd.*

Step 8: Adverse effects

Potential adverse consequences from this option. Potential beneficial effects can be considered 'negative' adverse effects.

1. How might the potential solution affect the structure *of agriculture or agricultural infrastructure*?
2. How might the potential solution *reinforce poor agricultural practices or disrupt useful practices*?
3. What are the *potential adverse effects* of these changes internally and externally to the production system?
4. How will its use affect other nearby *crop-production systems* and non-agricultural environments (can its use be restricted to a particular cropping system or geographical region)?
5. Are any of these changes *difficult to reverse*, once they occur?

Step 9: Recommendation

The multiple stakeholder group should present their problem formulation and option assessment to the appropriate decision-making body.

A science-driven PFOA must be a deliberative process (Forester, 1999) designed to provide for social reflection and discussion about transgenic organisms. A sound deliberative process is transparent, equitable, legitimate and data-driven when possible (Susskind *et al.*, 2000). Transparency allows for open communication of information between all parties and easily accessible reporting of decisions to the public (Hemmati, 2002). Providing an equitable PFOA process means that information from the broadest spectrum of society must be included with all stakeholders having the possibility to contribute. Civil society must perceive that there are sufficient avenues for input and consideration of diverse viewpoints and concerns. When transparency and equitable input are central to the process, the PFOA gains legitimacy in the public eye. This public legitimacy must be matched by traditional legitimacy or sanctioning by a formal political body that embeds the deliberative process. The deliberative process can be tied to a regulatory authority or legislative authority, but it must provide a means by which results from the PFOA inform government decision-making and action. Finally, the foundation of PFOA is a science-driven inquiry – in both environmental and social sciences. Questions are answered with data, impacts are assessed with valid indicators, and the limits of our understanding are clearly delineated by a research agenda or procedures for taking uncertainty into account.

Each country will need to develop a country-specific deliberative process that fits the particular structure and authority of the relevant decision-making bodies and implementing agencies. For many political systems in the world, the legitimating authority exists to incorporate PFOA in a legislative or regulatory context, but there are debates about necessary modifications of policies and regulation for transgenic organisms (Munson, 1993; Miller, 1994; Hallerman and Kapuscinski, 1995; Sagar *et al.*, 2000; NRC, 2002). For some legislative or regulatory situations, a PFOA can be incorporated into a public consultative process that is authorized by regulation or it may be added as an alternative process supported by civic society that informs the debate in traditional decision-making bodies.

The following sections describe the regulatory background in Brazil, in order to facilitate the consideration of how a PFOA would best fit in the existing system in Brazil, and document a trial run of the PFOA model proposed for this risk assessment approach. It is not a PFOA for Bt cotton. It is an evaluation by the authors of the concepts and protocols for this PFOA model.

Background on Brazilian Biosafety Legislation for GM Crops

In an effort to assure society that care and precaution will be taken to warrant both environmental and human health safety of all food/feed produced from GMOs, Brazil has been working, for nearly a decade, to enact regulatory legislation. A brief report of the accomplishments so far is as follows.

Past Brazilian biosafety legislation and the pressure for change

Law no. 8974 of 1995: The superceded Brazilian Biosafety Law for GM crops involved regulatory agencies under three ministries. Each agency granted both special temporary testing permits and final licensing for commercial products according to their own rules and protocols. In addition, in the case of plants containing biopesticide traits, specific pesticide-control legislation also applied (Law no. 7802, of 1989),[9] and Resolution 305 (June 2002) of the National Environmental Council (CONAMA) stated that GMO research and development are subject to the requirement of environmental impact studies – a preliminary step to any polluting activity.

The National Technical Committee for Biosafety (CTNBio) under the Ministry of Science and Technology, was given responsibility for assessment of health and environmental risks posed by GMOs. However, CTNBio's legal capacity to perform as stated in the Provisional Measure (PM)[10] in 2001 has been repeatedly challenged judicially (the case in point was a CTNBio decision concerning glyphosate-tolerant soybeans). This challenge ended in August 2004, with two judges voting to uphold and one voting to deny the CTNBio legal authority.

---

[9] Pesticide Law no. 802/1989 and Decree no. 4.074/2002 available at http://www.planalto.gov.br

[10] A Provisional Law Measure is a legal entity used by government when it needs any given law to go into force before it is submitted to the congress for approval. It is valid for 90 days and has to be voted and approved.

Difficulties of the regulatory agencies in defining their own rules and protocols, coupled with the legal entanglements concerning CTNBio's mandate, resulted in a nearly complete halt of GMO research projects of plants producing biopesticide products from 2001 up to 2003. Particularly affected was research on virus-resistant GM plants. Confronted with this research crisis, the agencies sped up their protocols, and in September 2003, Brazilian Agricultural Research Corporation (Embrapa) was licensed to plant a GM papaya field trial in the north-eastern state of Bahía. In 2004, Embrapa was licensed to field test GM beans and GM potatoes. In all three cases, the transgene was designed for virus resistance. After these, private companies obtained licences to conduct similar research.

Resolution 305 (June 2002) of the National Environmental Council (CONAMA)[11] stated that GMO research and development are subject to the requirement of environmental impact studies – a preliminary step to any polluting activity.

Labelling: Decree no. 4680 (April 2003) guarantees the right of information – as established in the Brazilian Consumers Defence Law (Law no. 8.078 of 1990) – compelling the labelling of every food that contains more than 1% of a GM ingredient. It also determines that animals that have been fed with transgenic grains, as well as the products prepared with the animals' meat, milk and eggs must be labelled. Practical difficulties in Brazil will make this labelling statute difficult to implement effectively.

The new biosafety law

Dissatisfaction with the ambiguities in the present legislation and the consequent difficulties to research led the government to submit a new bill to Congress, in November 2003, which was passed and signed by the president on March 24 2005.[12]

The new law covers licensing for laboratory, greenhouse, and field experiments, and applications in the areas of agriculture, environment and health. It assigns authority to CTNBio[13] for issuing technical opinions, case-by-case, on both research-related activities and the commercial use of GMOs and their by-products. The CTNBio opinion is forwarded to the registration and inspection agencies – SDA (Plant Protection Secretariat, Ministry of Agriculture), IBAMA (Brazilian Institute for the Environment, Ministry of Environment), and ANVISA (National Agency of Sanitary Control, Ministry of Health). The agencies are responsible for inspection and monitoring in their respective areas. They are

---

[11] Available at: http://www.mma.gov.br/port/conama/index.cfm
[12] An English translation of the law is available at: http://www.ctnbio.gor.br/index. php?action=/content/view&cod_objeto=1296
[13] The new CTNBio will be composed of 27 experts: 12 scientific experts in the areas of human health, animal health, plant health and environment, nine Ministry representatives, including the three regulatory Ministries, and six other experts chosen by specific ministries for specific areas of expertise.

bound to CTNBio's opinion, and in the case of research, CTNBio is expressly given final authority. However, whenever there is dissension regarding CTNBio's technical opinion on commercial use of a GMO, the registration and inspection agencies can appeal. The final decision on commercialization then lies with the new National Biosafety Committee (CNBS), made up of 11 ministers and subject to the Office of the President of the Republic. The CNBS can also, upon CTNBio's request, analyse the socio-economic convenience and opportunities and national interest entailed in the requests for commercial use.

At the time of writing, the rules governing the operation of the expanded CTNBio and the CNBS are still being discussed and it is unclear how the law will function. One recent decision taken by the old CTNBio related to importation of Bt maize varieties for animal feed[14] has been suspended by the CNBS, allowing just the three events analysed by CTNBio (gen *Cry1Ab*, *Cry1Ac* and *Cry9C*, for insect resistance; gen *pat* and *bar*, for glyphosinate herbicide tolerance; and gen *mEPSPS*, for glyphosate herbicide tolerance) to be imported.[15] Two other decisions of the old CTNBio that required analysis by CNBS have not yet been delivered: on commercialization of Bollgard Bt cotton and transgene contamination limit in cotton seed.[16]

Provisional measures

The increase in illegal planting of GM soybean and the delays in the judicial decision prompted the government to issue two PMs: PM 113 and PM 131. The first one was issued specifically to legalize marketing of GM soybean harvested in the early months of 2003, while the second was to legalize, for only 1 year, the 2003/04 planting and marketing of transgenic soybean seeds retained from the previous harvest. The production and marketing of HT soybean has been authorized in the new biosafety law, as well as the planting of saved seed in the season 2004/05. Since then, various HT soybean varieties have been brought onto the market in Brazil.

A consensus proposal

At another scale, the international forecast for the implementation of the Cartagena Protocol on Biosafety, which came into force in September 2003, but really gained momentum after the first meeting of the parties that took place in Malaysia, February 2004, implies that all signers will need to have risk-assessment regulations in place. Brazil became a party to the Cartagena Protocol in February 2004 and has not yet developed all the procedures required for export and import of GM products.

[14] Available at: http://www.ctnbio.gov.br/index.php?action=/content/view&cod_objeto= 1305 (English version) or http://www/ctnbio.gov.br/index.php?action=/content/view& cod_objeto=1299 (complete, Portuguese version)
[15] Source: Technical note from Dr Mônica Amâncio, lawyer.
[16] Source: http://www.prdf.mpf.gov.br

Bt cotton in the Brazil regulatory system

Regarding government decisions specific to the present case study, Bt cotton varieties owned by multinational companies were granted approval from the CTNBio for field trials beginning in 1997, in various regions of Brazil. These varieties produce the proteins Cry1Ab, Cry1Ac and Vip3A, and one variety contains both Cry1Ab and Cry1Ac. In 2000, a court injunction suspended these tests until they were granted approval by other Brazilian regulatory agencies. During the years 2001–2005, the companies did not receive approval for new field trials. On March 18 2005, the old CTNBio made a recommendation to commercialize Bt cotton containing Bollgard event 531. Their decision has been suspended by the CNBS and is being examined.

Test Run for the PFOA Model in the Brazil Case Study for Bt Cotton

The authors evaluated the PFOA model by discussing its purpose within an environmental risk assessment, testing a few questions from each step in the model to experience the type of discussion that might result from a multi-stakeholder exchange (Table 3.1), and deliberating over how a PFOA would best fit in the Brazilian regulatory system. The authors then summarized their findings about the PFOA content and process within the context of Brazil's deliberation over Bt cotton.

After discussing the questions in steps 1 and 2 (Table 3.1), the authors came to a consensus on the following problem statement for cotton pest problems.

Problem statement

During periods of high infestation, Lepidopteran pests cause yield reductions in cotton, increasing the number of insecticide applications and consequently increasing the cost of cotton production, possibly affecting the health of farmers and people who are involved with the crop, also causing environmental pollution in the soil and water systems. Reduced stability of yield makes planning more difficult and risky.

In the North-east, for small-scale farmers, the boll weevil damages cotton during the rainy season (March–May). Lepidopteran pests are not a major problem.

In the Midwest and Meridian regions, Lepidopteran pests are a moderate to major problem.

Continuing the test run of the PFOA model, we conducted a brainstorming session about possible options for addressing the problem. The options included Bt cotton, insecticide applications, biological control with *Trichogramma* wasps, integrated pest management (IPM) packages and organic packages. We selected Bt cotton, biological control (wasps) and insecticide applications as options used to evaluate questions in steps 6–8. Overall, our conversation jumped around a great deal, but participants developed a sense of the type of discussion produced by the questions and where the difficulties would emerge.

Evaluation of a Brazil PFOA

After reviewing the PFOA model and discussing the Kenya case study, the Brazilian authors agreed with their Kenyan counterparts on five key findings (Nelson *et al.*, 2004) and refined them in several aspects.

Finding #1: PFOA is a good idea for any agricultural technology but critical for GMOs. It should be done taking into consideration a precautionary approach on a case-by-case basis.

Finding #2: PFOA proved to be particularly useful for encouraging constructive dialogue and potential agreements.

Finding #3: For a successful PFOA, a nation should have in place a regulatory framework that would allow for the reduction of uncertainty about GMOs. Consideration of the PFOA could help to initiate discussions about possible regulatory options.

Finding #4: The discussion of a case study provides applied insights about key issues and consensus building.

Finding #5: Additional questions will strengthen the PFOA: what loss has occurred as a result of the problem in productivity as well as environmental, social and economic aspects? How will the technology's use affect the environment? How will use of the technology affect the conservation of genetic variability of the species and other related biodiversity?

We find there is a demand for a new way to host societal discussions, just as scholars have documented the need for creative approaches to critical natural resource-problem solving and risk assessment in other countries (O'Brien, 2000; Wondolleck and Yaffee, 2000). The use of a PFOA process in environmental risk assessment for GMOs provides a framing of the risk-assessment stages, allowing diverse stakeholders in the PFOA to propose issues that need to be evaluated in the risk-assessment stages, as well as serve as a multi-stakeholder discussion that can use the scientific results from the assessment to inform their evaluation of the options. The uncertainty associated with GMOs as an alternative make a PFOA process critical, so multiple stakeholders can contribute their views and science is used to inform the discussion.

Due to the nature of the GMO discussion, and the absence of sufficient scientific biosafety data in Brazil and monitoring data after release of new products or techniques in other countries, several authors believe it would be necessary to take a precautionary approach for each GMO technology. A precautionary approach, as represented in European regulation, balances between the need to take action and scientific uncertainty (Skorupinski, 2004), with the philosophy that if there is a 'reasonable assumption of possible harm' then the government should take measures for protection (CEC, 2000). Certainly, Brazil will need to develop its own understanding of the precautionary approach.

In particular, the options assessment would have to include more specific questions about environmental risks and a well-documented analysis of the crop losses a new technology would address. The adoption of a new technology might cause changes in environmental and societal aspects, which are not necessarily explicit. Purely economic questions should not obscure other aspects related to the problem. We must also consider the entire ecosystem involved, as well as the consequences for human health and well-being. For example, in the past some newly introduced technologies in a production system (e.g. chlorofluorocarbons (CFCs), pesticides) largely served to produce economic gains for multinational companies, but without regard for other consequences that could alter the components of the water–soil–atmosphere system and its relationship with living organisms.

The authors (a diverse group of ministry officials) believe a PFOA model would be useful for constructive dialogues, new insights and the consensus building necessary for possible agreements in Brazil, a finding supported by studies of multi-stakeholder dialogues in other countries (Irwin, 2001; McLean *et al.*, 2002; Glover *et al.*, 2003). Even in the highly politicized debates about Brazil's use of GMOs, we see some promise in a guided discussion about the range of issues in the societal consideration of a new genetically engineered product.

National governments are often critical actors for facilitating socially acceptable choices (Stern and Fineberg, 1996). Fundamental to the conceptualization of a Brazilian PFOA is the understanding that it has to be embedded in the regulatory systems in such a way that it reduces uncertainty for all members of society. At its best, it would serve as a forum for all stakeholders to understand the problem that needs to be addressed and the comparative impact of different options for solving the problem. In the Brazilian context, the PFOA would have to be organized by government authorities, but provide a forum that is viewed as legitimate by the multiple stakeholders. Representatives of these diverse stakeholders would have to have a voice and believe they had some influence in the assessment of options. This argument for transparency, equity, legitimacy and a data-driven process is well supported in the literature (Susskind *et al.*, 2000; Hemmati, 2002).

Many countries are considering modifications of policies and regulation for transgenic organisms (Munson, 1993; Miller, 1994; Hallerman and Kapuscinski, 1995; Sagar *et al.*, 2000; Burachik and Traynor, 2002; NRC, 2002; UNEP-GEF, 2003a,b). Because Brazil already has a complex system of laws and regulations, the authors suggest the following broad guidelines.

Finding #6: The PFOA should be organized by government authorities and discussed by a multi-stakeholder group.

Finding #7: In Brazil, the PFOA should be embedded in the policy and regulatory process.

Finding #8: In Brazil, PFOA staging can involve two meetings.

One option for incorporating the PFOA in the existing system would be to associate it with the CTNBio, but some modifications will be necessary (Figs 3.1 and 3.2). In 2003, the members of CTNBio were predominantly biotechnical scientists. In the future, this membership should be broadened

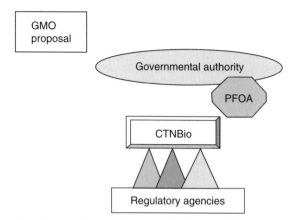

Fig. 3.1. Suggestions for embedding the problem formulation and option assessment (PFOA) in the Brazilian regulatory system. Three ministries (Environment, Agriculture and Health) and the National Technical Biosafety Commission (CTNBio, Science and Technology Ministry) play a role in the final decision for environmental release of a genetically modified organism (GMO).
Stage 1. Based on GMO proposal, the first PFOA meeting is facilitated by government authorities and carried out by a multi-stakeholder group.
Stage 2. CTNBio receives the registrant's proposal and the PFOA report. For a review of this report a composition of the CTNBio should be broadened in scientific expertise and stakeholders.
Stage 3. If the decision is favourable, it is sent to the ministries of Environment, Health, and Agriculture.
Stage 4. The proponent asks for respective licences from the ministries for laboratory, small- and large-field trials.
Stage 5. PFOA second meeting called to review the assessment in light of new results from testing.
Stage 6. The appropriate ministries review the PFOA report, promote public comment period, and may ultimately authorize for commercial use.

in scientific expertise in order to have the capacity to evaluate the results from a PFOA report that includes economic, social and environmental recommendations.

The CTNBio would receive the registrant's proposal of a transgenic organism; this may come from a private company or federal research institute such as Embrapa. CTNBio would decide if there is sufficient consideration for biosafety in the proposal to consider it for use in Brazil. If the proposal has been accepted for consideration, the CTNBio would convene the PFOA multi-stakeholder group for a first meeting to consider problem formulation (steps 1–2), and an initial review of options assessment including information needed for a full consideration of steps 6–8. The CTNBio would receive the registrant's proposal and the PFOA preliminary report and decide on whether to continue environmental risk assessment. If the decision is favourable, it is sent to the ministries of Environment, Health and Agriculture. The proponent asks for respective licences from the ministries for laboratory as well as small- and large-scale field trials. Near the end of field trials, the second PFOA meeting is called to review the assessment in light of new results from testing. The report will define the problem and its scope and compare the options. It will represent the degree of consensus among the stakeholders on important issues as well as the range of opinions expressed. Finally, the appropriate ministries review the PFOA report, conduct a public hearings session, and decide if the GMO should be authorized for commercial use. In all cases, the PFOA meetings would be guided by a pre-determined process, but sufficient flexibility needs to be allowed to respond to the unique characteristics of each case.

The staging of the PFOA process is particularly important because it must be designed to provide guidance for other sections of the environmental risk assessment, but at the same time the PFOA process needs to be informed by sci-entific results from the studies conducted by other sections (Figs 3.1 and 3.2). The first PFOA meeting could be held prior to risk-assessment studies to help understand the problem, provide a preliminary review of options and highlight critical societal concerns. Then studies can be conducted during the following stages in the development of a transgenic plant (Andow *et al.*, Chapter 1, this volume): (i) the design of transformation; (ii) focusing on set of events; (iii) characterization (laboratory and greenhouse); (iv) small-scale planned field trial; and (v) precommercialization. The second PFOA meeting should be con-ducted when laboratory and field environment biosafety test results can be included in the process.

The societal discussion about critical issues, specifically GMOs, is best served if it is driven by scientifically guided assessment and review (CBD, 1992; Gibbons, 1999; NRC, 2002), as well as consideration of the social, economic and ethical concerns. This is also true for any PFOA process. Brazilian work-shop participants emphasized that existing data and studies required by the environmental risk assessment would be necessary for a scientifically guided review. This also complies with the Cartagena Protocol on Biosafety.

Brazil has the scientific capacity to support a PFOA process and, if funds are allocated, the resources to conduct the multi-stakeholder dialogue. As the process is developed, there will always be a tension between efficiency in costs and time versus adequate coverage in representation and the extensive-ness of the deliberation. There will be start-up costs in developing the process and organizing adequate reporting systems for information. In Brazil, many agency staff and stakeholder representatives will require capacity-building assistance to facilitate or participate in a PFOA. Though Brazil may have the resources to contemplate a more elaborate process, other countries may need to develop more streamlined processes that still encourage discussion.

Finding #9: A database of existing studies should be organized and integrated to support PFOA discussion as needed.

Brazil has a wide variety of information sources to support steps 1–8 of the PFOA, but data about existing technologies in the agricultural sector are frequently spread among several institutions. This information would be very useful for questions about production problems and related issues, however, it is often hard to find or difficult to interpret. In other cases, the required

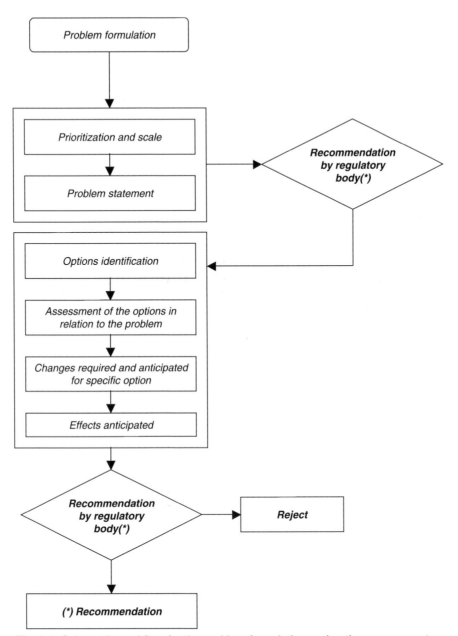

Fig. 3.2. Schematic workflow for the problem formulation and option assessment (PFOA) process. The regulatory body still needs to be defined by Brazilian authorities. (* indicates involvement of the recognized authority within the country. This authority receives the PFOA document and decides how to proceed.)

information is not available and it would be necessary to conduct additional surveys. Overall, for efficient problem formulation (steps 1 and 2) it would be necessary to establish a database where the information from several sources (census statistics, existing studies including economic, social, etc., papers and surveys) can be organized and integrated in a way that every person can understand it. When no studies exist, an inclusive research agenda should be developed to fill critical gaps using available resources, working to enhance the multi-disciplinary nature of data over time.

First, an inventory of all existing data sources will be required, such as the agricultural census, statistical yearbook, scientific database and botanical review. For example, the 1996 agricultural census produced by the Brazilian Institute for Geography and Statistics (IBGE) identified 93,688 rural establishments involved in cotton cropping in Brazil. In recent years, there has been a significant change in the location of the cultivated area in Brazil, with a shift from the North-east and Meridian regions to the Midwest and the western part of Bahía (Fontes *et al.*, Chapter 2, this volume). These changes had implications for the technological matrix and producer profiles, from the limited capital-intensive approaches of small farmers to the higher technological level of commercial farmers. Therefore, updated information is important for a PFOA analysis of Bt cotton.

In general, a literature review and new research efforts will be necessary at a regional level to provide information and analysis for consideration in steps 6–8. These studies will have to be conducted by a reliable scientific institution. A general producers' survey is recommended to identify needs prioritization and projections of demand. For Brazilian cotton, a recent survey in the Midwest and part of the North-east (west of Bahía state) will supply some data relevant for a problem formulation.

Some data can be obtained from the literature about contained environment research on GMOs. In addition, new research approaches and field data at the regional level have to be assessed to determine the barriers to use, environmental impacts (see Chapters 6 to 12, this volume), marketing issues and human health aspects of the GMOs, to name a few areas. For Brazilian Bt cotton, information about efficacy and cost are available from private companies, but these data will need to be reviewed by scientists in public institutions (Grossi *et al.*, Chapter 4, this volume). In the short term, published findings from abroad have to be assessed.

In the case of Bt cotton, gene flow is of particular concern for Brazil because there are three native cotton species in which cross hybridization and hybrid formation with cultivated varieties has already been documented. Further scientific studies will be necessary to evaluate introgression and genetic contamination effects (Johnston *et al.*, Chapter 11, this volume). In addition, Brazil hosts a rich biodiversity with more than a million insect species.

Finding #10: PFOA serves as a good foundation for future monitoring of environmental and societal impacts of the technology.

Once a particular option has been accepted by the decision makers, its implementation and impacts will have to be monitored to identify possible shifts in the environment and society. The participants recommended that the nation develop systems for monitoring new technologies, and consider the

PFOA model as an approach that can be adapted to postrelease monitoring as well as prerelease evaluation.

Next steps in the design of a Brazilian PFOA

In the 3-day workshop session the authors were introduced to the idea of a PFOA in environmental risk assessment, considered the types of questions and characteristics of the process, and made consensus recommendations about the value of a PFOA. The next step for a Brazilian team is to design the PFOA process based on a broader period of consultation and then implement the PFOA within environmental risk assessment. Brazil needs to develop its own PFOA process. There are many documented experiences of multi-stakeholder processes that can serve as a resource bank, as well as numerous techniques to facilitate discussion and option assessment (Crowfoot and Wondolleck, 1990; Susskind *et al.*, 1999; to name a few). More details about how multi-stakeholder dialogues have been applied in the consensus-building process in other countries would help the designers' understanding of how the PFOA might be applied. As a foundation, it would be very helpful to have a handbook that synthesizes this literature in light of the PFOA process for environmental risk assessment of GMOs. Many of the tools are publicly accessible on websites and through publications, but they are dispersed across organizations, agencies, countries and continents. A training handbook could guide the Brazilian design team through the major design issues and link them to ways others have addressed these issues. It would also serve as a reference as other countries develop their own PFOA for environmental risk assessment, or as a support tool for training regulators and stakeholders in the process. Overall, PFOA may be one component of a broader public-participation strategy for biosafety issues.

There are several critical issues that will need to be addressed while designing the Brazilian PFOA process. The keys to building a credible consensus on a science-intensive policy question are: (i) perceptions about the fairness of the method used to select stakeholder representatives who are invited to participate in a multi-stakeholder dialogue; (ii) the management (by a professional 'neutral') of the complex give and take involved in such a process; and (iii) the preparation of a written report – embodying the discussion and degree of consensus that will inform decision makers. The following represents a few issues and questions that will need to be considered, but this list is by no means exhaustive.

Independence and legitimacy
In the case of GMOs, PFOA is a needed tool for orienting scientific analysis of risks, and for addressing the conflict of questions of interest related to the subject. As with any other method, however, it can be manipulated depending on how different tendencies and expertise are represented. As stated in the findings section, a government authority, possibly CTNBio, should convene the PFOA, but there are still many decisions to be made about how to identify and invite stakeholders. The goal should be to balance the interests of all relevant stakeholders

with the need to generate a credible analysis of the parties involved (Crowfoot and Wondolleck, 1990). Also, its results and reports can be used to inform and help decision makers to make decisions based on scientific analysis of the environmental, social, ethical and economic risks posed by GMOs, or they may just be used to legitimate a priori decisions enforced by the lobby or pressure groups. It all depends on how independent and public the process will be.

One of the principal requirements for public legitimacy of the proposed PFOA process is independence. As pointed out in previous sections, if the host regulatory agency and the decision-making body are dominated by pro-biotech representatives or anti-biotech representatives, it could become one more tool for legitimating biased decisions. Designers of the process will have to carefully address:

• how to guarantee independence
• how to guarantee that conflicting interests will be represented
• how to use techniques for identifying relevant stakeholders
• how to assure legitimacy of selected representatives
• how to finance PFOA in environmental risk assessment
• how to prepare participants with information

Evaluating stakeholder contributions and comparing options
PFOA designers will have to consider how diverse stakeholder contributions can be made in an open discussion without judgement, but at the same time use agreed upon criteria for managing information, scientific findings and diverse values. For example:

• how to identify a neutral party to facilitate the process
• how to establish ground rules to govern the dialogue
• how to pay attention to disparities in power and the ability of different stakeholders to participate fully
• how to enforce that trustful risk-assessment analysis will be carried out when, and where necessary, and for the time it is necessary
• how to guarantee the transparency of the deliberation
• how to manage information so it is timely and accessible
• how to deal with a lack of consensus between experts
• how to guarantee that information produced will be released to the public
• how to assess the effectiveness of the participatory process

Linking recommendations to decision making
Finally, a written report about the PFOA deliberation will have to represent the critical issues, the degree of consensus and the range of opinions. Special attention will need to be paid to:

• how to present recommendations if there is no consensus
• how to take the results of the deliberation into the decision-making process in a timely manner
• how to recognize the conclusions of government authorities

Conclusion

The rapid advances in modern biotechnology, particularly genetically engineered products, will shape the coming decades of economic development. As more transgenic products are field tested and eventually approved for commercialization, scientists will also learn more about how to manage the possibility that something negative might happen, the risks and the socio-economic implications. In addition, a new array of possibilities is emerging from data obtained from functional genomics, and its potential impact on the sustainability of agricultural production could be enormous. As more developing countries begin to evaluate the local applications of the technology, each country will have to develop its own approach for societal discussion. Some members of society may begin to accept GM varieties as simply obtained by a different biological process, while others may take principled positions against technologies that have the potential for irreversible harm to the environment. Still others may use a holistic approach for the safe use of genetically engineered products, taking into account social, economic and ethical values.

For the last 15 years in Brazil, a range of other issues have been associated with the use of some biotechnology, mainly transgenic organisms and human and animal cloning, in such a way that political, economic, social, scientific and ethical issues are part of discussions throughout the country. Questions exist due to the lack of scientific, social, economic and environmental impact studies. Citizens also wonder about intellectual property rights, the build-up of gigantic transnational companies with focused influence around the world, and unresolved ethical, economic and social issues.

The coming years will be a challenging period for developing countries as they address issues such as the need for better training of scientists (including in the areas of risk assessment and monitoring of GM experiments) and in-house investments needed to continue the development of their capacity in the area of biosafety.

A societal multi-stakeholder discussion such as PFOA can strengthen the consideration of critical problems, the options we have for addressing them, and the role biotechnology and biosafety will play. We hope the coming years will give scientists greater confidence in developing appropriate methods to predict, or at least properly assess, the possible negative environmental impacts on a case-by-case basis and that a science-based discussion will inform public choices and allow these choices to be considered within the decision-making process.

References

Altieri, M.A. (2000) The ecological impacts of transgenic crops on agroecosystem health. *Ecosystem Health* 6, 13–23.

Biggs, S. and Matsaert, H. (1999) An actor-oriented approach for strengthening research and development capabilities in natural resource systems. *Public Administration and Development* 19, 231–262.

Bonalume Neto, R. (2003) GM confusion in Brazil. *Nature Biotechnology* 21, 1257–1258.

Brody, J. (2002) In a world of hazards, worries are often misplaced. *New York Times*, August 20.

Burachik, M. and Traynor, P.L. (2002) *Analysis of a National Biosafety System: Regulatory Procedures in Argentina.* ISNAR Country Report 63, International Service for National Agricultural Research, The Hague, The Netherlands.

CBD (Convention on Biodiversity) (1992) Convention on Biodiversity: convention text. Available at: http://www.biodiv.org/doc/legal/cbd-en.pdf (accessed February 2005).

CBD (Convention on Biodiversity) (2000) Cartagena Protocol on Biosafety: convention text. Available at: http://www.biodiv.org/biosafety/protocol.asp (accessed October 2003).

CBD (Convention on Biodiversity) (2004) Decision VII/11. Available at: http://www.biodiv.org (accessed February 2005).

CEC (Commission of the European Communities) (2000) Communication from the Commission on the Precautionary Principle, COM (2000) 1 final. Available at: http://europa.eu.int/comm/environment/docum/20001_en.htm, CEC, Brussels, Belgium (accessed October 2004).

Contini, E., Sampaio, M.J.A.M. and Ávila, F. (2003) The lack of clear GMO regulation and its costs for the Brazilian economy. *International Journal of Biotechnology* 5. Available at: http://www.environmental-expert.com/magazine/inderscience/ijbt/ (accessed October 2004).

Crowfoot, J.E. and Wondolleck, J.M. (1990) *Environmental Disputes: Community Involvement in Conflict Resolution.* Island Press, Washington, DC.

Department of Agriculture Western Australia (2004) GMO Trial Details. Available at: http://agspsrv34.agric.wa.gov.au/biotechnology/gmtrials.htm (accessed November 2004).

FAO (2004) The state of food and agriculture 2004. Agricultural biotechnology: meeting the needs of the poor? *FAO Agriculture Series*, No. 35. Rome, Italy.

Fontes, E.M.G. (2003) Legal and regulatory concerns of transgenic plants in Brazil. *Journal of Invertebrate Pathology* 83, 100–103.

Forester, J. (1999) *The Deliberative Practitioner: Encouraging Participatory Planning Processes.* MIT Press, Cambridge, Massachusetts.

Gibbons, M. (1999) Science's new social contract with society. *Nature* 402, C81–C84.

Glover, D., Keeley, J., McGee, R., Newell, P., Da Costa, P., Ortega, A.R., Loureiro, M. and Lin, L.L. (2003) *Public Participation in National Biosafety Frameworks: A Report for UNEP-GEF and DFID.* Institute of Development Studies, Brighton, UK.

Goldstein, I. (1993) *Training in Organizations*, 3rd edn. Brooks/Cole Publishing Company, Pacific Grove, California.

Grimble, R. and Wellard, K. (1997) Stakeholder methodologies in natural resource management: a review of principles, contexts, experiences and opportunities. *Agricultural Systems* 55(2), 173–193.

Hallerman, E.M. and Kapuscinski, A.R. (1995) Incorporating risk assessment and risk management into public policies on genetically modified finfish and shellfish. *Aquaculture* 137, 9–17.

Hemmati, M. (2002) *Multi-Stakeholder Processes for Governance and Sustainability: Beyond the Deadlock and Conflict.* Earthscan, London.

Irwin, A. (2001) Constructing the scientific citizen: science and democracy in the biosciences. *Public Understanding of Science* 10, 1–18.

Kapuscinski, A.R., Goodman, R.M., Hann, S.D., Jacobs, L.R., Pullins, E.E., Johnson, C.S., Kinsey, J.D., Krall, R.L., La Viña, A.G.M., Mellon, M.G. and Ruttan, V.W. (2003) Making safety first a reality for biotechnology products. *Nature Biotechnology* 21(6), 599–601.

Kessler, J. and Van Dorp, M. (1998) Structural adjustment and the environment: the need for an analytical methodology. *Ecological Economics* 27, 267–281.

Loevinsohn, M., Berdegué, J. and Guijt, I. (2002) Deepening the basis of rural resource management: learning processes and decision support. *Agricultural Systems* 73, 3–22.

McLean, M.A., Frederick, R.J., Traynor, P.L., Cohen, J.I. and Komen, J. (2002) A conceptual framework for implementing biosafety: linking policy, capacity and regulation. *International Service for National Agricultural Research Briefing Paper* 47, ISNAR, New York.

Miller, H.I. (1994) A need to reinvent biotechnology regulation at the EPA. *Science* 266, 1815–1818.

Munson, A. (1993) Genetically manipulated organisms: international policy-making and implications. *International Affairs* 69, 497–517.

NRC (National Research Council) (1983) *Risk Assessment in the Federal Government: Managing the Process*. National Academies Press, Washington, DC.

NRC (National Research Council) (1996) *Understanding Risk: Informing Decisions in a Democratic Society*. National Academies Press, Washington, DC.

NRC (National Research Council) (2002) *Environmental Effects of Transgenic Plants: The Scope and Adequacy of Regulation*. National Academies Press, Washington, DC.

Nelson, K.C., Kibata, G., Lutta, M., Okuro, J.O., Muyekho, F., Odindo, M., Ely, A. and Waquil, J. (2004) Chapter 3: Problem formulation and options assessment (PFOA) for genetically modified organisms: the Kenya case study. In: Hilbeck, A. and Andow, D.A. (eds) *Risk Assessment of Transgenic Crops: A Case Study of Bt Maize in Kenya*. CAB International, Wallingford, UK, pp. 57–82.

Nutti, M., Sampaio, M.J.A.M. and Watanabe, E. (2004) GMO research and agribusiness in Brazil: impact of the regulatory framework. In: Taylor, Iain E.P. and Barrett, K. (eds) *Genetically Engineered Plants: Decision-Making under Uncertainty*. Haworth's Food Products Press, Binghamton, New York.

O'Brien, M. (2000) *Making Better Environmental Decisions: An Alternative to Risk Assessment*, 3rd edn. MIT Press, Cambridge, Massachusetts.

Pelaez, V., Albergoni, L. and Guerra, M.P. (2004) Soja transgênica vs. soja convencional: uma análise comparativa de custos e benefícios. *Cadernos de Ciência e Tecnologia* 21(2), 279–309.

Sagar, A., Daemmrich, A. and Ashiya, M. (2000) The tragedy of the commoners: biotechnology and its publics. *Nature Biotechnology* 18(January), 2–4.

Schmoldt, D. and Peterson, D. (1998) Analytical group decision-making in natural resources: methodology and application. *Forest Science* 46(1), 62–75.

Skorupinski, B. (2001) Debating novel food: a comparison of public participation processes on genetically modified food in four European countries. In: *Third Congress of the European Society for Agricultural and Food Ethics*, Florence, 3–5.10.2001, pp. 381–384.

Skorupinski, B. (2004) Debating risks and hazards – about deliberation and deliberate release. In: Breckling, B. and Verhoeven, R. (eds) *Risk Hazard Damage: Specification of Criteria to Assess Environmental Impact of Genetically Modified Organisms*, Naturschutz und Biologische Vielfalt, Vol. 1. Federal Agency for Nature Conservation, Bonn, Germany, pp. 171–182.

Stern, P. and Fineberg, H. (eds) (1996) *Understanding Risk: Informing Decisions in a Democratic Society.* Committee on Risk Characterization. Commission on Behavioural and Social Sciences and Educations. National Research Council. National Academies Press, Washington, DC.

Susskind, L., McKearnan, S. and Thomas-Larmer, J. (eds) (1999) *The Consensus Building Handbook: A Comprehensive Guide to Reaching Agreement.* Sage Publications, Thousand Oaks, California.

Susskind, L., Levy, P.F. and Thomas-Larmer, J. (2000) *Negotiating Environmental Agreements: How to Avoid Escalating Confrontation, Needless Costs, and Unnecessary Litigation.* Island Press, Washington, DC.

Traynor, P.L., Fredrick, R.J. and Koch, M. (2002) *Biosafety and Risk Assessment in Agricultural Biotechnology. A Workbook for Technical Training.* The Agricultural Biotechnology Support Project. Institute of International Agriculture, Michigan State University, Michigan.

UNEP-GEF (2003a) Report of the subregional workshops for Asian countries on: Risk assessment and management and public awareness and participation. UNEP-GEF Project on Development of National Biosafety Frameworks, Asian Countries Subregional Workshop (January 21–24 2003, Kuala Lumpur, Malaysia). UNEP-GEF Biosafety Unit, Geneva, Switzerland.

UNEP-GEF (2003b) Sub-regional workshop for Latin American countries on: Development of a regulatory regime and administrative systems. UNEP-GEF Project on Development of National Biosafety Frameworks, Latin American Sub-Regional Workshop on Development of a Regulatory Regime and Administrative Systems (November 25–28 2003, Santiago, Chile). UNEP-GEF Biosafety Unit, Geneva, Switzerland.

Wondolleck, J.M. and Yaffee, S.L. (2000) *Making Collaboration Work: Lessons from Innovation in Natural Resource Management.* Island Press, Washington, DC.

4   Transgene Expression and Locus Structure of Bt Cotton

M.F. Grossi-de-Sa, W. Lucena, M.L. Souza, A.L. Nepomuceno, E.O. Osir, N. Amugune, Tran Thi Cuc Hoa, Truong Nam Hai, D.A. Somers and E. Romano

Corresponding author: Dr M. Fátima Grossi-de-Sa, Embrapa Recursos Genéticos e Biotecnologia, SAIN Parque Rural, Brasília, DF, 70700-900, Brazil. Fax: +55-61-4484673, e-mail: fatimasa@cenargen.embrapa.br

Introduction

New approaches to the regulation and safety assessment of genetically modified (GM) crops have been put forward due to public and scientific concerns (Kuiper *et al.*, 2001; Capalbo *et al.*, 2003; Conner *et al.*, 2003). Although a consensus has been accepted for the principles concerning the evaluation of food safety, no universal definition has been reached regarding environmental assessment. Knowledge of the introduced genes, their regulation and the integration site in the plant genome may provide important information for human health and environmental risk assessment (König, 2003). In fact, the integration of exogenous DNA into the plant genome can result in the disruption of host genes or the activation of silent genes, which may result in the formation of either new metabolites or altered levels of existing metabolites (Kuiper *et al.*, 2001). It should be emphasized that the occurrence of unintended effects is not specific for transgenic plants, but also occurs frequently in conventional breeding (Trewavas and Leaver, 2001; Dale *et al.*, 2002; Conner *et al.*, 2003; ICSU, 2003; Wilkinson *et al.*, 2003). This may be partly predictable on the basis of knowledge of the structure and function of the introduced gene, its involvement in metabolic pathways and its genome integration site. Thus, appropriate molecular characterization of a transgenic event intended for commercialization allows elimination of unwanted events with potential risks (Kuiper *et al.*, 2001). On the other hand, unexpected phenotypic effects in agronomic characteristics can be detected in field tests and in analysis of metabolic composition of GM crops, both required before the process of release for commercialization of GM plants.

It is important to emphasize that, while necessary, the molecular characterization of the introduced DNA is not sufficient on its own for prediction of all possible unintended effects, and not all molecular analyses will necessarily result in useful information for risk assessment (Andow et al., 2004). Therefore it is critical that the molecular characterization of transgenic plants should be carefully planned and restricted to those that are really necessary for a correct biosafety assessment.

The aim of this chapter is to describe the main conclusions of a public sector scientist working group on methodologies for the ecological risk assessment of transgenic crops with respect to transgene locus structure and expression (Andow et al., 2004). It also includes a scientific strategy for characterizing expression and transgene locus structure that can be applied to new *Bacillus thuringiensis* (Bt) cotton events in development in Brazil, and to the assessment of the Bt cotton events currently available for use in Brazil.

Scientific strategy for characterizing transgene locus structure and expression

The general principles of risk assessment for the characterization of transgene locus structure and expression should take into account interactions among the transgene(s), the crop and the environment in which the crop will be grown on a case-by-case basis (NRC, 1989, 2002; CBD, 2000; Kuiper et al., 2001). Each transgene confers a different trait with different phenotypes that will require different risk assessments. Even considering the same transgene in different transformation events in a single crop, the integration of the transgene into the genome occurs randomly, generating transgenes flanked by different genomic regions, which must be considered when evaluating each different transgenic event (Pawlowski and Somers, 1996; Salomon and Puchta, 1998; Gorbunova and Levy, 1999; Vergunst and Hooykass, 1999). For example, the characterization of one plant obtained by transformation with a *Bt* gene construct could reveal that the transgene was integrated in an innocuous region. However, this information cannot be utilized for all transgenic events since other integrations could occur in a region of an essential host gene. Hence, the strategy for characterizing transgene locus structure and expression involves four major issues (Andow et al., 2004):

Transgene design. During the process of development of transgenic events, several procedures that result in simpler transgene loci can be performed that facilitate or simplify subsequent risk assessment. If constructs were designed not to contain a selectable marker gene or to contain an appropriate selectable marker gene, risk assessment would be simplified. An example is the case of the antibiotic vancomycin that should be avoided in genetic constructs, due its extreme importance for medical purposes. Other selectable marker genes, such as the *nptII* gene, have been extensively studied and are considered safe by many scientists for use in construct design. In the process of transformation by biolistics, a fragment containing only the target gene and not the plasmid backbone can be used instead of the whole plasmid. This simple procedure ensures transgenic events without marker genes. Similarly, if the event is created via cotransformation, genetic segregation can eliminate the marker gene. These processes would substantially simplify the needs for further risk assessment associated with the marker gene.

Analysis of transgene locus structure. The characterization of a transgenic plant at molecular level is useful to provide information such as the composition and integrity of the inserted DNA, the number of copies of the inserted DNA, the number of transgene loci and possible modifications of the recipient genomic DNA. A detailed description of the transgene locus structure is important for several reasons:

1. It is possible to determine if the transgene would cause an insertional loss-of-function mutation by integrating into a recipient gene. Disturbance of the recipient-gene function could result in unwanted effects with possible environmental consequences.

2. It allows assessment of ectopic expression of the transgene or a recipient gene. Ectopic expression is the expression of a gene in a tissue in which it is normally not expressed. This can be caused by the juxtaposition of a transgene promoter into adjacent plant genes during transgene integration.

3. It allows an assessment of the likelihood of rearrangements during integration that can create spurious open reading frames (ORFs). Spurious ORF formation could result in the production of unintended gene products.

4. It allows the identification of repeated sequences in the transgene loci that could increase the probability of intra or interlocus instability by homologous recombination and possible rearrangements into the plant genome.

5. It allows assessment of the theoretical probability that the transgene has been incorporated into a transposable element, thereby allowing it to move independently on the genome.

While sequencing of transgene loci can provide information about the potential of unexpected effects, field tests and analysis of metabolic composition of GM crops can also be used to detect some unwanted phenotypic effects.

Analysis of transgene expression. Transgene expression can be measured in terms of the concentration of transgene product and in terms of the whole-organism phenotype. The characterization of phenotype expression should address what is essential for evaluating efficacy, non-target effects, resistance management and gene flow. Expression of transgenes frequently is not homogenous in the whole plant and can be affected by environmental conditions. Levels of Cry2Ab2 toxin in Bollgard II vary among plant parts. Studies have also demonstrated that the pattern of expression can change in different years, showing that environmental conditions and even geographical location can affect transgene expression among plant tissues (Akin *et al.*, 2002). The reduced toxin levels in some plant parts may allow survival of target Lepidopteran pests, and therefore has clear implications for the development of insect resistance. For instance, the Cry1Ac level in flowers of Bollgard cotton can be less than half that of other plant parts; consequently, flowers may be a food source that allows survival of insect pests (Akin *et al.*, 2002). Therefore, analyses of the level of transgene expression in several tissues should be conducted prior the commercialization of a transgenic crop. On the other hand, since changes in environmental conditions are unpredictable, studies to detect variations in expression in different environments should be conducted after the commercialization as a monitoring strategy to guide the development of new transgenic crops in order to avoid appearance of resistant insects.

Transgene transmission. The inheritance and stability of each introduced trait that is functional in the transformed plant must be determined. The transgene should be inherited as a normal Mendelian character for each novel trait. The pattern and stability of inheritance must be demonstrated as well as its level of expression.

Assessment Scope of Available Bt Cotton Events for Use in Brazil

A number of transgenic cotton events are being considered for introduction into Brazil (see Andow *et al.*, Chapter 1, this volume). Here, we will review the current publicly available information on two of the main events under consideration, Bollgard cotton (transformation event 531) and Bollgard II cotton (transformation event 15985).

Description of Bollgard (transformation event 531)

Plasmid vector construction
Line 531 was produced by *Agrobacterium*-mediated transformation of the cotton (*Gossypium hirsutum*) line L. cv Coker C312 with plasmid PV-GHBKO4 (Zambryski, 1992). This vector is 11.4 Kb in size (Fig. 4.1) and contains the origins of replication for *Agrobacterium* (*oriV*) and *Escherichia coli* (*ori322*) fused to the 360 bp DNA fragment from pTiT37 plasmid that contains the nopaline T-DNA right border. The plasmid vector contains two plant gene-expression cassettes. One cassette codes for the *Bt* insecticidal protein, Cry1Ac, and the second cassette codes for the selectable marker neo gene (*nptII*) under the control of the cauliflower mosaic virus (CaMV) 35S promoter and the non-translated region of the 3′ region of the nopaline synthase gene (*nos*) from the pTiT37 plasmid of *Agrobacterium tumefaciens* strain T37. The *cry1Ac* gene was modified for optimal expression in plants, which is composed of part of the 5′ end of the *cry1Ab* gene (Fischhoff *et al.*, 1987) with a portion of the *cry1Ac* gene (Adang *et al.*, 1985). This modified *cry1Ac* gene is under the control of the CaMV 35S promoter with a duplicated enhancer region (Odell *et al.*, 1985; Kay *et al.*, 1987) and the non-translated region of the soybean alpha subunit of the beta-conglycinin gene that provides the mRNA polyadenylation signals (7S 3′ terminator sequence) (Schuler *et al.*, 1982). The vector also contains a fragment with the *aad* gene isolated from *E. coli* bacterial transposon Tn7, which encodes the enzyme aminoglycoside

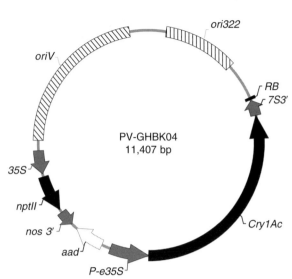

Fig. 4.1. Plasmid map of PV-GHBK04 used in the development of transgenic cotton Bollgard (adapted from Agbios, 2002).

adenyltransferase (AAD) that confers resistance to the antibiotics spectino-mycin and streptomycin. The *aad* gene cis under the control of a bacterial promoter and terminator and was used as a marker in bacteria. The *aad* gene has no plant-regulatory sequences and was not expressed in plant tissues.

Structure of the transgenic locus
The transgenic locus of event 531 contains two copies of the T-DNA insert integrated in a head-to-tail arrangement. One T-DNA insert contains a full-length *cry1Ac* gene and the *nptII* gene, and the second insert contains an inactive 3' portion of the *cry1Ac* gene. The two inserts are linked and seem to segregate as a single locus. In a second locus there is a third insert of a 242 bp fragment. This fragment is a portion of the 7S 3' genetic element that finalizes transcription of the *cry1Ac* gene construct in event 531. Further analysis demonstrate that the 242 bp fragment is located in a different locus from the other two inserts and is non-functional. The *ori322* region, present in the plasmid vector PV-GHBK04, was not found in the plant genome. The *aad* gene was found but was not expressed as it was under the control of a bacterial promoter (Agbios, 2002).

Transgene expression
The Cry1Ac protein expression levels in event 531 were quantified by enzyme-linked immunosorbent assay (ELISA) during a few cotton-develop-mental stages. Cry1Ac expression in line 531 was 1.56 µg protein/g and 0.86 µg protein/g fresh weight in leaf and seed tissues, respectively. In the whole plant, the Cry1Ac protein averaged 0.044 µg protein/g fresh weight (25 µg/plant) for a total of about 0.58 g/ha. Expression levels varied

approximately threefold over the growing season. Whole plant amounts were 1.1 µg protein/g fresh weight (200 µg/plant) for a total of 4.94 g/ha. The expression of the Cry1Ac protein in pollen of event 531 in laboratory studies was reported to be 11.5 ng/g fresh weight (Monsanto, 2002; OGTR, 2003). Bollgard cotton pollen has also been reported as containing 0.6 µg/g pollen fresh weight (Greenplate, 1997). For NptII protein, the levels in line 531 were 3.14 µg protein/g fresh weight and 2.45 µg protein/g fresh weight in leaf and seed tissues, respectively. Expression levels varied about twofold over the season (Agbios, 2002).

Transgene transmission
The stability of the inserted genes for line 531 was demonstrated over four generations (Agbios, 2002).

Description of Bollgard II (transformation event 15985)

Plasmid vector construction
The biolistic method was used to insert the *cry2Ab2* insect control-coding sequence and the *uidA* marker-coding sequence into the Bollgard cotton genome (McCabe and Martinell, 1993). The plasmid vector, PV-GHBK11 (Fig. 4.2), is 8.7 Kb in size and contains two plant gene-expression cassettes and a bacterial kanamycin resistance gene and origin of replication. The *KpnI* fragment of this plasmid containing only the two plant gene expression cassettes and no bacterial or plant selectable marker gene or origin of replication was used in the process of transformation, avoiding analysis of the removed sequences. The first cassette contains the *cry2Ab2* gene encoding the Bt insecticidal protein Cry2Ab2 and the second cassette contains the *uidA* gene encoding the β-D-glucuronidase (GUS) reporter protein (Jefferson et al., 1986; Gilissen et al., 1998) used only to facilitate selection of Cry2Ab2-producing plants (Agbios, 2002). Both genes, *cry2Ab2* and *uidA*, are under the control of the enhanced CaMV 35S promoter (e35S) and the 3′ region of the nopaline synthase gene (*nos* 3′) from *A. tumefaciens* (Depicker et al., 1982). The e35S promoter (Kay et al., 1987) controlling the *cry2Ab2* gene is also fused to the 5′ untranslated leader sequence from the petunia heat-shock protein 70 (HSP70) and the chloroplast transit peptide (CTP) from the *Arabidopsis thaliana* 5-enolpyruvyl shikimate-3-phosphate synthase (EPSPS) gene that is used to direct the protein to the chloroplasts (Shah et al., 1986).

Structure of the transgenic locus
The transgenic locus of Bollgard II contains a single DNA insert into the Bollgard I genome. This single insert contains the complete cassette of the *cry2Ab* gene, without the restriction site following the NOS 3′ polyadenylation sequence and the complete cassette of *uidA* gene, except that 260 bp of the 5′ end of the e35S promoter is not present. This truncated promoter is functional as demonstrated by production of the GUS protein. This information

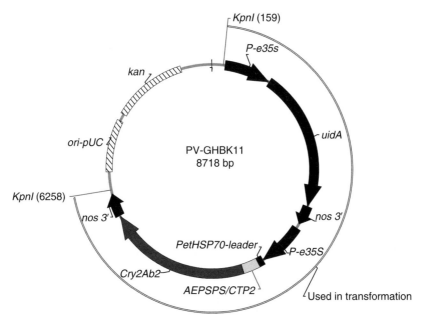

Fig. 4.2. Plasmid map of PV-GHBK11 used in the development of transgenic cotton Bollgard II (adapted from Agbios, 2002).

was obtained by DNA sequencing of the trangenic locus of Bollgard II. DNA sequencing of polymerase chain reaction (PCR)-amplified fragments corresponding to the 5′ and 3′ ends of the insert confirmed that the DNA flanking the insert was genomic cotton DNA (Agbios, 2002).

Transgene expression
Production of the full-length Cry2Ab2 and GUS proteins was confirmed by Western blot analysis during a few cotton-developmental stages. Levels of the Cry2Ab2 protein were measured in whole plants collected in different sites for 4 years, showing expression in leaves and cottonseeds. The Bollgard II cotton plants contained an estimated 8.8 µg Cry2Ab2 protein/g fresh weight. The expression of the Cry1Ac protein in pollen was detected; on the other hand, the Cry2Ab2 protein was not found in the same tissue (OGTR, 2003). The absence of Cry2Ab2 protein in the pollen was probably a consequence of the signal peptide used in the *cry* cassette that directs the protein to the chloroplasts (Agbios, 2002).

Transgene transmission
The transgene locus was inherited as a single genetic locus conforming to the expected Mendelian segregation pattern. The analyses were performed by Southern blot over four generations of selfing and two generations of backcrossing. The stability of the transgene locus structure and the expression was confirmed by several field tests undertaken at multiple sites in the

US since 1998 (Agbios, 2002). All elements present in the events 531 and 15985 are summarized in Table 4.1.

Now we will provide some general strategies for characterizing transgene expression and locus structure to highlight our approach.

Strategies for Locus-Structure Characterization

The locus structure should be characterized in different steps. Firstly, it is necessary to determine the subcellular localization of the transgene, which can be in the nuclear genome or in one of the organelle genomes (Daniell *et al.*, 1998). This localization should be determined by segregation analyses. In the case of integration in organelles the transgene will be maternally inherited (Daniell *et al.*, 1998), while in the events in which the integration occurs in the nucleus, the transgene inheritance should occur following Mendelian segregation (Srivastava *et al.*, 1996; Feldmann *et al.*, 1997; Howden *et al.*, 1998). Secondly, the number of transgenes should be determined. Southern genomic analysis can provide a good estimate of the copy number of the transgenes (Pawlowski and Somers, 1998). For this purpose, the genomic DNA is treated with restriction enzymes that cut near the transgene and the target gene should be used as probe. Thus, one hybridizing band indicates integration of one copy, while multiple bands suggest multiple copies. One example is shown in Fig. 4.3. In the case where more than one transgene is integrated in the same locus and linked, the number of hybridizing bands does not correspond to the number of transgene copies (Joersbo *et al.*, 1999; Makarevitch *et al.*, 2003). Southern analysis performed on segregating progeny of a transgenic plant provides a good indication of transgene locus numbers.

A more precise way to determine the number of transgene copies and to determine if transgene integration adversely affected the structure of the transgene locus and flanking genomic sequences is nucleotide sequencing of all transgene loci including flanking sequences. Several approaches may be used to isolate transgene locus fragments for sequencing. One approach is to construct a genomic library and perform screening of the wanted clone with a radioactive transgene probe. PCR can also be used for isolating transgene locus fragments. However, direct PCR cannot be used to isolate the regions flanking the exogenous DNA. For this purpose, more recently developed techniques such as inverse PCR and thermal asymmetric interlaced (TAIL)-PCR can be an alternative way to isolate genomic flanking regions. Sequencing flanking regions will also reveal if the transgene was integrated into a host gene, which results in loss-of-function of this recipient gene. Knockout of an important host gene could result in unwanted effects with possible environmental consequences (König *et al.*, 2004).

Some of these techniques can be very effective in determining transgene copy number, but at the same time they can be quite costly in terms of reagents, labour and time. Real-time PCR, despite the initial cost of the equipment, could in the long term be a more cost-effective technique to determine

Table 4.1. Genetic elements present in cotton line 531 (Bollgard) and 15985 (Bollgard II). Since Bollgard II was obtained by transformation of cotton line 531, all the genetic elements described for cotton line 531 are also present in Bollgard II (Agbios, 2002).

Gene	Protein	Promoter	Terminator	Copies	Form	Expressed in plant tissue
Genetic elements in cotton line 531 (Bollgard)						
cry1Ac	Cry1Ac δ-endotoxin (B. thuringiensis subsp. kurstaki (Btk))	Double-enhanced CaMV 35S	3' poly(A) signal from soybean alpha subunit of the beta-conglycinin gene	2	One complete and one truncated	Yes
Aad	3'(9)-O-aminoglycoside adenylyltransferase	Bacterial promoter	Bacterial terminator	1	Complete	No
neo	Neomycin phospho-transferase II (E. coli)	Nopaline synthase (nos) from A. tumefaciens	Nopaline synthase (nos) from A. tumefaciens	1	Complete	Yes
Genetic elements in cotton line 15985 (Bollgard II)						
Cry2Ab2	Cry2Ab2Cry1Ac delta-endotoxin (Bacillus thuringiensis)	Enhanced CaMV 35S plus 5' untranslated leader sequence from HSP70 and the chloroplast transit peptide from the A. thaliana 5-EPSPS gene	Nopaline synthase gene (nos 3') from A. tumefaciens	1	Complete	Yes
uidA	β-D-glucuronidase (GUS) reporter protein	Enhanced CaMV 35S plus	Nopaline synthase gene (nos 3') from A. tumefaciens	1	Complete (260 bp of the 5' end of the CaMV 35S promoter is not present)	Yes

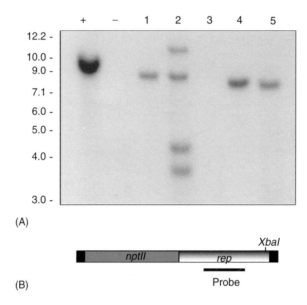

(A)

(B)   Probe

Fig. 4.3. Example of determination of the copy number of exogenous DNA by Southern blot in transgenic potato plants. (A) (+), plasmid positive control; (–), non-transformed plant; lanes 1, 4 and 5, transgenic plants presenting one copy of exogenous T-DNA; lane 2, transgenic plant presenting four copies of exogenous T-DNA. (B) Scheme of the T-DNA of the construct used in plant transformation showing the restriction site (*XbaI*) and the probe used in the Southern blot (E. Romano *et al.*, unpublished results).

transgene copy number and transgene expression levels (Schmidt and Parrott, 2001; Mason *et al.*, 2002).

Strategies for Transgene Expression Analysis

Analysis of transgene expression can be conducted on the transcript (RNA) and gene product (protein) levels. For Bt transgenic plants, the phenotype is related to the protein level; therefore the expression analysis should be focused on assays of the protein content. Since the regulation of gene expression can occur at different levels (transcriptional and translational), a high expression of a specific mRNA does not necessarily imply a high expression of the corresponding protein. Thus, it is not possible to consider levels of Bt toxins based on RNA-analysis methods. The main techniques used to analyse transgene expression at protein level are ELISA and Western blot. Ideally, the Cry protein should be expressed only in the tissue attacked by the target insect, and for this purpose the transgene can be designed to direct expression to the target tissue. Expression in other tissues can affect non-target insects and so in the case of Bollgard II the *cry2Ab2* gene was fused to the CTP from the *A. thaliana* EPSPS gene to direct the protein to the chloroplast (Shah *et al.*, 1986), therefore avoiding expression in pollen grains (Daniell *et al.*, 1998). While the design of the transgene could be optimized to avoid expression in unwanted tissues, this procedure is not sufficient to ensure the correct spatial expression. For example, a transgene fused to a leaf-specific promoter can be integrated adjacent to a constitutively expressed host gene causing expression of the transgene to not be restricted to leaf tissue.

Transgene expression restricted to the target tissue would be the ideal situation to avoid undesirable activity against non-target beneficial insects, which in practical terms is very difficult to achieve. However, the difficulty of developing transgenic events expressing insecticide proteins exclusively in target tissues does not necessarily constitute a biosafety problem. The environmental and ecological impact of GM crops should be judged to the impact of the practices that the GM crops are designed to replace (Conner *et al.*, 2003). GM cotton resistant against pests was designed to be a substitute for chemical pesticides; therefore the impact of these crops on non-target insects should be compared against the effect of non-transformed cotton with the normally associated pesticides that frequently present a broad activity range. For example, Bollgard cotton expresses the Cry1Ac protein in several non-target tissues; however, the use of this GM crop allows a reduction in the volume and frequency of chemical-insecticide applications, and consequently less damage to non-target insects has been achieved (FAO, 2004).

The analyses of gene expression should include the target and selectable marker genes. Several marker genes are considered safe and recently the European Food Safety Authority recognized that some antibiotic-resistant genes, such as the *nptII* and *hph* genes, are extremely unlikely (if at all) to have significant impacts on the environment or human or animal health (EFSA, 2004). On the other hand, if the marker genes are not present in the transgenic event, less analysis would need to be performed; thus, events without marker genes would save considerable time, expense and effort in characterizing transgene expression.

Final Considerations

As stated in an FAO (2004) report, appropriate biosafety regulations are essential, but excessive, expensive and unnecessary regulations can constitute a major barrier to the development of transgenic crops by the public sector. If the regulatory biosafety process is very expensive, only large transnational companies will be able to commercialize a transgenic crop (FAO, 2004). Therefore, analyses of transgene expression and locus structure for risk assessment should be limited to those that ensure a good basis for health and environmental biosafety assessment. For the case under consideration here, these analyses include: the sequence information of the inserted genetic elements; the sequence information of the genome regions flanking the exogenous DNA; and the information about Cry protein expression.

We consider that the sequencing of genome flanking regions of transgenes is advisable because insertional mutation of important host genes can
be detected allowing the elimination of unwanted events with potential risks.
However, it is difficult to know how much of the flanking regions should be
sequenced to identify a mutated host gene. For instance, the transgene could
integrate in a large intron without significant homology with other previously
published genes. On the other hand, field tests of agronomic traits and analysis of metabolic composition of GM crops can also be used to detect and eliminate some unwanted effects.

Molecular characterization methods of Southern and Western blot, as well
as sequencing of the transgenic loci, are enough to provide effective evaluation
of the structure of the introduced recombinant DNA, insertional mutations of
endogenous genes and the possible presence of spurious ORFs. Taking into
account that changes in plant gene-expression levels are at least as likely to
result from natural processes affecting genome rearrangements and the use of
other (non-transgenic) breeding technologies, and that very few such changes
may adversely impact human health and environment (König *et al.*, 2004),
information beyond the molecular methods mentioned above for characterizing transgene locus structure, determination of transgene expression levels and
determination of stable Mendelian inheritance is not at present essential for
characterizing transgenic events for prerelease risk assessment.

References

Adang, M.J., Staver, M.J., Rocheleau, T.A., Leighton, J., Barker, R.F. and Thompson,
D.V. (1985) Characterised full-length and truncated plasmid clones of the crystal
protein of *Bacillus thuringiensis* subsp. *kurstaki* HD-73 and toxicity to *Manduca
sexta. Gene* 36, 289–300.

Agbios (2002) Database query: MON-ØØ531-6, MON-ØØ757 (MON531/757/1076).
Essential Biosafety Crop Database, Agbios, Merrickville, Canada. Available at:
http://www.essentialbiosafety.info/main.php (accessed June 2003).

Akin, D.S., Steward, S.D., Knighten, K.S. and Adamczyk, J.J. Jr (2002) Quantification
of toxin levels in cottons expressing one and two insecticidal proteins of *Bacillus
thurigiensis*. Proceedings of the 2002 Beltwide Cotton Conferences. National
Cotton Council, Memphis, Tennessee. Available at: http://www.cotton.org/
beltwide/proceedings/2002/abstracts/H019.cfm (accessed 3 February 2005).

Andow, D.A., Somers, D.A., Amugune, N., Aragao, F.J.L., Ghosh, K., Gudu, S.,
Magiri, E., Moar, W.J. and Osir, E. (2004) Transgene locus structure and expres-
sion of Bt maize. In: Hilbeck, A. and Andow, D.A. (eds) *Environmental Risk
Assessment of Genetically Modified Organisms*, Vol. 1: *A Case Study of Bt
Maize in Kenya*. CAB International, Wallingford, UK, pp. 83–118.

Capalbo, D.M.F., Hilbeck, A., Andow, D.A., Snow, A., Bong, B.B., Wan, F.-H.,
Fontes, E.M.G., Osir, E.O., Fitt, G.P., Johnston, J., Songa, J., Heong, K.L. and
Birch, A.N.E. (2003) Brazil and the development of international scientific
biosafety testing guidelines for transgenic plants. *Journal of Invertebrate
Pathology* 83, 104–106.

CBD (Secretariat of the Convention on Biological Diversity) (2000) Cartagena Protocol
on Biosafety to the Convention on Biological Diversity: Text and Annexes.

Secretariat of the Convention on Biological Diversity, Montreal, Canada. Available at: http://www.biodiv.org/doc/legal/cartagena-protocol-en-pdf (accessed 1 December 2003).

Conner, A.J., Glare, T.R. and Nap, J.P. (2003) The release of genetically modified crops into the environment. Part II. Overview of ecological risk assessment. *The Plant Journal* 33(1), 19–46.

Dale, P., Clarke, B. and Fontes, E. (2002) Potential for environmental impact of transgenic crops. *Nature Biotechnology* 20, 567–574.

Daniell, H., Datta, R., Varma, S., Gray, S. and Lee, S.B. (1998) Containment of herbicide resistance through genetic engineering of the chloroplast genome. *Nature Biotechnology* 16, 345–348.

Depicker, A., Stachel, S., Dhaese, P., Zambryski, P. and Goodman, H.M. (1982) Nopaline synthase: transcript mapping and DNA sequence. *Journal of Molecular and Applied Genetics* 1, 561–573.

EFSA (European Food Safety Authority) (2004) Opinion of the scientific panel on genetically modified organisms on the use of antibiotic resistance genes as marker genes in genetically modified plants 1 (Question N° EFSA-Q-2003-109). *EFSA Journal* 48, 1–18. Available at: http://www.efsa.eu.int/science/gmo/gmo_opinions/384_en.html (accessed June 2004).

FAO (2004) The state of food and agriculture 2003–2004. Agricultural biotechnology: meeting the needs of the poor? Food and Agriculture Organization, Rome, Italy. Available at: http://www.fao.org/DOCREP/006/Y5160E/Y5160E00.HTM (accessed June 2004).

Feldmann, K.A., Coury, D.A. and Christianson, M.L. (1997) Exceptional segregation of a selectable marker (KanR) in *Arabidopsis* identifies genes important for gametophytic growth and development. *Genetics* 147, 1411–1422.

Fischhoff, D.A., Bowdish, K.S., Perlak, F.J., Marrone, P.G., McCormick, S.M., Niedermeyer, J.G., Dean, D.A., Kusano-Kretzmer, K., Mayer, E.J., Rochester, D.E., Rogers, S.G. and Fraley, R.T. (1987) Insect tolerant transgenic tomato plants. *Biotechnology* 5, 807–813.

Gilissen, L.J., Metz, P.L., Stiekema, W.J. and Nap, J.P. (1998). Biosafety of *E. coli* beta-glucuronidase (GUS) in plants. *Transgenic Research* 7(3), 157–163.

Gorbunova, V. and Levy, A.A. (1999) How plants make ends meet: DNA double-strand break repair. *Trends in Plant Science* 4, 263–269.

Greenplate, J. (1997) Response to reports of early damage in 1996 commercial Bt-transgenic cotton (Bollgard) plantings. *Society for Invertebrate Pathology Newsletter* 29(2), 15–18.

Howden, R., Park, S.K., Moore, J.M., Orme, J., Grossniklaus, U. and Twell, D. (1998) Selection of T-DNA-tagged male and female gametophytic mutants by segregation distortion in *Arabidopsis*. *Genetics* 149, 621–631.

ICSU (International Council for Science) (2003) *New Genetics, Food and Agriculture: Scientific Discoveries—Societal Dilemmas.* International Council for Science, Paris, France. Available at: http://www.icsu.org (accessed June 2004).

Jefferson, R.A., Burgess, S.M. and Hirsch, D. (1986) β-glucuronidase from *Escherichia coli* as a gene-fusion marker. *Proceedings of the National Academy of Sciences USA* 83, 8447–8451.

Joersbo, M., Brunstedt, J., Marcussen, J. and Okkels, F.T. (1999) Transformation of the endospermous legume guar (*Cyamopsis tetragonoloba* L.) and analysis of transgene transmission. *Molecular Breeding* 5, 521–529.

Kay, R., Chan, A., Daly, M. and McPherson, J. (1987) Duplication of CaMV 35S promoter sequences creates a strong enhancer for plant genes. *Science* 236, 1299–1302.

König, A. (2003) A framework for designing transgenic crops – science, safety, and citizens' concerns. *Nature Biotechnology* 21, 1274–1279.

König, A., Cockburn, A., Crevel, R.W.R., Debruyne, E., Grafstroem, R., Hammerling, U., Kimber, I., Knudsen, I., Kuiper, H.A., Peijnenburg, A.A.C.M., Penninks, A.H., Poulsen, M., Schauzu, M. and Wal, J.M. (2004) Assessment of the safety of foods derived from genetically modified (GM) crops. *Food and Chemical Toxicology* 42(7), 1047–1088.

Kuiper, H.A., Kleter, G.A., Noteborn, H.P.J. and Kok, E.J. (2001) Assessment of the food safety issues related to genetically modified foods. *The Plant Journal* 27, 503–528.

Makarevitch, I., Svitashev, S.K. and Somers, D.A. (2003) Complete sequence analysis of transgene loci from plants transformed via microprojectile bombardment. *Plant Molecular Biology* 52, 421–432.

Mason, G., Provero, P., Vaira, A.M. and Accotto, P. (2002) Estimating the number of integrations in transformed plants by quantitative real-time PCR. *BMC Biotechnology* 2. Available at: http://www.biomedcentral.com/1472-6750/2/20 (accessed February 2005).

McCabe, D.E. and Martinell, B.J. (1993) Transformation of elite cotton cultivars via particle bombardment of meristems. *BioTechnology* 11, 596–598.

Monsanto (2002) Safety assessment of Bollgard® cotton event 531. Available at: http://www.essentialbiosafety.com/docroot/decdocs/02-269-001.pdf (accessed July 2003).

NRC (National Research Council) (1989) *Field Testing Genetically Modified Organisms: Framework for Decisions.* National Academies Press, Washington, DC.

NRC (National Research Council) (2002) *Environmental Effects of Transgenic Plants: The Scope and Adequacy of Regulation.* National Academies Press, Washington, DC.

Odell, J.T., Nagy, F. and Chua, N.H. (1985) Identification of DNA sequences required for activity of the cauliflower mosaic virus 35S promoter. *Nature* 313, 810–812.

OGTR (Office of the Gene Technology Regulator) (2003) *Risk Assessment and Risk Management Plan Consultation Version: Commercial Release of Insecticidal (INGARD® event 531) Cotton.* Office of the Gene Technology Regulator, Woden, Australia.

Pawlowski, W.P. and Somers, D.A. (1996) Transgene inheritance in plants genetically engineered by microprojectile bombardment. *Molecular Biotechnology* 6, 17–30.

Pawlowski, W.P. and Somers, D.A. (1998) Transgenic DNA integrated into the oat genome is frequently interspersed by host DNA. *Proceedings of the National Academy of Sciences USA* 95, 12106–12110.

Salomon, S. and Puchta, H. (1998) Capture of genomic and T-DNA sequences during double-strand break repair in somatic plant cells. *EMBO Journal* 17, 6086–6095.

Schmidt, M.A. and Parrott, W.A. (2001) Quantitative detection of transgenes in soybean [*Glycine max* (L.) Merrill] and peanut (*Arachis hypogeaea* L.) by real-time polymerase chain reaction. *Plant Cell Reports* 20(5), 422–428.

Schuler, M.A., Schmitt, E.S. and Beachy, R.N. (1982) Closely related families of genes code for the alpha and alpha' subunits of the soybean 7S storage protein complex. *Nucleic Acid Research* 10, 8225–8261.

Shah, D., Horsch, R., Klee, H., Kishore, G., Winter, J., Turner, N., Hironaka, C., Sanders, P., Gasser, C., Aykent, S., Siegel, N., Rogers, S. and Fraley, R. (1986) Engineering herbicide tolerance in transgenic plants. *Science* 233, 478–481.

Srivastava, V., Vasil, V. and Vasil, I.K. (1996) Molecular characterization of the fate of transgenes in transformed wheat (*Triticum aestivum* L.). *Theoretical and Applied Genetics* 92, 1031–1037.

Trewavas, A. and Leaver, C. (2001) Is opposition to GM crops science or politics? An investigation into the arguments that GM crops pose a particular threat to the environment. *EMBO Reports* 2, 455–459.

Vergunst, A.C. and Hooykaas, P.J.J. (1999) Recombination in the plant genome and its application in biotechnology. *Critical Reviews in Plant Sciences* 18(1), 1–31.

Wilkinson, M.J., Sweet, J. and Poppy, G.M. (2003) Risk assessment of GM plants: avoiding gridlock? *Trends in Plant Science* 8, 208–212.

Zambryski, P. (1992) Chronicles from the *Agrobacterium*-plant cell DNA transfer story. *Annual Review Plant Physiology and Plant Molecular Biology* 43, 465–490.

5

Methodology to Support Non-Target and Biodiversity Risk Assessment

A. Hɪʟʙᴇᴄᴋ, D.A. Aɴᴅᴏᴡ, S. Aʀᴘᴀɪᴀ, A.N.E. Bɪʀᴄʜ, E.M.G. Fᴏɴᴛᴇs, G.L. Löᴠᴇɪ, E. R. Sᴜᴊɪɪ, R.E. Wʜᴇᴀᴛʟᴇʏ ᴀɴᴅ E. Uɴᴅᴇʀᴡᴏᴏᴅ

Corresponding author: Dr Angelika Hilbeck, Geobotanical Institute, Swiss Federal Institute of Technology, Zurich, Switzerland. Fax:+41 44 6321215, e-mail: angelika.hilbeck@env.ethz.ch

Introduction

The biodiversity of an agroecosystem is important for its intrinsic value, but also because it influences ecological functions that are vital for crop production in sustainable agricultural systems, and for wildlife and the environment surrounding the agroecosystem. Changes in biodiversity could possibly alter these functions and harm the agroecosystem, the farmer and the surrounding natural ecosystems. Therefore, it is essential to assess potential risks of transgenic crop plants to non-target organisms prior to their release into the field.

Most risk-assessment frameworks for transgenic plants require a stepwise, case-specific assessment of non-target risks. In previous publications, we introduced methods to support scientific risk assessment of biodiversity and non-target risks (Andow and Hilbeck, 2004; Birch *et al.*, 2004). This chapter provides an elaboration of these methods (Box 5.1), and in addition, provides a systematic, scientifically justifiable precautionary approach for dealing with scientific uncertainty.

Box 5.1. Case-specific methodology to support risk assessment of non-target organisms applied to the Brazil case example.

Step 1. Identification of functional categories of biodiversity.*

Step 2. List and prioritize non-target species and processes (species-selection matrix):*

    (a) Association with crop: non-target × crop plant coincidence;

    (b) Association with crop: non-target species × crop plant trophic relationship;

    (c) Functional significance of non-target species in the cropping system, in the surrounding agroecosystem and in the natural areas.

Step 3. Trophically mediated exposure pathways to transgenic plant and transgene products.

Step 4. Adverse-effect scenarios for trophically mediated and other ecological effects.

Step 5. Testing hypotheses and experimental designs to test causal chains of events associated with specific exposure pathways and adverse-effect scenarios.

*Steps 1 and 2 require no information about the transgenic plant except the crop species being considered.

Methodology to Support Non-Target and Biodiversity Risk Assessment

There are thousands of species present in and near any agricultural ecosystem, and it is impossible to evaluate the risk to every species. An important step in our methodology is the scientific and transparent selection of the most relevant species or ecological processes for risk assessment in the target regions of release. The objective is to screen the spectrum of known non-target species or processes and the theoretically possible risks associated with an effect on them, in order to focus risk assessment on the most important risks. The stepwise screening progressively reduces both the number of species or processes and possible risks considered, in order to end up with the most critical ones (Fig. 5.1).

We begin by identifying and screening appropriate functional groups of biodiversity. Functional analysis of ecological communities, ecosystems and biodiversity has been important for understanding ecological processes. Functional classifications can be defined ecologically or anthropocentrically (Andow and Hilbeck, 2004), but for the purposes of risk assessment it is most important that the functional groups encompass a broad range of possible environmental services they deliver and that might be affected by the introduction of the transgenic crop (Lövei, 2001). It is possible to screen these groups to select those functions and identify those associated risks that are of greatest concern.

The next step is to compile, classify and screen the species or processes known to exist in the particular agroecosystem where a transgenic crop is

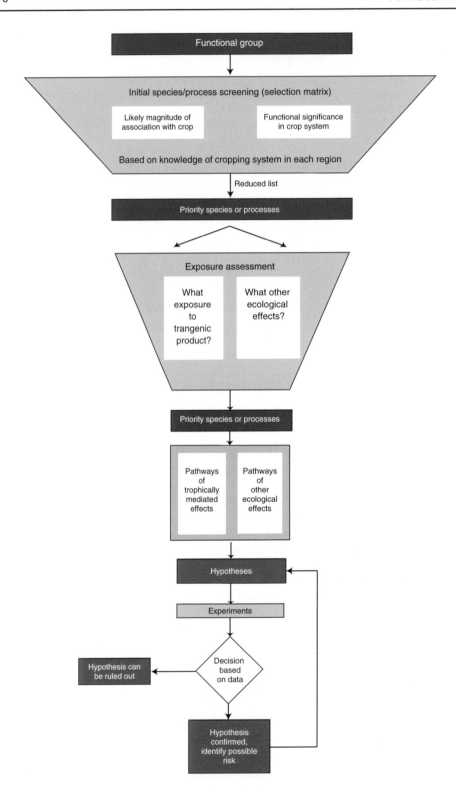

Fig. 5.1. (*opposite page*) Outline of the methodology to support non-target and biodiversity risk assessment. The objective is to screen the spectrum of known non-target species and the theoretically possible risks the introduction of the transgenic crop poses to these non-target species, in order to focus risk assessment on the most important possible risks. This stepwise screening progressively reduces both the number of species and possible risks considered, in order to end up with the most critical ones.

expected to be released. Possible adverse effects are then identified by determining the likely magnitude of association of these species or processes with the crop of concern (in this case, cotton) and their functional significance for the cropping system and nearby environment (Fig. 5.1). We assume that a greater adverse effect to the environment will result from a change in the population of a species, or a change in the rate of a process, that has greater exposure to the crop and a more significant ecological function. Depending on the ecological function being considered, the adverse effect may result from a reduction in the species population (e.g. pollinators) or an increase in the species population (e.g. herbivorous pests). This step screens all possible species or processes to produce a smaller list that should be candidates for further assessment.

After this initial screening procedure, the set of species and possible adverse effects can be further evaluated using information specific to the transgenic crop (in this case, *Bacillus thuringiensis* (Bt) cotton). This involves identifying and characterizing the possible pathways of exposure to the transgenic crop plant and its transgene products, and evaluating the likelihood and magnitude of any adverse effect that could stem from such exposure (see Grossi *et al.*, Chapter 4, this volume, for a discussion of transgene expression). By systematically considering scenarios for trophically mediated and other ecological adverse effects, we can identify one or more scenarios by which a specific transgenic crop could cause an adverse effect and formulate specific hypotheses by which this scenario can be tested (Fig. 5.1).

Finally, we develop experiments designed either to confirm or to refute the specific hypotheses. A confirmed hypothesis would provide evidence for the occurrence of the adverse-effect scenario and should be followed up with additional experiments to further substantiate the scenario and provide data for risk assessment (Fig. 5.1). If a hypothesis is refuted then the entire adverse-effect scenario can be ruled out. It might take one or several experiments to refute or confirm an adverse-effect scenario with satisfactory certainty.

The following describes in detail the main elements of this methodology for prerelease evaluations of transgenic crops. Postrelease methods can follow logically from these methods; the details will be developed in later publications.

Step 1. Identification of functional groups of biodiversity

An effective way to understand the role of biodiversity is through ecological function. Functional approaches are common in ecological theory and are also effective when knowledge about species is limited, for example in soil ecosystems. Soils are biologically, chemically and physically complex environments, and contain a large number of little-known species. Under these constraints, an operative approach is to aggregate species according to their ecosystem functions, and then evaluate risks associated with these functions. When species information is sufficient (such as for ground beetles or earthworms), species can be evaluated within these ecological functions as described below. A functional approach therefore underlies the subsequent species classification and selection.

Determining the relevant ecological functions is the critical first step in the methodology. This step involves the identification of the important functions that could be considered in the risk assessment for a given case. These functions may vary among different crop types and crop-production regions. Functions can be related primarily to human goals (referred to as anthropocentric functions), such as the maintenance of crop production or conservation goals. Functions can include ecological processes independent of specific human goals, such as nutrient cycling or species dispersal (Andow and Hilbeck, 2004). Species contributing to multiple functions can be considered under each of these functions.

In Brazil, six functional groups were identified for cotton: non-target pest herbivores, pollinators and species of conservation concern, predators, parasitoids, weeds and soil ecosystem functions. The selected groups reflect a focus on non-target risks that could adversely affect cotton production. Secondary pests, pest resurgences, reduction of pollination, loss of biological control by natural enemies, increased weed infestations and loss of soil fertility were the main potentially adverse effects addressed. These functional groups (except weeds) form the basis of the next five chapters, which report methods for the detailed assessments of the potential risks of Bt cotton on non-target organisms and soil ecological processes.

Cotton production has rapidly expanded into the Cerrado region, an important Brazilian biome with unique biodiversity. Therefore, rare, endemic or endangered species may also become potentially at risk. A full risk assessment of Bt cotton in Brazil may need to address possible risks associated with the surrounding natural habitats. Policy decisions, results of a problem formulation and option assessment (PFOA) (see Capalbo et al., Chapter 3, this volume) or heightened interest in conservation priorities may drive this need in Brazil. Arpaia et al. (Chapter 7, this volume) takes some initial steps in this direction by considering whether Lepidopteran species on an endangered species list might be associated with cotton.

Step 2. List and prioritize non-target species and processes (selection matrix)

Listing non-target species

Mobilizing and using all information and expertise available in Brazil, the known non-target species that actually occur in the cotton ecosystems are listed in the selected functional groups. It is important to list all species that occur anywhere in the entire range of production systems for the given crop. However, the species that occur in production systems where insecticides are heavily used are likely to be restricted to those that can best withstand such chemicals. The existing species community in such cropping systems is likely to be heavily biased towards 'survivors' of the 'pesticide treadmill'. Therefore, the range of species found in low-input, small-scale intercropping systems should also be considered. This needs to be taken into account, in particular, for those cases where the release of a transgenic crop targets low-input or subsistence cropping systems. Here, the value of a carefully conducted PFOA process (see Capalbo *et al.*, Chapter 3, this volume) becomes apparent in helping to shape the risk assessment process and identify relevant comparisons.

Many species found in an agricultural field may not be classifiable into a functional group because there is insufficient knowledge of the biology of these species. It is critical to consider these species with unknown function in the risk-assessment process, so that they are not overlooked inadvertently. These species can be grouped as species of unknown function. We suggest that species with unknown function or significance that have a high-standing biomass or are found in frequent association with the transgenic crop habitat should be prioritized for further evaluation, even though their significance is unknown.

Some species may be listed in more than one functional category. This should be expected, because individual species may have multiple functions in an ecosystem. For example, *Coleomegilla maculata* (Coleoptera: Coccinellidae) functions as a predaceous natural enemy of potential cotton pests as well as a pollen feeder (see Arpaia *et al.*, Chapter 7, this volume; Faria *et al.*, Chapter 8, this volume). Such species may be key species and would be examined carefully in the subsequent steps of the methodology.

Prioritization using the selection matrix

The species or processes are prioritized according to their potential association with the crop plant and the crop habitat or agroecosystem, and the significance of a possible adverse effect on their ecological functions (Box 5.2). The goal is to identify those species/processes with the greatest likelihood of being associated with the crop plant and therefore potentially affected, and the species/processes most likely to have a significant role in the crop ecosystem, which, if perturbed, would have the greatest adverse environmental effect. This approach goes beyond the simplistic notion that species abundance is a measure of ecological significance.

In this step, we consider the overall association of the species population or ecological process with the crop, without regard to the transgene phenotype. Presumably, populations that are not highly associated with the crop are unlikely to be highly exposed to any transgene product from the crop. Similarly, the significance of the non-target species is determined by considering all its ecological functions, without regard to the transgene phenotype. Again, species known to have a 'minor' ecological function in relation to crop production have a lower possibility of causing an adverse environmental effect than those that are known to have an important function. Similarly, ecological processes are considered by looking at the association of the biota responsible for the process with the crop, and the ecological significance of the process for crop production and sustainability. The criteria (Box 5.2) are systematically aligned in a selection matrix that allows an efficient and transparent prioritization process.

Box 5.2. Selection matrix.

Part I: Association with crop: non-target species × crop coincidence
Geographic distribution
The degree of overlap in the geographical distribution of the crop and the non-target species at the country or region or agroecological-zone scale (depending on what spatial scale has been chosen for the analysis).

Habitat specialization
The degree of association between the non-target species and the crop habitat. The crop habitat is defined as the crop field and its margins and includes all of the species associated with the field and its margins, including the crop, any intercrop and weeds. A habitat specialist occurs only in the crop habitat; a habitat generalist occurs in many other habitats.

Prevalence
The proportion of suitable crop habitat in which the non-target species can be reliably found. For all the suitable crop habitats (fields) in the area, in what proportion would the species be present?

Abundance
The average or typical density where the species is present. Assessment of abundance requires good field expertise with the sampling methods used to measure density, and knowledge of the typical population fluctuations of the species. Density measures can be difficult to compare across species when different sampling methods are used. Moreover, species may be difficult to compare because of vast differences in size and biology. For example, cotton aphids are small and can occur in high numbers while *Heliothis virescens* is larger and occurs in smaller numbers. Field expertise is needed to compare the relative densities of such species.

Part II: Association with crop: non-target species × crop trophic relationship
Phenology
Degree of temporal overlap of non-target species with the crop plant.

(a) From the non-target species perspective: What proportion of the non-target species life cycle takes place while the crop is alive?
(b) From the crop perspective: What proportion of the crop growing cycle is covered by the non-target species life cycle?

Linkage
For species: degree of specialization to a particular food. For herbivores this would be the degree of feeding specialization to the crop (host range) and/or for higher trophic level species this would be the degree of feeding specialization to the prey/host associated with the crop. Linkage might also be called feeding specialization and focuses on trophic relations. It can also specify what life stage of the non-target species feeds on the crop plant.

For processes: degree of association of biota responsible for the process with relevant crop plant tissues, parts, residues and secretions. For soil processes, this should include association with roots, plant parts that fall onto the soil (pollen, flowers, residue), plant residue incorporated into the soil and root exudates.

Box 5.2. Selection matrix – *cont'd.*

Part III: Functional significance for cropping system (relates to ecological function(s) of non-target species)

For species: functional significance in relation to the functional group (see step 1) in the cropping system. Examples: for predators and parasitoids rank importance as a biological control agent; for herbivores rank importance as potential secondary pest.

For species: other significance in the agroecosystem, specify functional role and rank its significance. The questions may vary according to the aim of the selection process. Example: for herbivores, consider the role as a disease vector, seed disperser, decomposer, plant biological control agent or other possible functions (see step 1).

For soil ecosystem processes: importance as an indicator of soil health. If possible, indicator organisms that are appropriate to the case study should be specified.

For processes: the significance in the functioning of the cropping system (specifying the system(s) of crop rotation, monocropping, intercropping and multiple cropping).

For processes: how directly does the ecological process affect crop development (do other ecosystem processes intervene to mitigate such effects)? This requires consideration of the coincidence between the seasonal pattern of variation in the rate of the process and the development of the crop.

As the last step in this process, a rank is assigned to each candidate species or process that summarizes the entire evaluation process of association with the crop, and another rank is assigned to the significance of the species/process in the agroecosystem. If there is sufficient information available, this ranking should be done for each region, agroecological zone or cropping system being considered. The ranks for likelihood of association with the crop and for its significance can be summed to give a final rank for each species/process. This final ranking reflects the consensus of the expert group, which underlines the importance of mobilizing and engaging all available expertise. It can be quantitative or qualitative, and is a relative rank; the species or processes are compared for each criterion, using published information, supplemented with available expert knowledge. By requiring expert consensus for each criterion, the process is transparent and the evaluations are more readily defendable.

It is important to identify existing knowledge gaps. These can be illustrated with a question mark, indicating insufficient information and uncertainty regarding that particular criterion. A highly-ranked species or process is one that may result in a greater adverse effect to the environment because it has both high maximum potential association with the crop and crop habitat and high functional significance. The result of the ranking process is a crop- and region-specific list of non-target species or processes that is prioritized for consideration in subsequent steps in a risk-assessment process.

At this point in the methodology, more species should be retained for further consideration than would be considered feasible to test. Subsequent steps in the methodology, during which the assessment is further tailored to the case, will reduce the number of candidate species. For the Brazil case study, we used a cut-off criterion that included the species of highest priority simply to reduce the number of species to a number for which the process could be completed as an illustrative case example. The number of candidate species might be expanded to include some of the lower-priority species in a full risk assessment.

Uncertainty and precaution

The gaps in knowledge in the selection matrix are identified by question marks. This uncertainty can be assessed in a precautionary manner by using a worst-case scenario. One can substitute high rankings for the question marks, and compute a final ranking under these worst-case assumptions. We have found that even when applying the worst-case scenario, it does not always result in a high final ranking for species or processes with uncertainty. Species or processes that end up with a high final ranking under the worst-case scenario usually had serious knowledge gaps that might merit research to calculate a more realistic ranking. By using this approach, we can better understand the limitations of the assessment and how significant any identified gaps in knowledge are for the risk-assessment process.

Step 3. Trophically mediated exposure to transgenic plant and transgene products

For the species or functions identified in the previous step, an exposure assessment is conducted to determine whether and to what degree the species or the biota responsible for the function come into contact with the transgene products and their metabolites. This assessment is case-specific to the transgenic crop and requires information on the phenotypic pattern of transgene expression in the various parts of the transgenic plant over the whole growing season (see Grossi *et al.*, Chapter 4, this volume). The purpose of this assessment is to differentiate candidate species or processes into those that are possibly exposed and those unlikely to be exposed to the transgene products and their metabolites in the transgenic plant.

The exposure assessment is formulated as a series of questions, allowing efficient and transparent ranking focused on potential routes of exposure to the transgene product and metabolites (Box 5.3). The analysis is carried out separately for each life cycle stage of the candidate species. It is conducted separately for each proposed growing region, when necessary. The ranking is a relative ranking of all the candidate species/processes for each of the questions. Slight variations to this approach (Box 5.3) were needed for each functional group, as illustrated in Chapters 6–9 (this volume).

Exposure can be bitrophic via exposure to the transgenic plant or plant parts, including residues and secretions that contain the transgene product,

Box 5.3. Trophically mediated exposure pathways to transgenic plant and transgene products.

1. Bitrophic exposure to transgene product and its metabolites
   (a) Exposure to plant tissues or secretions.

   (i) What plant tissues or secretions does each life stage of the species feed on?
   (ii) Do these tissues or secretions contain the transgene product or metabolites?
   (iii) Is this plant tissue or secretion an important part of its diet?

Is bitrophic exposure possible and important? (yes/no)
If yes: Is transgene product or metabolites detectable in the organism after feeding on plant tissue or secretion?
Rank. To what extent is bitrophic exposure likely to occur?
Does the plant show any other alterations that might affect its palatability for the non-target species? (e.g. changes in plant secondary chemistry, nutritional value, C:N ratio).
   (b) Exposure to plant tissues, transgene products or metabolites after they have moved.

   (i) Are there any transport processes that would move plant parts, transgene products or metabolites? Specify the process.
   (ii) Where are the plant parts, transgene products or metabolites moved to?
   (iii) How long do the products or metabolites persist?
   (iv) Does the species interact with the plant parts or transgene products or metabolites after movement?

Rank. To what extent is exposure to transported transgenic-plant tissues or the transgene product and its metabolites likely to occur?

2. Tritrophic exposure to the transgene product and/or its metabolites
   (a) Tritrophic exposure via feeding on herbivore products.

   (i) Does the species life cycle stage feed on herbivore products? (honeydew, frass, other excretions)
   (ii) Do any of these products have detectable transgene products or metabolites?
   (iii) Are herbivore products an important part of its diet?

Is tritrophic exposure via feeding on herbivore products possible and important? (yes/no)
If yes: Are the transgene products or metabolites detectable in the species life cycle stage after feeding on herbivore products?
Rank. To what extent does tritrophic exposure via herbivore products occur?
   (b) For predators and parasitoids (natural enemies) only: tritrophic exposure via prey or host.

   (i) Does the natural enemy eat a prey or parasitize a host that has fed on the transgenic plant? Which species?
   (ii) Are these prey/hosts likely to be exposed to transgene product or metabolites during the feeding period of the natural enemy?
   (iii) Are these prey or host a major or important part of its diet?

Is indirect (tritrophic) exposure via prey or hosts possible and important? (yes/no)

Continued

Box 5.3. Trophically mediated exposure pathways to transgenic plant and transgene products – *cont'd.*

If yes: Are the transgene products or metabolites detectable in the species life
cycle stage after feeding on the prey or host?
Rank. To what extent does tritrophic exposure occur via the prey or host?

3. **Higher trophic level exposure to transgene product and/or its metabolites**
   (a) Exposure via cannibalism or other intraguild feeding (for parasitoids this is
       called superparasitism)

       (i)   Does the species life cycle stage cannibalize other members of its own
             species or does it feed on other species in its own feeding guild?
       (ii) Is the species possibly exposed (from 1.a)?
       (iii) Is cannibalism an important part of its diet?

   Is indirect exposure via cannibalism possible and important? (yes/no)
   If yes: Is transgene product or metabolites detectable in the organism after canni-
   balism?
   Rank. To what extent does higher trophic level exposure via cannibalism occur?

4. **Exposure after gene flow**
   (a) Can exposure via gene flow occur? (see Johnston *et al.*, Chapter 11, this
       volume for methodology to answer this question).

       (i)   What are the likely recipients of gene flow (plants and recipients of
             horizontal gene flow)?
       (ii) Where are they located in the landscape (what habitat, landscape ele-
             ment, location relative to crop)?
       (iii) Do these recipient organisms express the transgene?
       (iv) Is the non-target species likely to encounter the recipient?
       (v)   For natural enemies: Is the non-target species likely to encounter a
             prey/host that has fed on the recipient?

   Is bitrophic or tritrophic exposure via gene-flow recipients possible? (yes/no)

5. **Behavioural modification of exposure**
   (a) Will behaviour modify exposure (e.g. searching behaviour of parasitoids
       and pollinators, oviposition behaviour, food choice behaviour such as
       avoidance of sublethally affected prey)?

       (i) What behaviours might increase exposure?
       (ii) What behaviours might decrease exposure?

   Rank. To what extent are behaviours likely to increase exposure?

or exposure can occur through higher trophic level exposure to the transgene product or metabolites in organisms that have been exposed to the
transgene product or its metabolites. Moreover, the plant parts and transgene
products can move separately from the transgenic crop, leading to exposure
in other parts of the environment, e.g. pollen, nectar, seeds or plant residues.
The transgene may move via gene flow (pollen, seed, horizontal gene flow) to
other related plants that may then express the transgene, thereby causing
exposure. For a Bt toxin that acts as a gut toxin, exposure must occur via con-

sumption to have any effect. Additionally, it must be anticipated that the transgene products or their metabolites might interact with existing plant compounds and that the result of this interaction might affect the non-target organism (Birch *et al.*, 2002; Andow and Hilbeck, 2004) or might affect the quality of the transgenic plant tissue, thereby affecting the non-target organism (Saxena and Stotzky, 2001; Birch *et al.*, 2002).

Many of the questions inquire about the importance of various foods in the diet of the species, emphasizing the role of the food for the organism rather than the quantity consumed. For example, nectar feeding in parasitoids and plant feeding in *Orius* spp. or *Chrysoperla* spp. could be considered important to adult longevity in the absence of hosts or prey. Further, organisms may be exposed via multiple routes within a single life stage. For example, *Chrysoperla carnea* larvae that are generalist predators feed on various sources of plant liquid including nectar, exudates from trichomes and plant sap leaking from wounds (Limburg and Rosenheim, 2001). Some coccinellid larvae are known to feed on pollen in addition to their typical herbivore prey, and could therefore experience combined exposure through bitrophic and tritrophic routes.

Some species will be considered unlikely to be exposed. Here, as in step 2, it might be advisable to use a cut-off criterion to select only the highest-ranked species, i.e. those most likely to be exposed. This procedure will eliminate some of the candidate species identified at step 2. Any pragmatic 'cut-off' criterion should be flexible and enable inclusion of species of lower ranking to increase the degree of precaution or to address economic or political considerations. For the Brazil case study, the cut-off criterion was the species of highest rank. This was decided simply to reduce the number of species to a number for which the process could be completed as an illustrative case example. The prioritized species or processes are therefore ecologically relevant (as determined in step 2) and likely to be exposed to the transgene products and/or its metabolites (as determined in step 3) on a specific transgenic crop in the location of the case study.

Uncertainty and precaution

It is likely that there will be significant knowledge gaps for some of the candidate species that will create a degree of uncertainty in the consequent summary exposure analysis. Similar to step 2, the following procedure may be used to address uncertainty. For each knowledge gap, we assumed a worst-case scenario (this assumes that the greatest level of exposure is possible, in the absence of published data). It is possible that even under the worst-case scenario, the expected exposure may not be considered to be significant, so it can be concluded that even with such a precautionary approach, the exposure is not significant and there is no need for additional research to clarify this uncertainty. It is also possible that the worst-case scenario would imply significant exposure that under any other assumption would be considered insignificant. By this means, important uncertainties can be identified and this straightforward and transparent assessment helps to identify and prioritize the knowledge gaps that should be filled through necessary research.

Step 4. Adverse-effect scenarios

The first three steps identify the focal species and processes for non-target risk assessment, and evaluate exposure pathways including the identification of important knowledge gaps related to exposure. The goal of step 4 is to identify all possible adverse-effect scenarios. An adverse-effect scenario is a causal chain of events that starts with exposure and ends with an identified adverse effect (Hayes *et al.*, 2004), without concern for their probability of occurrence or their magnitude. The purpose is to consider many potential scenarios and to assure a regulatory agency that all conceivable adverse effects have been considered. It is important to keep in mind that this step does not prove the occurrence of an adverse effect or the causal pathway leading to an adverse effect. Our process for identifying adverse-effect scenarios ends with a series of transparent, testable scientific hypotheses for which experiments can be developed and undertaken to confirm or refute the scenario.

The bi- and multi-trophic exposure pathways from step 3 are used to guide the development of adverse-effect scenarios for each selected species or process. For species assessments, an adverse-effect scenario could begin either with a hypothetical change in a population parameter, population density or with a behaviour resulting from the possible exposure. These changes would consequently result in a change in the ecological function or functions of the population. Such changes would then be evaluated as adverse or beneficial. In Box 5.4, guidance to identification of adverse-effect scenarios for species populations is provided in the form of a catalogue of questions addressing several generally important ecological criteria. For scenarios of ecological processes, for example in soil ecosystems, an environmental effect would be a change in the timing, rate or magnitude of that ecological process. Such changes would then be evaluated as adverse or beneficial.

These adverse-effect scenarios are possible causal pathways by which a significant adverse effect could occur. The knowledge gap(s) associated with these scenarios should be identified and, in step 5, experiments to address these gaps can be proposed.

Step 5. Hypotheses and experiments to test adverse-effect scenarios

Formulation of transparent, testable hypotheses
This step uses the information from steps 3 and 4 to construct hypotheses that can be supported or refuted by appropriate laboratory, greenhouse or field experiments. These hypotheses build on the hypothetical causal chain of events, starting with the transgenic plant and proceeding as a series of sequential causes and effects through the identified potential exposure pathway(s) and the identified adverse-effect scenario(s) for the high-priority non-target species or processes. By considering the entire hypothetical causal chain of events, it should be possible to identify specific hypotheses

---

Box 5.4. Guide to identification of pathways of potential adverse effects (adverse-effect scenarios).

Population-level effects (potential exposure affects population density or behaviour). This is the first link in a causal chain of an adverse-effect scenario.

Population components
• Is the life history or fitness parameters of the species, such as development time, survival and reproduction, affected by the potential exposure pathway? What is the scientific evidence supporting the occurrence or lack of occurrence of the effect?
• For colonial species such as bees or ants: is colony survival, reproduction or other fitness parameter affected by the potential exposure pathway? What is the scientific evidence supporting the occurrence or lack of occurrence of the effect?

Population density
• If there is a hypothesized effect on a population component, would this effect result in a higher or lower population density?
• Would the potential exposure pathway affect population density? What is the scientific evidence supporting the occurrence or lack of occurrence of a change in population density?

Behaviour
• Would the potential exposure pathway affect a behavioural parameter such as preferential host, or host plant and oviposition choices?
• Would the presence of the transgenic plant affect a behaviour?

Functional effects or consequences from population-level effects. This is the second link in a causal chain of an adverse-effect scenario.
• For each identified potential population effect that originates with a particular potential exposure pathway, what is the likely effect on the ecological functions of the species (functions can be found in the species selection matrix under significance).
• Would the potential exposure pathway affect one of the significant functions of the species? What is the scientific evidence supporting the occurrence or lack of occurrence of a change in any of the functions?

Assessment of adversity (or benefit). This is the third and final link in a causal chain of an adverse-effect scenario.
• For each functional effect identified above, determine if the effect would be considered adverse (or beneficial). It is possible that it would be considered adverse under some conditions, but not under other conditions. If so, these conditions should be specified, if possible. It is also possible that the effect would be considered adverse to some people, but not so for others. In this case, the people who would consider the effect to be adverse should be specified, if possible.

---

that can be falsified relatively easily by experiment and to efficiently test the entire hypothetical chain of events. These specific hypotheses are not null hypotheses in the statistical sense and so are not formulated as such (Andow, 2003). They are hypothetical statements about the likely occurrence of a specific causal link that could result in an adverse effect on the environment. Thus, it may be necessary to conduct a series of experiments before any particular hypothesis can be sufficiently supported or refuted. Also, the same species or process can have more than one associated possible adverse effect, which itself could be caused by more than one hypothetical causal chain of events. The end product from this step should be a set of exposure pathways and adverse-effect scenarios that have been investigated experimentally with appropriately designed testing methods. We call these 'possible risks' because they have been confirmed to be possible but have not been confirmed to occur in the actual environment.

Development of ecologically meaningful experiments

From the specific hypotheses, ecologically meaningful experimental methods and protocols can be developed. If the causal chain of events (combined exposure pathway and adverse-effect scenario) cannot be confirmed or refuted, the hypotheses may need to be revised and alternative experiments developed. This iterative process continues until the causal chain can be confirmed or refuted based on the available scientific information and expert knowledge. The first experiments are designed for the prerelease stage of transgenic plant development (see Andow *et al.*, Chapter 1, this volume), to be carried out in closed environments (laboratory or greenhouse). Additionally, field experiments that do not involve the transgenic plant can be conducted, e.g. by simulation of Bt expression using Bt insecticides. If a causal chain has been confirmed during the prerelease stage, it will be necessary to conduct experiments in the field, to determine if a risk actually occurs under normal cultivation conditions. These experiments will be carried out at the small-scale field trial stage and would typically include the transgenic plant. When developing proper experiments, a number of scientific criteria for good experimental designs are important (Box 5.5).

Depending on the hypothetical causal chain of events and gaps of knowledge, experiments should be designed that accomplish either one or both of the following objectives:

- Experiments that confirm or refute suspected exposure pathways.
- Experiments that confirm or refute an adverse-effect scenario.

Experiments that verify or refute suspected exposure pathways

These experiments verify or refute suspected exposure pathways. For example, for phloem sap-sucking insects, a suspected exposure route would be that the transgene product, e.g. a Bt toxin, is present in the phloem of the transgenic plant and ingested by aphids when feeding on the phloem. A lack of Bt toxin in the phloem of maize plants implies that no sap-sucking insect is exposed to Bt toxin on the transgenic maize events tested (Raps *et al.*, 2001). Hence, there can be no risks that extend from this pathway, and any other experiments related to this exposure adverse-effect pathway are superfluous. In contrast, Bernal *et al.* (2002) found Bt toxin in the honeydew of the brown plant hopper feeding on Bt rice that would confirm this as a suspected exposure pathway. Good experimental practice here would include testing for the

---

Box 5.5. Important aspects to follow when developing proper experiments for testing hypotheses related to possible environmental risks of transgenic plants.

If the experiment is a *test of a potential exposure pathway*, it minimally should include:

1. Methods for detecting the transgene product and if possible to detect transgene metabolites;
2. Have a proper scientific control or controls; and
3. Have sufficient replication and sufficient numbers of individuals screened, so that statistical power of the experimental design is not an issue for interpretation of results.

If the experiment is a *test of an adverse-effect scenario*, it minimally should include:

1. Food (e.g. ecologically relevant plant and or prey species) that is used by test species in their relevant habitat should be used in laboratory tests. If transgene product is used, it should be identical to what is produced in the transgenic plant;
2. Verification that the food offered to the species actually contained the administered material at the intended concentration or dose throughout the investigation;
3. Verification that all life stages of the species are exposed appropriately to the transgene product and actually contact the product in relevant ways;
4. Either use intact plants or plant parts in the experimental system with verification that the plant parts used contain the transgene product or use the transgene product at concentrations or doses much higher than normally expressed in the plant as a worst-case scenario for short-term exposure;
5. A proper scientific control or controls; and
6. Sufficient replication and sufficient numbers of individuals screened, so that statistical power of the experimental design is not an issue for interpretation of results.

---

presence of the transgene product in the relevant food/prey items of the predator or parasitoid (Zhang *et al.*, 2004).

Experiments that confirm or refute adverse-effect scenarios
Several issues are critical to the sound design of experiments to evaluate adverse-effect scenarios: exposure methodologies, measurement endpoints and ecological realism.

Exposure methodologies. Two kinds of exposure methodologies are necessary (Andow and Hilbeck, 2004; Birch *et al.*, 2004). Firstly, conventional ecotoxicology methodologies can be adapted to allow an assessment of the effects of exposure to the transgene products (Hilbeck, 2002). Secondly, a 'whole plant' methodology is required; the whole transgenic plant, not just the transgene product, has to be evaluated. This is the only scientifically practical way to include the great majority of possible exposure pathways allowing the detection of any potential effects due to altered secondary metabolism, pleiotropic effects and the various possible interactions between these individual components in the transgenic plant when growing in its normal environment.

Endpoints for species assessments. An endpoint for an experiment is the important experimental parameter(s) that provides a measure of the adverse effect on the population. An appropriate endpoint for prerelease experiments is generational relative fitness or some component of relative fitness (Andow and Hilbeck, 2004; Birch *et al.*, 2004). Generational relative fitness is the relative lifetime survival and reproduction of the non-target population throughout one entire generation of a test species, often including physiologically very distinct life stages. For example, for a holometabolous insect, this includes the egg, larva, pupa and adult. Generational relative fitness is a particularly useful endpoint, because it relates directly to the possible adverse effect. If the transgenic plant were to adversely affect a non-target species population, its effects would come through some component of relative fitness. Hence, survival experiments should estimate survival through all of the developmental stages of the non-target species, and adult life stage parameters should be measured, including age-specific mortality and female fecundity.

Mimicking ecological reality. In principle, the duration of the experiment should correspond to the time the non-target species would be exposed to the transgenic plant, plant parts and residues in relation to the temporal pattern of expression and persistence of the transgene product and its metabolites in the cropping situation. These factors must also be considered when designing surrogate experiments with microbially produced transgene products. While an integrated testing programme following this methodology can start with simple ecotoxicological tests using microbially produced transgene products (termed 'surrogate' proteins; Freese and Schubert, 2004), it is imperative that the whole plant is also presented to the test species to mimic the way the species would come into contact with the plant under field conditions. Raising plants and test species in controlled environments requires carefully designed growing conditions. If appropriate care is not taken, the experimental plants could become etiolated (elongated) with low specific leaf weights and not represent typical primary and secondary plant metabolism. The test species could become inbred, physiologically stressed, physiologically variable or even diseased. Excised plant tissues quickly change metabolically, so laboratory bioassays using excised plant material should be either short (24–48 h maximum) or newly excised plant material must be supplied every 24–48 h. However, such experiments are only an approximation of the growing transgenic crop in the field. A more detailed discussion on the whole-plant approach to assess transgenic plants is provided by Andow and Hilbeck (2004) and Birch *et al.* (2004). Suggestions on improving the ecological realism of tests on natural enemies can be found in Lövei and Arpaia (2005).

Staging the Assessment of Potential Non-Target and Biodiversity Risks

As indicated in Andow *et al.* (Chapter 1, this volume), it is important to specify how an environmental risk-assessment process can be staged to correspond with the development of a transgenic plant. Through this, risk assessment and biosafety issues actually become a constructive component of transgenic plant development and help to improve the product. Here, we suggest how the science supporting non-target risk assessment can be staged by indicating the stage during which specific scientific risk information should be collected. In addition, we describe some of the scientific requirements necessary to collect this information. This information has been compiled for flower-visiting species, herbivores, natural enemies and soil ecosystem processes in Table 5.1.

During the transformation of the plant cells, which is the earliest stage of transgenic plant development (stage 1, Table 5.1), the target plant, transgenic trait and likely target market will have been determined. At this stage, functional groups should be selected (step 1) and the selection matrix completed (step 2). Neither of these steps requires specific information about the transgene or transgene expression, and both steps are needed to plan the future risk assessment. In addition, for transgene products of microbial origin such as Bt, simple tests in closed systems using a surrogate protein product can be conducted. However, such tests can only be an initial step of an integrated testing programme, as they only provide toxicological information under artificial conditions, and are inadequate for a complete risk assessment of the transgenic plant by themselves. Plant-expressed transgene products often differ from those produced by microbes (Freese and Schubert, 2004) and interaction effects with other plant compounds and effects due to genetic engineering not necessarily related to the transgene product cannot be evaluated. However, they do help in guiding the prioritization process and the design of follow-up tests with whole-plant material.

During the stages when transgene events are screened, regenerated, selected and characterized (stages 2 and 3, Table 5.1), the risk-assessment steps 3, 4 and 5 should be completed. The initial exposure assessment (step 3) should be conducted as early as possible. Completing this step does not require detailed measurements of transgene expression on the actual transgenic plant. It can be completed using worst-case scenarios in place of definite knowledge. By doing so, it should be possible to determine the phenotypic information on transgene expression that will be needed to conduct the risk assessment. Identification of adverse-effect scenarios (step 4) and generating specific testable hypotheses (step 5) should be initiated immediately thereafter. Once these steps are completed, the necessary experiments can be designed and conducted. Some of these experiments may require characterization of the fate and transport of transgene products. These experiments may also require that purified transgene product is available for experimentation, is sufficiently characterized and that analytical detection methods are ready to use. For more sophisticated exposure pathways it may be necessary to have a sufficient supply of transgenic plant material, stable expression of the transgene product in the plant, well characterized tissue- and time-specific expression patterns and reliable *in planta* transgene product-detection methods. The experiments testing the adverse-effect scenarios will need to be adapted specifically to the specific adverse-effect scenario identified. Each of these experiments will have essential

Table 5.1. Possible staging of testing for environmental effects related to the development of a transgenic plant that expresses a bioactive protein.

Stage in transgenic-plant development	Flower-visiting species	Herbivores	Natural enemies	Soil ecology
1. Transformation of plant cells (crop and desired transgenic trait are known)	Collect information on flower-visiting species associated with crop Selection matrix procedure to determine high-priority species for risk assessment	Collect information on herbivores associated with crop Selection matrix procedure to determine high-priority species for risk	Collect information on natural enemies associated with crop Selection matrix procedure to determine high-priority species for risk assessment	Collect information on soil processes in the cropping system Exposure analysis and selection matrix procedure to prioritize soil processes assessment
Scientific requirement: Purified protein exists	Individual level ecotoxicity testing with purified (surrogate) protein[a]; measure survival, size, development time of high-priority species		Individual level ecotoxicity testing with purified (surrogate) protein[b]: measure survival, size, development time of high-priority species	Measure persistence, adsorption, ecotoxicity (micro-, meso- and macro-fauna) of purified (surrogate) proteins in main soil types of main production regions[c]
2. Event selection for multiplication of plants	Individual level eco-toxicity testing with purified (surrogate) protein: measure survival, size, development time of high-priority species	Individual level eco-toxicity testing with purified (surrogate) protein: measure survival, size, development time of high-priority species	Individual level eco-toxicity testing with purified (surrogate) protein: survival, size, development time of high-priority species	Measure persistence, adsorption, ecotoxicity (micro-, meso- and macro-fauna of purified (surrogate) proteins in main soil types of main production regions[c]

	Exposure analysis	Exposure analysis	Exposure analysis	Exposure analysis
Scientific requirements: Transgene product (protein) exists, transgene product (protein) characterized, transgene product (protein)-detection methods exist				
3. Event characterization, efficacy testing in the laboratory and greenhouse Scientific requirements: Sufficient supply of transgenic plants, stable expression of transgene product, transgenic plant well-characterized (including tissue- and time-specific expression), *in-planta* transgene product-detection methods exist	Individual-level tests and population/colony-level tests using both purified protein and transgenic plant material. Colony-level testing: verify/quantify exposure, colony development (% occupation, % brooding, queen fertility), flight activity, foraging behaviour in relation to flower(ing) characteristics	Individual-level testing using both purified protein and transgenic plant material: verify/quantify exposure, biological activity, bionomics, behavioural changes (oviposition, preference, acceptance), physiological testing [Some of the species will be part of efficacy screening in the R&D programme of the transgenic plant]	Individual-level tests and population level-tests using both purified protein and transgenic plant material: testing for direct and indirect effects, verify/quantify exposure, bitrophic tests, tritrophic tests, behaviour tests, (foraging, feeding, mating, oviposition, choice)	Transgenic plant material (laboratory, greenhouse): measure fate of plant materials and plant-produced toxin, persistence, effects on soil processes in different soil types Individual-level testing with meso- and macro-fauna: verify/quantify exposure

Continued

Table 5.1. Possible staging of testing for environmental effects related to the development of a transgenic plant that expresses a bioactive protein – *cont'd.*

Stage in transgenic-plant development	Flower-visiting species	Herbivores	Natural enemies	Soil ecology
4. Small-scale field tests	Population (colony) and community tests using transgenic plants Biodiversity (species richness, assemblages) Population dynamics (temporal abundance, etc.)	Community-level testing using transgenic plants Population dynamics Damage levels	Population and community tests using transgenic plants Biodiversity (species richness, assemblages) Population dynamics (temporal abundance, etc.)	Tests using transgenic plants and different soil types Fate and persistence of transgenic plant material and transgene product Effects on soil processes, e.g. mycorrhizae, rhizosphere functions
Scientific requirements: Access to sufficient transgenic plant material, reliable transgene product-detection methods for routine detection of large sample sizes				
Legal requirement: Ease of permission	Foraging behaviour in relation to transgenic plant characteristics (attractiveness of flowers)	Multi-species-level assemblages (biodiversity)	Foraging behaviour in relation to transgenic plant and host characteristics	Meso- and macro-fauna tests (analogous to above-ground non-target herbivores)
5. Large-scale field trials	Population (colony) and community tests using transgenic plants	Population (colony) and community tests using transgenic plants	Population and community tests using transgenic plants	Using transgenic plants in large-scale field plots
Scientific requirement: Access to sufficient transgenic plant material, reliable transgene product-detection methods for routine detection of large sample sizes	Biodiversity (species richness, assemblages)	Biodiversity (multi-species assemblages)	Biodiversity (species richness, assemblages)	Soil health/fertility

Legal requirement: Ease of permission	Population dynamics (temporal abundance, etc.) Foraging behaviour in relation to transgenic plant characteristics (attractiveness of flowers, etc.) Travel distances of pollen related to bee size	Population dynamics Damage levels on transgenic crop	Population dynamics (temporal abundance, etc.) Foraging behaviour in relation to transgenic plant characteristics Attractiveness of transgenic-plant flowers	Degradation rates and soil biomass Persistence of protein in soil in crop rotation Long-term effects on macro-invertebrates
6. Commercial release and production Scientific requirement: Reliable transgene product-detection methods for routine detection of large sample sizes	Post-commercial release monitoring[d] Biodiversity Functions	Post-commercial release monitoring Functional biodiversity Functions	Post-commercial release monitoring Biodiversity Functions	Post-commercial release monitoring Biodiversity

[a]Results of such experiments must be interpreted with caution – see Arpaia *et al.*, Chapter 7, this volume.
[b]Results of such experiments must be interpreted with caution – see Faria *et al.*, Chapter 8, this volume.
[c]Results of such experiments must be interpreted with caution – see Mendonça *et al.*, Chapter 10, this volume.
[d]Methodologies will be developed in future publications.

scientific prerequisites, but the prerequisites will be case-specific and are difficult to specify in general. Steps 3–5 can be conducted iteratively, revising the hypothesized exposure pathways and adverse-effect scenarios as experimental information is obtained. At the end of this process, it is likely that some possible risks have been supported by prerelease experiments.

During small-scale field trials (stage 4, Table 5.1), experiments addressing these possible risks can be carried out to determine if the risk actually can occur in the field. Importantly, it is likely that some experiments will give false negative results, i.e. that a risk actually occurs but the experiment cannot detect it (Marvier, 2002; Andow, 2003; Snow et al., 2005). Hence, small-scale field experiments should be aimed at evaluating the ecological mechanisms underlying the possible occurrence of the risk (e.g. Flores et al., 2005). Such experiments may require larger amounts of transgenic plant material and reliable transgene product-detection methods for routine analysis of large sample sizes. Most importantly, they require legal permits for field trials. During the large-scale field trials (stage 5, Table 5.1), the remaining risks can be examined more directly and biodiversity studies on assemblages of species can be carried out. Differences between testing schemes at stage 4 and 5 are largely related to scale, such as population versus community effects and spatial scale. After commercial release, postrelease experiments help in continued oversight and monitoring efforts, and allow verification or falsification of previously determined risks (or the absence thereof) and identification of unanticipated problems (NRC, 2002). We envision this process to be linked through feedbacks of the results gained at one level to inform further testing at the laboratory level, in particular if certain possible risks are confirmed or unanticipated effects were observed that merit further investigation. In the Brazil case study (Chapters 6–10, this volume), experiments were designed for prerelease assessments, with consideration of small-scale field trials. Large-scale field tests may also need to be planned and undertaken, based on information from small-scale field trials.

Conclusion

We have developed a scientific, case-specific, step-by-step methodology to support non-target environmental risk assessment that aims to evaluate the actual potential environmental risks of a transgenic plant rather than rely on indicator species. This methodology is a screening process that considers all possible non-target species and adverse effects and eliminates those that are less likely to result in an adverse effect on the environment. It starts by using specific information about the crop and geographical region to develop a list of non-target species and ecological functions that could be most at risk. It proceeds through a species-selection process based on exposure to the crop and the ecological significance of the non-target species. It continues by screening species through potential exposure to the transgene products and identifying scenarios resulting in a possible adverse environmental effect. This process provides not only a list of the most exposed species for testing, but also allows for estimates of the potential impact and identifies causal chains leading to possible adverse effects that lend themselves to specific testable hypotheses. By incorporating a method for dealing with uncertainty, our methodology also allows for a prioritizing of data needs. In the long run, we expect that this focus on testable hypotheses and prioritization of data needs will result in the optimum allocation of restricted resources for risk assessment, so that the greatest efforts to assess those risks are expended on transgenic plants that pose the most serious environmental risks.

In the present chapter, we focus on the methodology to support risk assessment of a transgenic crop prior to its release into the environment. We have proposed how the steps in the methodology can be staged in relation to the development of the transgenic crop. By staging the process, we can ensure that efforts are not wasted by conducting parts of an assessment too early, and that efforts yield the highest return, by conducting the appropriate parts of a risk assessment at the appropriate time. In addition, we have suggested how the methodology could be extended to small-scale field testing, large-scale field testing and post-commercial evaluation.

In the following five chapters, this methodology is illustrated using Bt cotton and non-target herbivores, pollinators and pollen visitors, including some species of conservation concern, predators, parasitoids and soil ecosystem functions. In these chapters, steps 1–3 are tested and steps 4 and 5 are illustrated by example. The functional groups were chosen by Brazilian cotton experts because they reflected a concern that adverse effects on non-targets could damage cotton production. We expect, however, that the case study presented in this book will need to be expanded before a full risk assessment of Bt cotton for Brazil is accomplished.

Acknowledgements

Table 5.1 was developed at the Brazil workshop with the participation of Sonja Righetti, Mamoudou Sétamou, Leda C. Mendonça-Hagler and Gaetan Dubois, as well as co-authors of this chapter.

References

Andow, D.A. (2003) Negative and positive data, statistical power, and confidence intervals. *Environmental Biosafety Research* 2, 75–80.

Andow, D.A. and Hilbeck, A. (2004) Science-based risk assessment for non-target effects of transgenic crops. *BioScience* 54, 637–649.

Bernal, C.C., Aguda, R.M. and Cohen, M.B. (2002) Effect of rice lines transformed with *Bacillus thuringiensis* toxin genes on the brown planthopper and its predator *Cyrtorhinus lividipennis*. *Entomologia Experimentalis et Applicata* 102, 21–28.

Birch, A.N.E., Geoghegan, I.E., Griffiths, D.W. and McNicol, J.W. (2002) The effect of genetic transformations for pest resistance on foliar solanidine-based glycoalkaloids of potato. *Annals of Applied Biology* 140, 143–149.

Birch, A.N.E., Wheatley, R., Anyango, B., Arpaia, S., Capalbo, D., Getu Degaga, E., Fontes, E., Kalama, P., Lelmen, E., Lövei, G., Melo, I.S., Muyekho, F., Ngi-Song, A., Ochiendo, D., Ogwang, J., Pitelli, R., Sétamou, M., Sithanantham, S., Smith, J., Nguyen Van Son, Songa, J., Sujii, E., Tran Quang Tan, Wan, F.-H. and Hilbeck, A. (2004) Biodiversity and non-target impacts: a case study of Bt maize in Kenya. In: Hilbeck, A. and Andow, D.A. (eds) *Environmental Risk Assessment of Transgenic Organisms: A Case Study of Bt Maize in Kenya.* CAB International Publishing, Wallingford, UK.

Flores, S., Saxena, D. and Stotzky, G. (2005). Transgenic Bt plants decompose less in soil than non-Bt plants. *Soil Biology and Biochemistry* 37(6), 1073–1082.

Freese, W. and Schubert, D. (2004). Safety testing and regulation of genetically engineered foods. *Biotechnology and Genetic Engineering Reviews* 21, 1–24.

Hayes, K.R., Gregg, P.C., Gupta, V.V.S.R., Jessop, R., Lonsdale, M., Sindel, B., Stanley, J. and Williams, C.K. (2004) Identifying hazards in complex ecological systems. Part 3: Hierarchical holographic model for herbicide tolerant oilseed rape. *Environmental Biosafety Research* 3, 109–128.

Hilbeck, A. (2002) Transgenic host plant resistance and nontarget effects. In: Letourneau, D. and Burrows, B. (eds) *Assessing Environmental and Human Health Effects of Transgenic Organisms.* CRC Press LLC, Boca Raton, USA, pp. 167–185.

Limburg, D.D. and Rosenheim, J.A. (2001) Extrafloral nectar consumption and its influence on survival and development of an omnivorous predator, larval *Chrysoperla plorabunda* (Neuroptera: Chrysopidae). *Environmental Entomology* 30, 595–603.

Lövei, G. (2001) Ecological risks and benefits of transgenic plants. *New Zealand Plant Protection* 54, 93–100.

Lövei, G. and Arpaia, S. (2005) The impact of transgenic plants on natural enemies: a critical review of laboratory studies. *Entomologia Experimentalis et Applicata* 114, 1–14.

Marvier, M. (2002) Improving risk assessment for nontarget safety of transgenic crops. *Ecological Applications* 12(4), 1119–1124.

NRC (2002) *Environmental Effects of Transgenic Plants: The Scope and Adequacy of Regulation.* National Academy Press, Washington, DC.

Raps, A., Kehr, J., Gugerli, P., Moar, W.J., Bigler F. and Hilbeck, A. (2001) Immunological analysis of phloem sap of *Bacillus thuringiensis* corn and of the nontarget herbivore *Rhopalosiphum padi* (Homoptera: Aphididae) for the presence of Cry1Ab. *Molecular Ecology* 10, 525–533.

Saxena, D. and Stotzky, G. (2001) Bt corn has higher lignin content than non-Bt corn. *American Journal of Botany* 88, 1704–1706.

Snow, A.A., Andow, D.A., Gepts, P., Hallerman, E.M., Power, A., Tiedje, J.M. and Wolfenbarger, L.L. (2005) Genetically modified organisms and the environment: current status and recommendation. *Ecological Applications* 15(2), 377–404.

Zhang, G.-F., Wan, F.-H., Guo, J.-Y. and Hou, M.-L. (2004) Expression of Bt toxin in transgenic Bt cotton and its transmission through pests *Helicoverpa armigera* and *Aphis gossypii* to natural enemy *Propylea japonica* in cotton plots. *Acta Entomologica Sinica* 47, 334–341.

6 Non-Target and Biodiversity Impacts on Non-Target Herbivorous Pests

E.R. Sujii, G.L. Lövei, M. Sétamou, P. Silvie, M.G. Fernandes, G.S.J. Dubois and R.P. Almeida

Corresponding author: Edison R. Sujii, Embrapa Genetic Resources and Biotechnology (CENARGEN), Parque Estacão Biológica, Av. W5 Norte – Final, 70770-900 Brasília DF, Brazil. Fax: +55 61 34484673, e-mail: sujii@cenargen.embrapa.br

Introduction

Herbivore species present in crop fields can potentially reach damaging densities, depending on a number of factors that are strongly influenced by the traits of the plants that the animal community exploits. Changes in crop traits that aim to suppress the occurrence of dominant pests, such as the expression of toxins in transgenic plants that confer resistance to particular pests (e.g. *Bacillus thuringiensis* (Bt) toxin against Lepidoptera) can cause guild rearrangement. These changes in the community structure can lead to unexpected outbreaks of some non-target herbivore species because of: (i) direct effects on the fitness and behaviour of these species; or (ii) indirect effects as these herbivores escape from intraguild competition, insecticides and/or predation. For example, in US Bt-cotton fields, pentatomids such as *Nezara viridula*, *Acrosternum hilare* and *Euschistus servus* are secondary pests with known potential to become major pests (Greene and Turnipseed, 1996; Greene and Capps, 2002; Turnipseed *et al.*, 2002). According to Lutrell *et al.* (1994), herbivorous insects on cotton are an important production-limiting factor globally, due to the high losses they can cause, both in yield and fibre quality. There could be more than 1000 insect species in cotton agro-ecosystems (Luttrell *et al.*, 1994; Matthews, 1994); however, the number of economically important pests is relatively small. Since very little is known about herbivore species of cultural or conservation concern in cotton fields in Brazil, we restricted our evaluations in this chapter to herbivores that have a known potential to become an agronomic pest problem. However, we do recommend that a full risk assessment of Bt cotton in Brazil should also include other herbivores in the assessment that have not yet been recorded as a pest but might have the potential to become one. Pollen-feeders and pollinators are considered in Arpaia *et al.* (Chapter 7, this volume).

Lepidoptera are among the most important pests worldwide (Lutrell *et al.*, 1994). The available options for reducing the population density of these pests on cotton in Brazil include insecticide application, biological control (using predators and parasitoids, also known as natural enemies) and organic production programmes (see Fontes *et al.*, Chapter 2, this volume). Transgenic Bt cotton is considered a viable alternative in both small and big production systems (Benedict *et al.*, 2000; Pray *et al.*, 2001).

Cotton-growing in Brazil started in the North-east and Meridian regions, but gradually shifted to the current cotton belt of Brazil, situated in the Cerrado biome in the Midwest region (see Fontes *et al.*, Chapter 2, this volume). Currently 76.5% of the national cotton-production area is in this region (CONAB, 2003). As a consequence of this change, many of the historical records of insects on cotton in Brazil do not come from the currently most important region. Therefore, information on cotton pests and other herbivores from the literature was compared and complemented with field knowledge of entomologists working in the different regions. Due to substantial differences in environmental conditions, cotton-growing practices, landscape structure and cotton growing history, adverse-effect scenarios should be developed region-by-region. A general 'Brazilian' approach would be too general and less effective.

There is a substantial amount of information on cotton-associated arthropods in Brazil, often published in proceedings of cotton growers' meetings, publications from the national entomological and agricultural scientific journals (e.g. *Neotropical Entomology*, *Pesquisa Agropecuária Brasileira*, *O Biológico*), and technical bulletins of research centres and papers in non-peer-reviewed publications. This information comes from the different regions, reflecting the historical development of the crop in the country, and it is mainly oriented towards the identification of pests and their management. Thus, this information is fragmented, was collected using different methods, and has never been synthesized. A full risk assessment of Bt cotton in Brazil would require a comprehensive synthesis of the existing information.

Species Selection: Listing and Prioritizing Species

Our methodology provides a screening procedure for adverse effects to non-target species (Box 5.1 and Fig. 5.1, Hilbeck *et al.*, Chapter 5, this volume). We identify the non-target herbivores as an important functional group, and list the species for this group found on cotton in Brazil. For each species, we evaluate the likelihood of association with the crop and crop habitat and possible adverse effects on the cropping system caused by an effect on the species (i.e. the significance of the species for the cropping system). We combine the likelihood of association with the crop and significance in the cropping system to prioritize species for subsequent evaluation. We develop exposure and adverse-effect scenarios for the highest priority species by identifying trophic-exposure pathways and other ecological adverse-effect pathways. We then formulate specific hypotheses to test these scenarios, and describe experiments that can test these hypotheses in the laboratory and field. This case-example analysis does not constitute a complete risk assessment of the effects of Bt cotton on the functional group of herbivores in Brazil; a full characterization of risk would need to be carried out after doing initial laboratory experiments and include field experiments of exposure.

We compiled a detailed list of economically important arthropod pests attacking cotton plants in Brazil (see Table 6.1), and this formed the basis of the following assessment as outlined in Hilbeck *et al.* (Chapter 5, this volume). Table 6.1 contains 35 species of herbivorous insects and three species of Acari (Gondim *et al.*, 2001; Silvie *et al.*, 2001; Gallo *et al.*, 2002). In the USA, only five to ten key pests cause significant damage in productive cotton systems (Bachelor and Bradley, 1989). However, surveys of the species-rich fauna associated with cotton in Brazil are rare and were conducted earlier (e.g. Silva *et al.*, 1968, estimated around 259 insect species associated with the cotton agroecosystem). Most recent publications report on data collected in other countries and regions of South America. For example, a study conducted in Paraguay, a neighbouring country south of Brazil, reports more than 300 arthropod species on cotton (Michel and Prudent, 1985; Michel, 1989). From this, we estimated that cotton in Brazil likely harbours potential pests from six orders and 14 families, many of which are not targeted by the Bt toxin incorporated in currently available transgenic cotton.

One important question in risk assessment is to know the potential direct and indirect impacts of transgenic Bt crops on non-target pest species, both at individual and community levels. Will the introduction of Bt cotton targeting Lepidopteran pests change the pest-management programme and favour vectors of blue disease or the boll weevil? Will a possible reduction of pesticide use against the target pest species in Bt cotton be associated with an increase in natural enemy populations, and consequently a reduction in population levels of non-target species? Or could natural enemy populations be reduced because the disappearance of the target species means they lack sufficient prey in Bt cotton? Could natural enemy populations be reduced due to a negative impact of the Bt crop on them? Reductions in natural enemy populations may favour secondary-pest outbreaks. Or could the disappearance of the target species release non-target pests from intra- and interspecific competition? These are questions that should be addressed in the assessment of potential impacts on non-target pests before the introduction of Bt cotton.

To address these questions systematically, it is important to prioritize the non-target pest species to include in a risk assessment. It would be impossible to test all of the species listed in Table 6.1. This process is facilitated by determining the 'target pest species'. We identified a group of susceptible Lepidoptera as target pest species in Brazil of the Bt cotton considered in this case study, including *Alabama argillacea* (Hübner), *Heliothis virescens* (Fabricius), *Pectinophora*

Table 6.1. List of pest herbivore taxa known to occur on cotton in Brazil and their feeding-guild categorization.

Feeding guild	Description of pest taxa		
	Species or species group	Order and family	Common name
Leaf sucker	*Tetranychus urticae* (Koch), *T. ludeni* (Zacher)	Acarina: Tetranychidae	Two-spotted spider mite and red spider mite
Leaf sucker	*Polyphagotarsonemus latus* (Banks)	Acarina: Tetranychidae	Broad mite
Defoliator	*Alabama argillacea* (Hüebner)	Lepidoptera: Noctuidae	Cotton leafworm
Stem cutter	*Agrotis ipsilon* (Hufnagel) and *Agrotis* sp.	Lepidoptera: Noctuidae	Black cutworm
Square, boll borer	*Spodoptera frugiperda* (J.E. Smith)	Lepidoptera: Noctuidae	Fall armyworm
Defoliator	*Spodoptera eridania* (Crotch), *S. exígua* (Hüebner), *S. ornithogalli* (Gueneé), *S. cosmioides* (Walker) and *S. sunia* (Guenné)	Lepidoptera: Noctuidae	Leafworm
Defoliator	*Trichoplusia ni* (Hübner)	Lepidoptera: Noctuidae	Cotton looper
Defoliator, square	*Pseudoplusia includens* (Walker)	Lepidoptera: Noctuidae	Semi-looper
Square, boll borer	*Pectinophora gossypiella* (Saunders)	Lepidoptera: Gelechiidae	Pink bollworm
Square, boll borer	*Heliothis virescens* (Fabricius)	Lepidoptera: Noctuidae	Tobacco budworm
Square, boll borer	*Helicoverpa zea* (Boddie)	Lepidoptera: Noctuidae	Budworm
Square, boll borer	*Anthonomus grandis* (Boheman)	Coleoptera: Curculionidae	Boll weevil
Root borer	*Chalcodermus niger* (Hustache)	Coleoptera: Curculionidae	Cotton root borer
Stem borer	*Eutinobothrus brasiliensis* (Hambleton)	Coleoptera: Curculionidae	Cotton root borer
Stem borer	*Conotrachelus denieri* (Hustache)	Coleoptera: Curculionidae	Cotton stem borer
Defoliator	*Typophorus nigritus* (Crotch)	Coleoptera: Chrysomelidae	Leaf beetle

Table 6.1. List of pest herbivore taxa known to occur on cotton in Brazil and their feeding-guild categorization – *cont'd.*

Defoliator	*Diabrotica speciosa* (Genmar)	Coleoptera: Chrysomelidae	Leaf beetle
Defoliator	*Costalimaita ferruginea vulgata* (Fabricius)	Coleoptera: Chrysomelidae	Leaf beetle
Leaf scraper	*Thrips tabaci* (Lindeman)	Thysanoptera: Thripidae	Onion thrips
Leaf scraper	*Franklinella schultzei* (Trybom)	Thysanoptera: Thripidae	Thrips
Leaf scraper	*Caliothrips brasiliensis* (Morgan)	Thysanoptera: Thripidae	Thrips
Root sucker	*Atarsocoris brachiariae* (Becker), *Scaptocoris castanea* (Perty)	Hemiptera: Cydnidae	Brown bug
Stem, leaf sucker	*Horciasoides nobilellus* (Bergston)	Hemiptera: Miridae	Plantbug
Stem, leaf sucker	*Aphis gossypii* (Glover)	Hemiptera: Aphididae	Cotton aphid
Leaf sucker	*Bemisia tabaci* (Gennaduis)	Hemiptera: Aleyrodidae	Whitefly
Stem, leaf sucker	*Agallia albidula* (Uhler), *Agallia* sp., *Sonesimia grossa* (Signoret), *Xenophloeae viridis* F.	Hemiptera: Cicadellidae	Cicads
Stem, leaf sucker	*Lygus* sp.	Hemiptera: Miridae	Lygus bug
Stem, leaf sucker	*Gargaphia torresi* (Lima)	Hemiptera: Tingidae	Lacebug
Stem, leaf sucker	*Nezara viridula* L.	Hemiptera: Pentatomidae	Green stink bug
Seed sucker	*Dysdercus* spp.	Hemiptera: Pyrrhocoridae	Cotton stainer

gossypiella (Saunders) and *Agrotis ipsilon* (Hufnagel). These target pest species were therefore removed from the initial list (Table 6.1). We considered the *Spodoptera* species as non-target pests, because they will not be well controlled by Cry1Ac Bt cotton (Moore and Navon, 1973; Höfte and Whiteley, 1989; Müller-Cohn *et al.*, 1996). Less common pest species were also removed from the initial list (Table 6.1), because they were considered economically unimportant due to their low damage and low population-increase potential. The species-selection procedure was conducted with the remaining 20 pest species (see Tables 6.2–6.4). The matrix presents the results of our consensus evaluation of the potential association between these selected arthropods and cotton, and the potential environmental significance of each remaining species. Population distribution in space and time of each herbivore species, and its trophic interaction with cotton and other plants were evaluated in the assessment of potential association with the crop and crop habitat (Table 6.2A and 6.2B). At this stage, no information about expression of the Bt transgenes in cotton was required. The environmental significance of each species in various ecological roles was a second set of criteria used to compare the species (Table 6.3). We scored each species for each criterion ranging from one (strong association, high importance) to three (weak association, low importance; see the footnote of Table 6.4 for further detail). Each score was the result of a consensus expert opinion of the chapter authors, considering the three regions and different crop systems used throughout the country based on available literature on pests in different regions and experience of the authors or consulted experts. The two scores were added for each species to give the overall rank.

From these assessments, it was possible to prioritize a group of four important non-target pest species for their close association with cotton, general occurrence in the cotton-growing regions of Brazil and their potential for economic damage if they increase their populations. These herbivores are:

1. Cotton aphid, *Aphis gossypii* (Glover) (Hemiptera: Aphididae).
2. Fall armyworm, *Spodoptera frugiperda* (J.E. Smith) (Lepidoptera: Noctuidae).
3. Boll weevil, *Anthonomus grandis* (Boheman) (Coleoptera: Curculionidae).
4. Whitefly, *Bemisia tabaci* (Gennadius) (Hemiptera: Aleyrodidae).

We note that for judging the importance of phytophagous species, some modification of the criteria of the selection matrix would be useful (Table 6.2A). Among all the factors that we analysed, 'abundance' and 'damage potential' (considered in Tables 6.2A and 6.2B) were particularly important for species ranking and determination of pest species for further evaluation.

Exposure Assessment

The four identified non-target herbivore species were further evaluated to determine if they could be exposed to the toxin expressed by the Bt cotton. All of these species have direct and indirect potential exposure to the transgene, toxin and secondary metabolites (Andow and Hilbeck, 2004). Different pathways leading to exposure to the transgene product or its metabolites were evaluated in a second matrix for exposure analysis (Table 6.5; Hilbeck *et al.*, Chapter 5, this volume). This second evaluation round concluded that all four species as key non-target pests are likely to be directly exposed to Bt containing plant tissues. Furthermore, the cotton aphid, fall armyworm and boll weevil each have a different feeding mode and consequently need a different set of experiments to evaluate the possible ecological risks associated with the

Table 6.2A. Selection matrix – association with the crop and crop habitat.

(A) Non-target pest species × crop coincidence
Geographical distribution: degree of overlap of cotton crop and non-target taxon.
Habitat specialization: degree of association between taxon and cotton habitat.
Prevalence: proportion of cotton habitat occupied by taxon.
Abundance on cotton crop and on other plants occurring in the cotton habitat: average or typical densities.
For definitions see Box 5.2 in Hilbeck et al., Chapter 5, this volume.

Selection matrix	Association with crop – coincidence				
Non-target pest taxon	Geographical distribution	Habitat specialization	Prevalence in the suitable cotton habitat	Abundance on cotton crop	Abundance on other plants
Tetranychus urticae, T. ludeni	All three regions	Generalist	Low	Variable	Variable
Polyphagotarsonemus latus	Restricted	Generalist	Low	Variable	Rare
Spodoptera frugiperda	All three regions	Generalist	High	Variable, high in MW	High
Anthonomus grandis	All three regions	Specialist	High	High	Low, adults eat pollen
Chalcodermus niger	Midwest (MW)	Specialist	Variable	Variable	Low
Eutinobothrus brasiliensis	MW	Specialist	Variable	Variable	Low
Conotrachelus denieri	South	Specialist	Variable	Variable	No info
Diabrotica speciosa	All three regions	Generalist	Low	Low	Variable
Costalimaita ferruginea, Typophorus nigritus	All three regions	Generalist	Low	Low	High
Franklinella schultzei	All three regions	Generalist	High	Variable	Variable
Atarsocoris brachiariae, Scaptocoris castanea	South and MW	Restricted range	Variable	Variable	Variable
Horciasoides nobilellus	All three regions	Restricted	Variable	Low	Low
Aphis gossypii	All three regions	Generalist	High	High	High
Bemisia tabaci	All three regions	Generalist	High	High	High
Agallia albidula	All three regions	Restricted	Variable	Variable	Variable
Lygus sp.	All three regions	No info	Low	Variable	Low
Gargaphia torresi	South	No info	Low	No info	No info
Nezara viridula	South and MW	Restricted	Low	Low	Variable
Dysdercus spp.	All three regions	Restricted	Low	Variable	Low

Table 6.2B. Selection matrix – association with the crop.

(B) Non-target pest species × crop trophic relationship
Phenology: temporal overlap (i) from non-target taxon perspective, (ii) from crop perspective.
Linkage: specialization of trophic relationship and dependency on cotton.
For definitions see Box 5.2 in Hilbeck *et al.*, Chapter 5, this volume.

Selection matrix	Association with crop – trophic relationship		
Non-target pest taxon	Phenology of pest taxon and cotton crop		Linkage with crop
	Degree of overlap of pest-taxon life cycle with cotton	Proportion of cotton-growing season when taxon is present	Degree of feeding specialization and dependency
Tetranychus urticae, T. ludeni	High	80%	All stages: low, non-dependent
Polyphagotar-sonemus latus	High	80%	All stages: low, non-dependent
Spodoptera frugiperda	High	40%	All stages: low, non-dependent
Anthonomus grandis	High	70%	Larvae: obligatory on cotton species, rarely on other Malvaceae – *Cienfuegosia, Hibiscus, Abutilon* Adults: low, eat pollen of various plants, but if cotton is present they will feed on it
Chalcodermus niger	High	No info	No info
Eutinobothrus brasiliensis	High	No info	Larvae: low, non-dependent (mainly maize)
Conotrachelus denieri	High	50%	No info
Diabrotica speciosa	Partial	80%	Adults eat young leaves and pollen
Costalimaita ferruginea, Typophorus nigritus	Partial	30%	Adults eat young leaves
Franklinella schultzei	High	30%	All stages: low, non-dependent (vegetable crops – onions, garlic and other related plants and weeds)

Table 6.2B. Selection matrix – association with the crop – *cont'd.*

Atarsocoris brachiariae, Scaptocoris castanea	Partial	100%	No info
Horciasoides nobilellus	High	40%	No info
Aphis gossypii	High	100%	No info
Bemisia tabaci	High	100%	All stages: wider range, do like cotton
Agallia albidula	No info	70%	No info
Lygus sp.	No info	No info	No info
Gargaphia torresi	No info	No info	Nymphs and adults: low, non-dependent (mainly on Brassicaceae)
Nezara viridula	Partial	No info	Nymphs and adults: low, non-dependent
Dysdercus spp.	No info	100%	Nymphs and adults: Malvales, e.g. *Hibiscus*

Table 6.3. Selection matrix – functional significance for cropping system (related to significance of adverse effect).

Selection matrix	Significance as:			
Non-target pest taxon	Pest of cotton	Vector of disease	Food for natural enemies	Pest of other crops
Tetranychus urticae, T. ludeni	Variable	No	Restricted	Variable
Polyphagotarsonemus latus	Variable	No	Restricted	Variable
Spodoptera frugiperda	Primary	No	Broad	Main
Anthonomus grandis	Primary	No	Restricted	No
Chalcodermus niger	Variable	No	Restricted	No info
Eutinobrothus brasiliensis	Variable	No	Restricted	No info
Conotrachelus denieri	Variable	No	Restricted	No
Diabrotica speciosa	Secondary	No	Restricted	Variable
Costalimaita ferruginea, Typophorus nigritus	Secondary	No	Restricted	Variable

Continued

Table 6.3. Selection matrix – functional significance for cropping system (related to significance of adverse effect) – *cont'd.*

Selection matrix	Significance as:			
Non-target pest taxon	Pest of cotton	Vector of disease	Food for natural enemies	Pest of other crops
Franklinella schultzei	Variable	No	Restricted	Variable
Atarsocoris brachiariae, Scaptocoris castanea	Secondary	No	No info	Variable
Horciasoides nobilellus	Secondary	No	No info	No
Aphis gossypii	Primary	Yes	Broad	Main
Bemisia tabaci	Primary	Yes	Broad	Main
Agallia albidula	Secondary	No	Restricted	No
Lygus sp.	Secondary	No	Restricted	No
Gargaphia torresi	Secondary	No	No info	No
Nezara viridula	Secondary	No	Restricted	Main
Dysdercus spp.	Variable	No	Restricted	No

commercialization of Bt cotton. The whitefly, *B. tabaci,* has a feeding mechanism very similar to that of aphids.

Because the authors have complementary expertise on cotton pests and their control strategies, knowledge gaps did not seriously affect the outcome of the selection process and exposure assessment. For example, in the case of several secondary, only locally important pests, such as lacebugs, lygus bugs and stem borers, we did not have all the ecological information to complete the assessment of potential exposure to the crop in the selection matrices. Similarities in the ecology of these potential pests to related taxa in other regions were used to address these gaps. However, in the case of complete absence of information 'the worst case scenario' was applied by assuming that the species has the potential to become an important pest (see Hilbeck *et al.,* Chapter 5, this volume).

The current Brazilian regulations (CTNBio[1] Guideline IN 3 published on 12 November 1996) require some results and justification for authorization to conduct small-scale field tests (this can be an area of up to 5 ha). Consequently, the selection and exposure assessment matrices should be completed prior to the first small-scale field test of a transgenic plant when an elite event is available and can be backcrossed to local varieties (see Andow *et al.,* Chapter 1, this volume).

---

[1] Source: http://www.mct.gov.br/legis/biosseguranca.htm

Table 6.4. Selection matrix: prioritized species.[a]

Combination of association with crop (average estimates from Tables 6.2A–B) and functional significance in the cropping system (estimate from Table 6.3). See Hilbeck *et al.*, Chapter 5, this volume for more detail on the procedure followed.

Selection matrix	Ranking of expert group		
Non-target pest taxon	Association with crop	Functional significance for cropping system	Prioritized rank for species[a]
Tetranychus urticae, T. ludeni	2	2	4
Polyphagotarsonemus latus	3	2	5
Spodoptera frugiperda	1	1	2
Anthonomus grandis	1	1	2
Chalcodermus niger	2	3	5
Eutinobothrus brasiliensis	2	3	5
Conotrachelus denieri	1	2	3
Diabrotica speciosa	3	3	6
Costalimaita ferruginea, Typophorus nigritus	3	2	5
Franklinella schultzei	2	2	4
Atarsocoris brachiariae, Scaptocoris castanea	2	3	5
Horciasoides nobilellus	3	3	6
Aphis gossypii	1	1	2
Bemisia tabaci	1	1	2
Agallia albidula	2	3	5
Lygus sp.	3	3	6
Gargaphia torresi	3	3	6
Nezara viridula	3	2	5
Dysdercus spp.	2	3	5

[a]1–2 = highest priority (widely distributed, abundant, mostly present, closely linked, environmental effects significant); 3–4 = intermediate priority; 5–6 = low priority (distribution, abundance and presence limited, loosely linked, environmental effect not significant).

Identification of Adverse-Effect Scenarios

A possible population outbreak of the selected non-target species leading to higher crop losses from pest damage was the adverse-effect scenario identified from our preliminary evaluation for all four identified non-target species. This could offset the expected pest-control benefits of Bt cotton.

The cotton aphid, *A. gossypii*, is an important cotton pest. It damages the crops by sucking plant phloem sap and causing injury on new leaves of young plants, producing honeydew and improving the substrate for the growth of fungi that lower the quality of the fibre. Aphids can also act as a vector of virus disease (e.g. blue disease) on susceptible cotton varieties. Outbreaks of aphids would lead to increased use of broad-spectrum chemical insecticides on cotton

Table 6.5. Exposure-assessment matrix: exposure to transgenic crop tissues and transgene products.

See Box 5.3 in Hilbeck *et al.* (Chapter 5, this volume) for more detail on the questions and procedure followed.

Species name	Spodoptera frugiperda	Anthonomus grandis	Aphis gossypii	Bemisia tabaci
Plant tissues or secretions on which the taxon feeds	Larvae: leaf tissues and squares	Larvae and adults: squares	All stages: phloem Adults: leaves and terminal buds of vegetative phase, pollen	All stages: phloem
Do these tissues/secretions express the transgene product?	Yes	Yes	Unknown	Unknown
Is this tissue/secretion an important part of its diet?	Yes	Yes	Yes	Yes
Is direct (bitrophic) exposure possible?	Highly possible	Highly possible – unknown	Highly possible – unknown	Highly possible – unknown
Does taxon feed on other spp. excretions or products (e.g. honeydew, frass, faeces?)	No (only unintentionally)	No	No	No
Do these excretions/products contain transgene product or metabolites?	—	—	—	—
Are these excretions/products an important part of its diet?	—	—	—	—
Is indirect (tritrophic) exposure possible via feeding on herbivore products?	+?	—	—	—
Any cannibalism or feeding on other species?	Yes	No	No	No
Does the species eaten feed on transgene-expressing tissues?	Yes	—	—	—
Is cannibalism or feeding on other species (intraguild feeding) an important part of its diet?	No	—	—	—

	Col 1	Col 2	Col 3	Col 4
Is exposure via cannibalism or other intraguild feeding possible?	Possible – unknown	—	—	—
What changes in plant characters could increase exposure?	Plant secondary metabolites and volatiles	Plant secondary metabolites and volatiles	Plant secondary metabolites and volatiles, nutrient content of phloem	Plant secondary metabolites and volatiles
What behaviours can modify exposure?	Larval host plant choice, adult oviposition behaviour	Larval host plant choice, adult oviposition behaviour	Larval host plant choice, adult oviposition behaviour	Larval host plant choice, adult oviposition behaviour
Are changes in plant characters and/or behaviours likely to increase exposure?	Possible – unknown	Possible – unknown	Possible – unknown	Possible – unknown
Ranking of probability of exposure to transgenic plant and transgene product	1	1	2	2

crops, which could cause a reduction of natural enemy populations and therefore less natural biological control.

Fall armyworm attacks leaves and squares of cotton producing direct
damage and reducing boll and fibre production. The low susceptibility of
S. frugiperda to Bt toxins expressed in Bt cotton (e.g. Bollgard and Bollgard II)
is an additional concern due to the potential for rapid resistance evolution (see
Fitt *et al.*, Chapter 12, this volume).

Boll weevil is a main pest in the cotton crops where it occurs due to the
heavy damage it causes. The adults puncture the squares and bolls to feed on
the tissues inside and to lay eggs. Damaged squares drop off or hang withered
and dry. The larvae feed inside the squares and bolls, destroying or reducing
the production. The yield losses caused by boll weevil range from 58% to
84% under high infestations (Ramalho, 1994). Increases in boll weevil populations due to changes in the plant as a nutritional source for the insect represent a possible adverse effect to be evaluated.

Whiteflies are also important pests of cotton, particularly because, like
aphids, the honeydew they produce makes the cotton 'sticky', which hinders
mechanical spinning and stimulates fungal infection. They also transmit disease, but as they do not transmit in a persistent manner, the diseases are not
considered significant constraints in cotton in Brazil.

Increases in non-target-pest populations, known as secondary pest outbreaks, may result from several processes:

• *Altered plant/insect interactions*

The Bt cotton varieties to be commercialized in Brazil may have a new value as
insect food through their production of nutritionally valuable resources or
through their lack of non-transgenic resistant traits (see Fontes *et al.*, Chapter 2,
this volume, for more information on morphological and physiological resistant
traits). These changes could increase the fitness of non-target pests on Bt cotton
and therefore their damage potential, if they are not negatively affected by the
Bt toxin.

• *Release from biological control*

Some herbivore populations are controlled by predation (predators and parasitoids). A reduction in the populations of predators could cause population
surges of these species on cotton, changing their pest status. For example, a
herbivore population can be released from control by heavy pesticide spraying, which decimates the natural enemies. Predators and parasitoids are often
more susceptible to pesticides than herbivores, and frequent sprayings can do
more damage to natural enemies than pests and the pest will quickly reach
high densities. This happened with 'Green Revolution' rice in the Philippines.
Frequent sprayings eliminated the natural enemies of the brown leafhopper
Nilaparvata lugens, which subsequently reached very high densities
(Kenmore *et al.*, 1984). In the case of Bt cotton, the profound re-arrange-
ment of the herbivore complex may drive some predators away from the cotton fields and thus decrease their control of some herbivore species. The
result can be a classical 'predator release'.

• *Release from intraguild competition*

Some herbivore populations are limited by competition with other herbivorous species for the same food resource. When the density of one or more herbivore species (e.g. Lepidoptera on Bt cotton) is selectively reduced, the species that were competitively inferior to this dominant species can increase rapidly. These 'secondary-pest outbreaks' have frequently been reported during the initial period of pesticide use (van Driesche and Bellows, 1996). Since Bt cotton was introduced in China, the density of the main pest, the cotton bollworm *Helicoverpa armigera*, has drastically decreased. The biomass that was earlier consumed by this pest has thus become available to other arthropods, and their densities have increased (Wu *et al.*, 2002).

These mechanisms are intertwined, and it is not always easy or possible to identify the precise cause of a secondary pest outbreak.

Testable Hypotheses and Experimental Designs

General considerations

Altered plant/insect interactions: To test the adverse-effect scenarios under this heading, we look at: (i) the biological activity of the Bt plant on the non-target pest; and (ii) its effect on the behaviour of the non-target pest. For the experiments under (i) the exposure of the non-target pests to the toxins produced by Bt cotton is a critical issue. In the absence of exposure of the insect to Bt toxin, further testing on this species may be unnecessary under these hypotheses. These experiments also supply information on Bt toxin accumulation in the non-target pest, which is important for assessing whether higher trophic-level organisms (such as predators and parasitoids) are exposed. One important assumption is that tests should start during laboratory characterization of selected events (stage 3, Andow *et al.*, Chapter 1, this volume), when an elite event – a Bt cotton transformed plant that is effective against target species and can be backcrossed with local commercial varieties – is available.

We suggest below the key examples of experiments for the selected non-target species that should be performed before field release, tailored to the specific feeding behaviour of each species. The whitefly, *B. tabaci*, is not further considered because their feeding mechanism is very similar to that of aphids. This way, the experiments to test the exposure hypotheses are the same for both species, and the aphid protocol could be used for whiteflies with minor adaptations.

Release from biological control and release from intraguild competition: The adverse-effect scenarios under this process can only be tested in field experiments evaluating populations of non-target pests and their natural enemies, at the small-scale and large-scale field trial stages. We make some recommendations on experimental methods to look at population dynamics of the selected non-target pests.

Biological activity of Bt plants on non-target pest

Testable hypothesis: Non-target pests feeding on Bt plants are exposed to the Bt toxin which increases their fitness (via changes of one or more fitness components such as mortality, population growth rate, length of individual development or longevity), thus increasing their outbreak potential.

Cotton aphid experiments

We do not know if the Bt toxin is present in the phloem sap. Expression of Cry1A toxins in the phloem of some Bt cottons is possible. The cotton aphid, an obligate phloem feeder, acquires Cry1A toxin after feeding on some Chinese varieties of Bt cotton (Zhang *et al.*, 2004). Following a 'worst-case scenario', we assumed that the Bt toxin is present in the phloem sap and proposed experiments to verify or refute exposure and effects on the aphids' biology and behaviour. A test for Bt toxin presence in the phloem sap by enzyme-linked immunosorbent assay (ELISA) could determine whether or not these experiments are necessary.

PLANT MATERIAL. Individually potted cotton plants, infested with aphids at the one-leaf stage. Start with at least 30 plants per treatment.

TREATMENTS. The elite event expressing Bt toxins active against target species (if possible backcrossed into the local commercial varieties) is considered the treatment and is compared against the untransformed parent line that does not express the Bt toxin(s). The use of a local cultivar as control should be considered.

METHOD. Place one wingless mature aphid female in a clip cage attached to a leaf. After 24 h, remove the clip cage and keep the infested potted plants in cages of $1 \times 1 \times 1$ m, with at least 9 pots/cage. The position of the pots should be periodically changed (pots in the first line can be moved to the back of the cage after every second inspection). Keep the pots in water-filled trays to prevent aphid migration between plants. Have one plant/cage aphid-free and inspect to assess migration between plants. Place the larger cages in different mini-glasshouses if possible (for statistical as well as logistical reasons). Create uniformly aged starting cohort of aphids; start with mature female, harvest young 24 h nymphs to start an even-aged cohort. All plants should be inspected every 3 days to record the growth and development of aphids.

Conduct the experiment during laboratory characterization of elite events (Andow *et al.*, Chapter 1, this volume) in a glasshouse or controlled-climate facility, such as a greenhouse with biosafety registration. Environmental conditions should be set to simulate/mimic local regional conditions. If possible, the effects of a stress factor, such as nutritional, water, nematode or shade stress, on the expression of Bt toxins should be examined. Perform the experiments in at least three regions. We suggest these for the initial tests because facilities will still be influenced by local conditions (e.g. photoperiod, sunshine), and it is much easier to perform the tests even under controlled conditions that match the regions' conditions. If regional tests do not indicate differences, further tests can be done in one region only.

Variables to be measured include intrinsic aphid population growth rate and toxin levels in plant and insects (and their products). Plant toxin levels should be checked on three occasions during the experiment. Aphid toxin levels should be estimated by pooling five to ten adult aphids/sample with at least ten such samples analysed by standard ELISA.

Fall armyworm experiment

PLANT MATERIAL. Use potted plants at the stage when the first bolls appear on the plant.

TREATMENTS. The elite event expressing Bt toxins active against target species (if possible backcrossed into the local commercial varieties) is considered the treatment and is compared against the untransformed parent line that does not express the Bt toxin(s).

METHOD. Infest plants with either one neonate (L1) or second instar (L2) per plant. In a preliminary experiment, test if neonates can be used (preferable) – and use L2 only if there is high L1 mortality due to failure of establishing a feeding site. Sample plants destructively, taking weekly samples for 3 weeks + pupation, with 15 replicates each. Dissect all the bolls for the presence of larvae. Record mortality, weigh larvae. Start with 75 plants to allow for plant mortality. When pupae are formed, determine pupal sex and weight. Keep them in Petri dishes with moist tissue paper at ambient temperature. Initially, perform these tests in at least three regions. If no regional differences appear, tests can be continued in one region only.

Variables to be measured include the biological variables (fitness components): larval and pupal mortality; developmental time of immature stages (larvae to adult); growth; fecundity (surrogate parameter – pupal mass is related to reproductive potential); mate females with males from a laboratory colony, measure realized fertility by measuring area occupied by eggs; and record hatching rate. Calculate fitness from these variables through the intrinsic rate of natural increase (r) (Begon *et al.*, 1996). Toxin should be identified using methods suggested for plant expression. Repeat the experiment over two generations at least, using plants of the same age.

Boll weevil experiment

PLANT MATERIAL. Raise potted plants until buds appear. Once buds are produced, select 20 × 5 buds/treatment, put singly in Petri dishes with one female boll weevil. Keep the buds at a temperature of 25°C. After 48 h, inspect buds for signs of feeding and oviposition.

TREATMENTS. The elite event expressing Bt toxins active against target species (if possible backcrossed into the local commercial varieties) is considered the treatment and is compared against the untransformed parent line that does not express the Bt toxin(s).

METHOD. Select females of the same age and allow them to oviposit for 2 days on diet (first eggs often are infertile), then expose them to experimental plant bolls. Record feeding and oviposition punctures after 48 h; record development time, sex ratio, mortality at adult emergence. After a gap of 7 days of emergence, dissect bolls to establish larval mortality/fate of larvae. Weigh females as a surrogate fitness parameter. Estimate lifetime fecundity: test 10 females/treatments by providing them with normal, non-manipulated as well as Bt cotton buds. Provide five buds every 2 days for every female. After 48 h, inspect the buds for oviposition punctures or dissect for recording eggs and continue the monitoring of females for record of longevity and fecundity. Calculate fitness as proposed above for fall armyworm. Repeat the experiment with females emerging from the first experiment to simulate long-term exposure. The reason for doing the experiment over two generations is that there could be cumulative effects that will not be detectable during exposure for a single generation only.

Effect on behaviour of non-target herbivores

Testable hypothesis: The adults and larvae of non-target pests will distinguish among Bt and non-Bt cotton thus affecting their host finding/acceptance (feeding) and oviposition preference. This may result in differential feeding and, consequently, non-random mixing of individuals within a population and may have important consequences for resistance management.

Cotton aphids, fall armyworm and boll weevil
For fall armyworm, perform leaf disc test in Petri dish, at ambient temperature with Bt plant and non-Bt plant. Release four armyworm larvae in a Petri dish and follow the insect behaviour for at least 12 h. For boll weevil, perform cotton bolls test in Petri dish, at ambient temperature with Bt plant and non-Bt plant for boll weevil adults. Release four adults in the dish. For aphids, set up a choice experiment using intact plants of Bt and non-Bt cotton. Measure different behaviours such as time to start feeding, time spent feeding and dispersion inside the dish. For general advice on identifying behavioural categories and how to measure them, consult Martin and Bateson (1999). Repeat the assay at least four times with 20 replications.

Fall armyworm
Tests should be done using both non-choice and choice situations.
    Use the same type of plant material as the previous experiments under point 1.0. Plants are arranged in a circle of 1.2 m diameter in a glasshouse or climate chamber. No-choice methods will indicate a potential nutritional effect, while choice experiments allow the registration of a behavioural-level effect.
    No-choice method: Use flowering plants. Release 4 gravid females of fall armyworm. Allow oviposition for 2 days. Measure plant height, number of lateral branches, number of flowers/plant and numbers of eggs on plants. Repeat the experiment three to four times.

Choice method: Use nine plants of three cultivars each (3 × 3). Be aware of the position effects – randomize block arrangement. (Do not disturb the experiment during its run – do not reposition plants). Measure the same variables as in the non-choice experiment.

Boll weevil and cotton aphid

Same experimental design as for fall armyworm, but collect buds and count egg-laying punctures. Count the number of buds affected. Count and compare feeding versus oviposition punctures. For aphids, count the number and rate of settling.

For the choice experiment, observe location of insects after 1 h, 2 h, 4 h, 8 h and 12 h. Score feeding marks for boll weevils. Use adult boll weevils and winged adult aphids. Let the number of insects = number of plants = 20.

Population dynamics of non-target pest

These field experiments should be initiated during small-scale field trials (stage 4) in three regions. Use only species that are present in the region. We do not suggest field monitoring of single species because such a situation (single herbivore species present) does not occur frequently, and it is difficult to maintain a single species field experiment without interference from other species.

Testable hypothesis: Multi-trophic and intraguild interactions in the field may affect herbivore population dynamics in unexpected ways compared to that observed in lab conditions. This can result in secondary pest outbreaks and yield loss.

Multi-species studies – field studies of non-target pest species

Allow natural infestations to occur; fall armyworm immigration can be assured by planting maize near the cotton plot.

TREATMENTS. Conventional practice, integrated pest management (IPM), Bt cotton. It is necessary to precisely define the 'conventional' control option and IPM according to each region, as the actual practices vary among regions.

METHODS. Plot size should be region-specific, but minimally 0.5 ha. Plot shape should be as close to square as possible (further considerations assume a 70 m × 70 m plot, with a 5 m bare strip around separating the individual plots). In addition, leave the outer 5 m as an edge-buffer strip. Select 10 positions within the central area of each plot, using a regular grid. Mark these fixed positions. Randomly allocate five of these as 'fixed' locations, and select and label one plant near this position. These will be the 'fixed plants'; always record data on these plants on any sampling occasion. At the other five locations, always select a different plant randomly to conduct the census. If necessary, this plant can be bagged and cut, to count the target insects at the edge of the field, in shade or in the laboratory. Ideally, plots should be separated as far as possible to avoid populations of boll weevil from immigrating into the neighbouring plots. Considering practical constraints, it is alternatively proposed to use randomized complete blocks with 30–50 m distance from each other and at least 10 m between treatments.

APHID CENSUS. Count the number of individuals and aphid colonies on plants on a weekly basis. If the colonies reach the size of > 200 aphids, make a rough estimate only. Try to make an estimate to two significant digits.

BOLL WEEVIL SAMPLING. Pick up fallen squares from the ground within 5 m of the focal plant. Take a total of 50 squares per plot. Dissect them in the laboratory. Take 100–200 squares from plants, from the set locations. Dissect them in the lab. Repeat the census weekly in one plot, fortnightly from the other three plots for all treatments.

FALL ARMYWORM CENSUS. Count the number of individuals on flowers and reproductive structures and consider each aggregated group of neonate larvae as one individual on leaves. Try to make an estimate to two significant digits.

Discussion and Conclusions

There is a substantial amount of information on economically important arthropods present in cotton agroecosystems in Brazil. However, this information has not been synthesized, is not organized in a systematic way and there are differences in the methods used for collecting these data in the various studies. We highly recommend carrying out a 'synthesis of existing knowledge' on cotton pests as a basis for non-target impact assessment.

A further complicating factor is that cotton growing started in the Northeast and Meridian regions of Brazil, and moved recently to the Midwest region, which is currently the most important region for cotton production. Consequently, much of the historical record on insects does not come from the currently most important region. This makes prediction difficult. The authors suggest that the identification of adverse effects should consider realistic future management regimes and provide informed hypotheses of potential adverse effects in and near the cotton agroecosystems. A purely species-based approach has limited power to do this. Building future scenarios based on previous experiences with commercial release of Bt plants in other countries could be the first step to understanding ecological effects of this technology. We recommend conducting a synthesis of effects in other countries and then relating lessons learned to the situation in Brazil.

Non-target risk assessment should take into account the expected changes in the production system. The selection matrix provides only the first step by indicating how to select species for laboratory testing. A logical next step, as yet not fully developed, is a process to provide a transition from the laboratory protocols emphasized here to the field tests needed to support risk assessment. When developing these different testing systems, predicted agricultural changes after adopting the transgenic crop should also be considered. Due to substantial differences in environmental conditions, cotton-growing practices, landscape structures and cotton-growing history, adverse-effect scenarios and research hypotheses should be developed region-by-region. A general 'Brazilian' risk assessment would be too general and less informative.

A step-by-step sequence of tests on selected species with gradually increasing levels of complexity can be useful to evaluate the effect of a large-scale release of Bt cotton on the pest herbivore function in the agroecosystem. Laboratory and field experiments are both important in this evaluation and should be part of an integrated risk assessment process. We stress that this is not the same as the 'tiered system' used in pesticide testing. Many potential impacts may never show up in the laboratory but will in the field under different conditions. The model employs a testing programme with stepwise increasing levels of complexity, which is continued until it fills key 'critical' gaps of information that make certain adverse-effect scenarios irrelevant or confirm the existence of a risk.

References

Andow, D.A. and Hilbeck, A. (2004) Science-based risk assessment for non-target effects of transgenic crops. *BioScience* 54, 637–649.

Bachelor, J.S. and Bradley, J.R. (1989) Evaluation of bollworm action thresholds in the absence of the boll weevil in North Carolina: the egg concept. In: *Proceedings of the Beltwide Cotton Production Research Conference 1989, Memphis.* National Cotton Council of America, Memphis, Tennessee, pp. 308–311.

Begon, M., Harper, J.L. and Townsend, C.R. (eds) (1996) *Ecology: Individuals, Populations and communities.* 3rd edn. Blackwell Science, Oxford, UK.

Benedict, J.H., Julie, M.E. and Carrol, J.C. (2000) *Bollgard Cotton: Assessment of Global Economic, Environmental, and Social Benefits.* Fleishmann-Hillard, Kansas.

CONAB (2003) *Algodão em pluma informativo especial Julho 2003.* Available at: http://www.conab.gov.br/download/cas/especiais/CONJ%20ESPECIAL%20JU LHO%2003.pdf (accessed September 2004)

Gallo, D., Nakano, O., Carvalho, R.P.L., Baptista, G.C. de, Berti Filho, E., Parra, J.R.P., Zucchi, R.A., Alves, S.B., Vendramin, J.D., Marchini, L.C., Lopes, J.R.S. and Omoto C. (2002) *Entomologia Agrícola.* Fealq, Piracicaba, Brazil.

Gondim, D.M.C., Bélot, J.L., Silvie P. and Petit, N. (2001) *Manual de identificação das pragas, doenças, deficiências minerais e injúrias do algodoeiro do Brasil.* Boletim Técnico, 33. 3ª edn. Codetec/CIRAD, Cascavel, Brazil.

Greene, J.K. and Capps, C.D. (2002) Management of 'secondary pests' in transgenic Bt cotton. In: *Proceedings of the Beltwide Cotton Conferences 2002.* National Cotton Research Council, Memphis, Tennessee. CD-ROM.

Greene, J.K. and Turnipseed, S.G. (1996) Stink bug thresholds in transgenic Bt cotton. In: *Proceedings of the Beltwide Cotton Conferences 1996.* National Cotton Research Council, Memphis, Tennessee, pp. 936–938.

Höfte, H. and Whiteley, H.R. (1989) Insecticidal crystal proteins of *Bacillus thuringiensis. Microbiological Review* 53, 242–255.

Kenmore, P.E., Cariño, F.O., Perez, C.A., Dyck, V.A. and Gutierrez, A.P. (1984) Population regulation of the rice brown planthopper (*Nilaparvata lugens* Stål) within rice fields in the Philippines. *Journal Plant Protection Tropics* 1, 19–37.

Luttrell, R.G., Fitt, G.P., Ramalho, F.S. and Sugonyaev, E.S. (1994) Cotton pest management: Part 1. A worldwide perspective. *Annual Review of Entomology* 39, 517–526.

Martin, P. and Bateson, P. (1999) *Measuring Behaviour*, 2nd edn. Cambridge University Press, Cambridge, UK.

Matthews, G.A. (1994) Insect and mite pests: general introduction. In: Matthews, G.A. and Tunstall, J.P. (eds) *Insect Pests of Cotton*. CAB International, Wallingford, UK, pp. 29–37.

Michel, B. (1989) Nouvelle contribution à la connaissance des insectes et arachnides rencontrés en culture cotonnière au Paraguay. *Coton et Fibres Tropicales* 44, 51–54.

Michel, B. and Prudent, P. (1985) Acariens et insectes déprédateurs du cotonnier (*Gossypium hirsutum* L.) au Paraguay. *Coton et Fibres Tropicales* 40, 219–224.

Moore, J. and Navon, A. (1973) Studies of the susceptibility of the cotton leaf worm *Spodoptera littoralis* (Boisd.) to various strains of *Bacillus thuringiensis*. *Phytoparasitica* 1, 23–32.

Müller-Cohn, J., Chaufaux, J., Buisson, C., Gilois, N., Sanchis, V. and Lereclus, D. (1996) *Spodoptera littoralis* (Lepidoptera: Noctuidae) resistance to CryIC and cross-resistance to other *Bacillus thuringiensis* crystal toxins. *Journal of Economic Entomology* 89, 791–797.

Pray, C.E., Ma, D., Huang, J. and Qiao, F. (2001) Impact of Bt Cotton in China. *World Development* 29, 813–825.

Ramalho, F.S. (1994) Cotton pest management: Part 4. A Brazilian perspective. *Annual Review of Entomology* 39, 563–578.

Silva, A.G.A., Gonçalves, C.R., Galvão, D.M., Gonçalves, A.J.L., Gomes, J., Silva, M. do N. and Simoni, L. de (1968) *Quarto catálogo dos insetos que vivem nas plantas do Brasil-Seus parasitos e predadores*. Ministério da Agricultura, Rio de Janeiro, Brazil, Vol. 1, p. 2.

Silvie, P., Leroy, T., Belot, J.-L. and Michel, B. (2001) *Manual de identificação das pragas e seus danos no algodoeiro*. Boletim Técnico No. 34, COODETEC/CIRAD-CA, Cascavel, Brazil.

Turnipseed, S.G., Sullivan, M.J., Hagerty, A.M., Jenkins, R.A. and Ridge, R. (2002) Predaceous arthropods and the Stinkbug/Plantbug complex as factors that may limit the potential of Bt cotton. *Proceedings of Beltwide Cotton Conferences 2002*. National Cotton Research Council, Memphis, Tennessee. CD-ROM.

Van Driesche, R.G. and Bellows, T.S. Jr (1996) *Biological Control*. Chapman & Hall, New York.

Wu, K., Li, W., Feng, H. and Guo, Y. (2002) Seasonal abundance of the mirids, *Lygus lucorum* and *Adelphocoris* spp. (Hemiptera: Miridae) on Bt cotton in northern China. *Environmental Entomology* 21, 997–1002.

Zhang, G.-F., Wan, F.-H., Guo J.-Y. and Hou M.-L. (2004) Expression of Bt toxin in transgenic Bt cotton and its transmission through pests *Helicoverpa armigera* and *Aphis gossypii* to natural enemy *Propylea japonica* in cotton plots. *Acta Entomologica Sinica* 47, 334–341.

7 Non-Target and Biodiversity Impacts on Pollinators and Flower-Visiting Insects

S. ARPAIA, V.L.I. FONSECA, C.S. PIRES AND F.A. SILVEIRA

Corresponding author: Dr Salvatore Arpaia, Biotec-Gen, ENEA, S.S.106 Jonica km 419.5, Rotondella (MT), I-75026, Italy. Fax: +39 0835 974515, e-mail: salvatore.arpaia@trisaia.enea.it

Introduction

There is an overwhelming abundance of scientific literature highlighting the important ecological service that pollinators furnish to the survival of wild plant species (e.g. Buchmann and Nabhan, 1996), to the productivity of many food crops (e.g. Roubik, 1995) and to the maintenance of whole ecological communities (e.g. Kearns and Inouye, 1997). In recent years, there has been an increasing accumulation of data indicating that seed and fruit yields of insect-pollinated plants may often be lower than expected, not because of climate, soil or cultural factors, but simply because of the decreasing abundance of certain insect species with pollination function (Kevan and Phillips, 2001).

Cotton is a partially cross-pollinated crop, although many breeders have treated it as a completely self-fertile and self-pollinating plant. Pollinator insects may assume, in several cases, a major role in increasing the rate of natural crossing in cotton (e.g. Simpson, 1954; McGregor, 1976) and they can increase cotton yield and quality (Pimentel *et al.*, 1992). In Brazil, there are three main cotton-growing regions and different varieties are cultivated in each of them (see Fontes *et al.*, Chapter 2, this volume). In this context, the exact determination of the cotton species/varieties in use is important information, since McGregor (1976) clearly indicated that the effect of bees on pollination depends on the variety.

The risk assessment model used for the Brazil case study is a multi-step model that starts with the identification of functional groups of biodiversity (see Box 5.1, Hilbeck *et al.*, Chapter 5, this volume). The implementation of step 1 led to the identification of flower-visiting species as one functional group to assess. This chapter illustrates the implementation of steps 2–5 for this important group of non-target organisms.

Species Selection

Only fragmentary data about pollinator activity and flower-visiting insects on cotton in the three Brazilian cotton-growing areas are available, as no specific surveys have been conducted for this purpose. A first specific field survey of bees was done in cotton fields from the Brasília area in January and February of 2003 (C.S. Pires and F.A. Silveira, unpublished results) and additional data are presently being collected in other regions of the country. Therefore, in order to compile a preliminary list of possible pollinating and flower-visiting species (Table 7.1), we largely relied on published literature on cotton in different areas of the world (e.g. McGregor, 1976) and on the expertise of the chapter co-authors from Brazil. Species were included in the list according to two main criteria: (i) the species has been recorded on flowers of cotton or other Malvaceae in Brazil; and (ii) the species is a member of the regional fauna and belongs to a genus that include species recorded on cotton flowers in other parts of the world (Silveira, 2003).

Some herbivores and predators also have flower-visiting habits and feed on pollen as a complementary food. Among the species known to be possibly linked to the crop, two groups were mainly considered: Coccinellidae (lady beetles) and Dermaptera (earwigs). It must be noted that the predators known to occur in cotton vary among the three major cotton-growing areas. Nevertheless, based on the available information, it was agreed that Coccinellidae and Dermaptera can be considered important in all the three areas, even if different species within these groups may prove to be relevant in different areas.

Table 7.1 lists the species that formed the basis for prioritization and further analysis as outlined in Hilbeck *et al.* (Chapter 5, this volume). Species entirely dependent on flower resources for survival are called anthophylous. Some of them are considered to be cotton pollinators; others may prove to be so; others not. The remaining species are predators, sap feeders or leaf eaters that occasionally eat pollen. Firstly, the species were ranked in order to prioritize the flower-visiting species according to their potential association with cotton crops and the cotton crop habitat. This step did not yet require specific information on the transgenic Bt cotton but did require good information on cotton and cotton production in Brazil. The following criteria were used for the selection matrix:

1. Geographical distribution in the cotton-cropping system in Brazil (cotton field and field margins, Table 7.2A).
2. Habitat specialization: preferred occurrence of the species (in field, margins, elsewhere, Table 7.2A).
3. Prevalence: proportion of cotton fields occupied by the species (ranked as high, medium or patchy).
4. Abundance in cotton fields and other environments around crop (Table 7.2A).

Table 7.1. List of flower-visiting taxa on cotton in Brazil. Species are listed according to assemblage or feeding guild.

Assemblage	Species or species group	Order	Common name
Anthophylous	*Bombus* spp.	Hymenoptera	Bumblebee
Anthophylous	*Apis mellifera* L.	Hymenoptera	Domesticated honeybee
Anthophylous	*Melissodes* spp.	Hymenoptera	Solitary bee
Anthophylous	*Paratrigona* spp.	Hymenoptera	Stingless bee
Anthophylous	*Oxaea flavescens* (Klug)	Hymenoptera	Solitary bee
Anthophylous	*Melissoptila* spp.	Hymenoptera	Solitary bee
Anthophylous	*Eufriesea* sp.	Hymenoptera	Orchid bee
Anthophylous	*Centris* spp.	Hymenoptera	Solitary bee
Anthophylous	*Trigona spinipes* (Fabricius)	Hymenoptera	Stingless bee
Pollen feeder	*Anthonomus grandis* (Boheman)	Coleoptera	Boll weevil
Pollen feeder	Coccinellidae[a]	Coleoptera	Lady beetles
Pollen feeder	*Diabrotica speciosa* (Genmar)	Coleoptera	Leaf beetle
Pollen feeder	Phytoseiidae	Acarina	Predatory mites
Pollen feeder	Chrysoperla complex	Neuroptera	Lacewings
Pollen feeder	Dermaptera[b]	Dermaptera	Earwigs
Pollen feeder	Lepidoptera larvae on associated plants	Lepidoptera	Caterpillars

[a]At least five different species of Coccinellidae (*Cycloneda sanguinea*, *Hippodamia* spp., *Coleomegilla maculata*, *Scymnus* spp., *Eriopsis* spp.). These vary in abundance regionally.
[b]At least four different species of Dermaptera, including several *Doru* spp.

5. Phenology: synchrony between life-cycle stages feeding on pollen and/or nectar and phenology of cotton crop during the growing season (Table 7.2B).
6. Linkage between the species and the crop pollen and/or nectar (a measure of dependency on crop, Table 7.2B).
7. Ecological significance (Table 7.3).

From each step of the outlined procedure (see above), it became increasingly clear that two bee taxa reached the highest overall ranking: honeybees and bumblebees (Table 7.4). This was due to the widespread distribution and high abundance recorded for their populations (Table 7.2A), their intimate relationships with their food sources (cotton plants in this case) throughout their life cycle (Table 7.2B) and their potential significance as pollinators (Table 7.3). A stingless bee, *Paratrigona lineata*, known to be rather abundant in the Brasília cotton area, was assumed to be rather inefficient as a pollinator, since

Table 7.2A. Selection matrix – association with the crop and crop habitat.

(A) Non-target flower-visiting species × crop coincidence
Geographical distribution: degree of overlap of cotton crop and non-target taxon.
Habitat specialization: degree of association between taxon and cotton habitat.
Prevalence: proportion of cotton habitat occupied by taxon.
Abundance on cotton crop and on other plants occurring in the cotton habitat: average or typical densities.
(For definitions see Box 5.2, Hilbeck *et al.*, Chapter 5, this volume.)

Selection matrix			Association with crop – coincidence		
Flower-visiting taxon	Geographical distribution	Habitat specialization	Prevalence in suitable cotton habitat	Abundance on cotton crop	Abundance on other plants
Bombus spp.	All three regions	On crop area, field margins, wild areas	No info	High	High
Apis mellifera	All three regions	On crop area, field margins, wild areas	No info	High	High
Melissodes spp.	All three regions	On crop area, field margins, wild areas	No info	Possibly high	Medium
Paratrigona spp.	All three regions	On crop area, field margins, wild areas	No info	High	High
Oxaea flavescens	All three regions	On crop area, field margins, wild areas	No info	Medium	Medium
Melissoptila spp.	All three regions	On crop area, field margins, wild areas	No info	Possibly high	Medium
Eufriesea sp.	All three regions	On crop area, field margins, wild areas	No info	Low	Low
Centris spp.	All three regions	On crop area, field margins, wild areas	No info	Medium	Medium
Trigona spinipes	All three regions	On crop area, field margins, wild areas	No info	Medium	High

Anthonomus grandis	All three regions	On crop plant, rarely on other Malvaceae	High in all three regions	High	Zero
Coccinellidae (five species)	All three regions, but not always the same species	On crop plant	High in all three regions	Medium to high	Medium to high
Diabrotica speciosa	All three regions	On crop plant, field margins and wild areas	High	High	High
Phytoseiidae	All three regions	On crop plant	Medium	High	Medium to high
Chrysoperla complex	All three regions	On crop plant	High in MD, NE; patchy in MW	Medium	Medium
Dermaptera (four species)	All three regions, but not always the same species	On crop plant	High in all three regions	Medium	Medium
Lepidoptera larvae on associated plants	All three regions	Field margins	No info.	Zero	Medium

NE, North-east; MD, Meridian; MW, Midwest.

Table 7.2B. Selection matrix – association with the crop.

(B) Non-target flower-visiting species × crop trophic relationship.
Phenology: temporal overlap (i) from non-target taxon perspective, (ii) from crop perspective.
Linkage: specialization of trophic relationship and dependency on cotton.
(For definitions see Box 5.2, Hilbeck *et al.*, Chapter 5, this volume.)

Selection matrix	Association with crop–trophic relationship		
	Phenology of flower-visiting taxon and cotton crop		Linkage with crop
Flower-visiting taxon	Life cycle stage that feeds on pollen/nectar	Phase of cotton-growing season when taxon is present	Degree of feeding specialization on cotton pollen/nectar and dependency on cotton pollen/nectar
Bombus spp.	Adults and larvae	Reproductive	Larvae and adults: low, non-dependent
Apis mellifera	Adults and larvae	Reproductive	Larvae and adults: low, non-dependent
Melissodes spp.	Adults and larvae	Reproductive	Larvae and adults: low, non-dependent
Paratrigona spp.	Adults and larvae	Reproductive	Larvae and adults: low, non-dependent
Oxaea flavescens	Adults and larvae	Reproductive	Larvae and adults: low, non-dependent
Melissoptila spp.	Adults and larvae	Reproductive	Larvae and adults: low, non-dependent
Eufriesea sp.	Adults and larvae	Reproductive	Larvae and adults: low, non-dependent
Centris spp.	Adults and larvae	Reproductive	Larvae and adults: low, non-dependent
Trigona spinipes	Adults and larvae	Reproductive	Larvae and adults: low, non-dependent
Anthonomus grandis	Adults	Throughout	Adult only: obligate on cotton, pollen constitutes only part of the diet
Coccinellidae (five species)	Adults	Throughout	Adult only: low, non-dependent
Diabrotica speciosa	Adults	Throughout	Adult: low, non-dependent
Phytoseiidae	Adults and larvae	Throughout	Nymphs and adults: medium dependence, pollen feeding important when prey is scarce
Chrysoperla complex	Adults	Throughout	Adults only: low, non-dependent
Dermaptera (four species)	Adults and nymphs	Throughout	Adults only: low, non-dependent
Lepidoptera larvae on associated plants	Larva	Reproductive	Larvae: none

Table 7.3. Selection matrix – functional significance in cropping system (related to significance of possible adverse effect).

Selection matrix

Flower-visiting taxon	Significance as:						
	Pollinator	Natural enemy	Food for natural enemies	Pest	Vector of disease	Pollinator on other crops	Pollinator in natural areas
Bombus spp.	Yes	No	No	No	No	Yes	Yes
Apis mellifera	Yes	No	No	No	No	Yes	Yes
Melissodes spp.	Likely	No	No	No	No	Likely	Yes
Paratrigona spp.	Unlikely	No	No	No	No	Yes	Yes
Oxaea flavescens	Likely	No	No	No	No	Likely	Yes
Melissoptila spp.	Likely	No	No	No	No	Likely	Yes
Eufriesea sp.	Unknown	No	No	No	No	No	Yes
Centris spp.	Likely	No	No	No	No	Likely	Yes
Trigona spinipes	Unlikely	No	No	No	No	Yes	Yes
Anthonomus grandis	No	No	No	Yes	No	No	No
Coccinellidae (five species)	No	Yes	No	No	No	Yes	Likely
Diabrotica speciosa	Unlikely	No	No	Yes	No	Yes	Unlikely
Phytoseiidae	No	Yes	No	No	No	Yes	Likely
Chrysoperla complex	No	Yes	No	No	No	Yes	Yes
Dermaptera (four species)	Unlikely	Yes	No	No	No	Yes	Likely
Lepidoptera larvae on associated plants	No	No	Yes	No	No	Yes	?

Table 7.4. Selection matrix – prioritized species.[a]

Combination of association with crop (average estimates from Tables 7.2A–B) and functional significance in the cropping system (estimate from Table 7.3). See Hilbeck *et al.*, Chapter 5, this volume for more detail on the procedure followed.

	Ranking of expert group		
Flower-visiting taxon	Likelihood of association with crop	Significance for cropping system	Prioritized species rank[a]
Bombus spp.	1	1	2
Apis mellifera	1	1	2
Melissodes spp.	1 ?	2 ?	3 ?
Paratrigona spp.	1	2	3
Oxaea flavescens	3	2	5
Melissoptila spp.	2	2	4
Eufriesea sp.	3	2 ?	5 ?
Centris spp.	3	2	5
Trigona spinipes	3	2	5
Anthonomus grandis	2	1	3
Coccinellidae (five species)	1	2	3
Diabrotica speciosa	3	3	6
Phytoseiidae	2	2	4
Chrysoperla complex	2	2	4
Dermaptera (four species)	2	1	3
Lepidoptera larvae on associated plants	3	3	5

[a]1–2 = highest priority (widely distributed, abundant, mostly present, closely linked, environmental effects significant); 3–4 = intermediate priority; 5–6 = low priority (distribution, abundance and presence limited, loosely linked, environmental effect not significant)

the bee seems to be too small to touch the stigma with the parts of its body dusted with pollen. Other bee species in the list that are also potentially important pollinators for this crop received a lower rank mainly due to two reasons: either a more limited geographical distribution and abundance (e.g. *Eufriesea* spp.) or because they were reported as visitors for the same plant family as cotton but were not recorded in Brazilian cotton fields. Among the latter, those in the genus *Melissoptila* are known as common visitors to the flowers of *Sida* spp. plants belonging to the Malvaceae, but were listed only as potential visitors to cotton flowers (Silveira, 2003). Recently, one species in this genus was recorded in a bee survey of cotton in the Brasília area (C.S. Pires and F.A. Silveira, unpublished results). Other genera present in Brazil, such as *Melissodes*, are considered to be effective pollinators of cotton flowers elsewhere in the world (McGregor, 1976), but have not yet been recorded on cotton flowers in Brazil. The species rankings based on our preliminary list (Table 7.1) will likely need to be modified as more information is collected from specific areas of Brazil. In fact, different groups of bees have been recorded from recent surveys that are being conducted in the semiarid north-eastern region of the country (C.S. Pires, F.A. Silveira and P. Barroso, unpublished results). Among these, *Lithurgus huberi* (Megachilidae) and unidentified species of *Ptilothrix* and *Melitoma* (Apidae, Emphorini) and *Euglossa* (Apidae, Euglossina) genera may prove to be effective pollinators (due to their body size and relative abundance). On the other hand, species of *Ceratina* genus (Apidae, Xylocopinae), although too small to be effective pollinators, are extremely common on the cotton flowers and might sustain heavy losses if they prove to be susceptible to Cry proteins. All these groups should then be included in a future full risk assessment for that region.

Not all the criteria for ranking the ecological role of each species listed in Box 5.2 in Hilbeck *et al.* (Chapter 5, this volume) were equally important in this assessment. For instance, none of the species are known as plant-disease vectors and their role as prey for natural enemies is not generally very important in natural or man-managed environments. In contrast, the role of the species as pollinators was obviously considered the major issue.

Uncertainty was a significant issue in the selection matrix. In order to cope with this problem, a 'worst case' scenario approach was used, which resulted in a higher ranking for some of the pollinator species analysed. Nevertheless, no other species reached the highest rank given to honeybees and bumblebees, for which there exists less uncertainty. For example, *Trigona spinipes* has an abundance specified as 'medium'. These bees form huge colonies with tens of thousands of workers and are very efficient recruiters. Where cotton is a main source of food, close-by nests might likely crowd portions of the fields and therefore their abundance can be occasionally rather high. However, even in this case, other features of the species (i.e. dependency on crop, significance as pollinator) maintained it at a lower ranking compared to other taxa. There are also other flower-visiting species that are not significant as pollinators, but nevertheless are significantly associated with the crop (e.g. ladybugs, earwigs, leaf beetle) that therefore received a high rank during the selection process.

Table 7.4 contains an overall ranking for each species. The authors reached a unanimous consensus on the ranking, while it was acknowledged that lack of information was somewhat limiting the analysis of the Brazilian cotton system.

Some flower-visiting species may have a high charismatic value and therefore the potential impact of transgenic plants on them can cause great concern, for example as a result of the study by Losey *et al.* (1999), which could only be partly reduced after further studies (e.g. Sears *et al.*, 2001). In Brazil, no insect species of such cultural significance can be identified as a visitor of cotton flowers or consumer of cotton pollen. Further, some flower-visiting species may be endangered or threatened in their existence, even if elsewhere, and therefore enjoy a national or international protection status. Therefore, a list of endangered species in Brazil including butterflies, beetles and wild bees (MMA, 2004) was matched with a preliminary survey of butterflies (E.M.G. Fontes, Brasília, June 2003, personal communication) and bees in a cotton field in the Brasília region. None of the endangered species were found to be a visitor of cotton fields. However, information from a single survey is not sufficient to enable us to determine which species of butterflies could be potentially exposed to transgenic cotton in Brazil. As a first step in selecting endangered species potentially exposed to Bt toxin in cotton, detailed information about their distributions in today's main cotton-growing areas is needed. Furthermore, a preliminary analysis of their feeding habits and food range is necessary to refine the list to those species likely to occur in cotton fields.

Assessment of Potential Exposure Pathways to Transgenic Crop and Transgene Products

The exposure analysis to transgenic product or its metabolites in transgenic plants was considered for those species that had received the highest rank in the previous selection matrix (Table 7.4): honeybees, bumblebees, lady beetles and earwig groups. To assess the potential trophically mediated and indirect exposure, the following questions were answered (see Box 5.3, Hilbeck *et al.*, Chapter 5, this volume):

1. Which of the tissues/secretions that the species feeds upon express the transgenic products?
2. Could the plant–insect relationships lead to exposure?
3. What change in plant characteristics could increase exposure?
4. Does the organism show avoidance behaviour of the host plant?
5. Are behaviours likely to modify exposure to increase or decrease it?

The assessment of the potential exposure pathways led to the highest ranking for the two taxa of bumblebees and honeybees (Table 7.5). Several of their biological characteristics generated this result: both groups visit flowers to collect pollen and nectar, they can equally be considered as generalists in terms of preferred food sources and they have been observed visiting cotton plants regularly. Cotton flowers have floral and extrafloral nectaries, as do cotton leaves of some cotton varieties (see Fontes *et al.*, Chapter 2, this volume). Field observations confirm that nectaries are the most preferred food source for bees, even if they also collect pollen from flowers.

The two cotton events that are most likely to be field-tested soon in Brazilian varieties (event 531 expressing the Cry1Ac protein (Bollgard®) and event 15985 expressing Cry1Ac and Cry2Ab (Bollgard II®)) express the Bt protein(s) in pollen (Grossi *et al.*, Chapter 4, this volume). The expression of the Cry1Ac protein in pollen of event 531 in laboratory studies in the USA was reported to be 11.5 ng/g fresh weight (Monsanto, 2002; OGTR, 2003). Bollgard-cotton pollen has also been reported to contain 0.6 μg/g pollen fresh weight (Greenplate, 1997). Event 15985 also expresses Cry1Ac in pollen, but Cry2Ab was not detected in pollen (OGTR, 2003), although one study found the highest concentrations of Cry2Ab in white stamens, at 26.2 ppm (Akin *et al.*, 2002). Expression may also change during the development of the Bt cotton plant or in response to environmental conditions, and both Bollgard and Bollgard II have been reported to have variable expression levels in stamens (Greenplate *et al.*, 1998; Akin *et al.*, 2002). Bt toxin presence in cotton nectar or other plant exudates has never been assessed for transgene

Table 7.5. Exposure assessment: prioritization of flower-visiting species based on maximum potential exposure to transgene product or metabolites in the transgenic plant.

Taxon	Plant tissues or secretions on which the taxon needs	Which of these tissues/secretions express the transgene product?	Could the plant–insect relationship lead to exposure?	What changes in plant characters could increase exposure?	Avoidance behaviour of host plant?	Are behaviours likely to increase or decrease exposure?	Rank: probability of exposure
Apis mellifera	Pollen, floral and extrafloral nectaries	Pollen	+++	Any change in plant metabolism (e.g. volatiles, sugar content) that affect plant suitability as food source	Likely	Yes	1
Bombus spp.	Pollen, floral and extrafloral nectaries	Pollen	+++		Likely	Yes	1
Coccinellid group	Pollen	Pollen	+		Unlikely	No	2
Earwigs group	Pollen	Pollen	+		Unlikely	No	2

+++ pollen represents the main food source for the species and therefore exposure to transgene product occurs;
+ pollen represents a secondary food source for the species and therefore exposure to transgene may occur;
1 = highest risk of exposure to transgene product;
2 = intermediate risk of exposure to transgene product.

events likely to be used in Brazil (refer to Grossi *et al.*, Chapter 4, this volume). Fluid from damaged leaf or stem tissue will probably contain Bt toxin as these tissues express high levels of the protein during the whole cotton growth cycle.

Both honeybees and bumblebees carry food to the colony, thus potentially exposing both larvae and other adults to Bt toxin. Possible modifications in the feeding habits of these Hymenoptera due to the Bt plant may alter their exposure to the transgenic toxin. Namely, any change in plant metabolism that possibly interferes with flower/plant characters (e.g. appearance, odour, nectar or pollen composition) might change their attractiveness for the bees and consequently change toxin exposure for better or worse (Wright *et al.*, 2002).

A number of predators that are frequent in cotton fields can alternatively feed on pollen as adults and sometimes as larvae as well. Hence, as for the bee species above, also for Coccinellidae and Dermaptera predators, a direct bitrophic exposure is therefore considered to be likely as long as toxin expression in pollen is confirmed. Specific considerations about insect–plant biology kept these groups at a lower level of prioritization compared with the two Hymenoptera; plant tissues (i.e. pollen) only represent a secondary part of the predators' diet, whereas exposure through the plant pollen is a proven fact for *Apis mellifera* and *Bombus* spp. because pollen is a major primary food source for these species. While there is imperfect knowledge about the biology of each of these predator species, these gaps in knowledge were considered not to restrict the conclusions of this second prioritization process based on exposure to the transgenic crop and its transgene products. Four non-target flower-visiting insects were therefore selected for further analysis of adverse effects (Table 7.5).

Identification of Adverse-Effect Scenarios

In order to develop testable hypotheses for each selected non-target organism, the next logical step requires identifying adverse-effect scenarios for the species being exposed to the transgene.

Pollinator species

When considering pollinator–plant relationships, we hypothesize that the potential exposure will be direct. The scientific studies produced so far with the use of several *Bacillus thuringiensis*-derived Cry toxins have not identified acute toxic effects on European honeybees and bumblebees (reviewed in Malone and Pham-Delègue, 2001). None the less, any new risk assessment is specific to the transgenic plants that are proposed for cultivation so that the effect of the toxins and metabolites produced can be evaluated on the local flower-visiting species. For example, in Brazil, Africanized honeybees predominate and the bumblebee species are not the same as in Europe or in the USA. To assess the potential adverse effects for the pollinator species selected, we suggest a stepwise approach at different levels of complexity from individual to community.

Potential adverse effects:

1. Mortality and morbidity effects on immatures (either directly and/or indirectly through effects on microorganisms involved in food processing in the bee nests).

2. Impaired colony development (through direct toxicity to adults and/or immatures or indirectly through lowering of the queens' fecundity).

3. Avoidance of transgenic plants by foragers and overall reduced pollination activity.

Organism level

ADVERSE-EFFECT SCENARIO. One mode of action of the Cry toxins involves a specific binding to brush border membranes in the midgut of larvae after feeding, which eventually leads to cell swelling and lysis. The specificity of several *B. thuringiensis* proteins is due in part to the presence of binding sites in the target insects' midgut cells. While this specificity is known for several insect species, no such studies on honeybees exist. If the larvae of a colony are affected by the presence of a Cry toxin in their diet, the effects can be lethal or can cause an impaired development due to malnutrition. Therefore, feeding studies with young larvae may constitute a logical first step for a biosafety assessment. The toxin expressed in plants may not be identical in structure, and thus in its biological activity, to the natural protoxin produced by a given *B. thuringiensis* strain. To consider this possibility, this type of experiment can be run in the laboratory, initially with artificial diet added with the purified toxin and then with controlled amounts of Bt pollen as soon as it is available, in order to have complete control over food source and quality, toxin concentration and distribution and environment.

Colony level

It is well known that many aspects of honeybee biology have evolved for the benefit of the colony. Monitoring colony 'wellness' is therefore a reasonable indicator for non-optimal environmental conditions for social bees.

ADVERSE-EFFECT SCENARIO. The hypothesis is that bee colonies react to the environment in which transgenic plants have been introduced, compromising normal colony development. These effects could also be detected in other non-transgenic cotton fields if transgenes move beyond their intended destinations through gene flow (see Johnston *et al.*, Chapter 11, this volume). There are several social behaviours that are commonly used for judging the 'wellness' of a bee colony (e.g. communication, hygienic behaviour). Some of these may clearly reflect reactions of the colony to adverse conditions (e.g. mite infestation in the hive), but we felt that the complexity of such behaviours, the incomplete knowledge of the genetic bases that are involved and the little experience developed so far in this area could produce ambiguous results. As discussed below, we suggest that other parameters, such as comb occupation, percentage of brooding, colony activity, etc., be considered for study.

Attractiveness

Unintended effects of genetic modifications might affect floral attractiveness via altered volatile production, flower appearance or chemical composition in plants.

ADVERSE-EFFECT SCENARIO. Bee foragers actively look for flowers that are the main source of rewards. If their ability to find food is impaired by the presence of transgenic plants, the fate of the whole colony is at risk. Therefore, bee–flower relationships need to be studied in order to identify any possible behavioural abnormalities. In addition, if bees avoid Bt flowers, pollination levels may decline, adversely affecting cotton yield.

Pollen-feeding predators

The analysis of the two other pollen-feeding taxa (ladybeetles and earwigs) is restricted to the situation of direct exposure due to feeding on transgenic-cotton pollen as a complement in the predators' diet. Trophically mediated exposure is considered in Faria *et al.* (Chapter 8, this volume).

ADVERSE-EFFECT SCENARIO. These carnivorous species complement their diet with the consumption of pollen at certain life stages. The Bt toxins expressed in plant pollen could therefore produce toxic effects on larvae and/or adults upon ingestion, as these species might have binding sites for Cry toxins. Toxicity tests with pollen-based diet in the laboratory are needed. The use of pollen is preferred over pure toxins, as the transgenic plant tissue as a whole should be tested in order to include the possible effects of alterations in secondary metabolites as a consequence of pleiotropic effects or because of metabolic interactions between the novel proteins.

Testing Hypothesis Formulation and Experimental Designs

In order to correctly address all of the questions arising from the adverse-effect scenarios described above, testing hypotheses were derived for each scenario. Then an accurate experimental design (e.g. with meaningful routes of exposure that mimic reality) was proposed to find an answer to each question (Birch *et al.*, 2004).

Pollinator species

Testing hypothesis: mortality or morbidity is higher for immatures after transgenic Bt food has been carried to the colony by foragers.

Organism level

Bioassays can be conducted effectively using artificial diet that allows honeybee larvae to develop until maturity (Brødsgaard *et al.*, 1998). This provides a reliable method for testing the effects of Bt toxins at the concentrations that larvae are expected to be exposed to in the field and at much higher levels. To conduct such a bioassay, larvae are reared in sterile tissue culture multi-wells and grafted daily to new wells with food; handling is thereby reduced to one time per day with no additional feeding. Toxins can be mixed into the standard food at the appropriate concentrations. Larvae will be monitored once a day until emerging as adults, recording the length of the larval and pupal stage as well as survival. Larval developmental time is calculated to the LS stage (when larvae stop feeding and defaecate to begin pupation) and to adult emergence. Both pupae (at the 'black eyes' stage; Arpaia, 1996) and unfed newly emerged adults can be weighed to investigate possible differences in body mass. Young individuals are likely to be the most sensitive ones to a large array of proteins. It is therefore advisable to use neonates to start the experiment.

A similar reliable protocol for rearing Brazilian species of bumblebees on a meridic diet is not available. A possibility for assessing toxic effects on immature bumblebees is the preparation of small size colonies (ten larvae and three adult workers) in cups and feeding them with sugar solutions and pollen until larvae pupate (Gretenkord, 1997). These tests should be the first ones to be conducted in a biosafety programme, and it is suggested that they may be carried out using the purified toxin during the first phases of transgenic crop development at the stage of plant-cell transformation and transgene event selection, and also later on when transgenic pollen is available at the stage at which the transgenic event in the plant is being characterized in laboratory and enclosed greenhouse conditions (see Andow *et al.*, Chapter 1, this volume).

Colony level

Colony development can be studied and several indicators of colony reaction to the environment can be measured by hive examination (as opposed to adult life span). Colony development can be monitored under semi-field conditions. Smaller size colonies can be prepared to reduce the work needed for experimental observations. This also allows more replications to increase the reliability of the results. To create semi-field conditions, a screen house has to be set up starting from a field crop of appropriate size, covered with meshing nets. Each treatment should be in a different unit so that the bees are not given the choice between Bt cotton and isogenic food; in this way, colonies are only exposed to the appropriate treatment (either transgenic or control cotton plants). Colonies will then develop and freely feed on monocultures as the main food source in the screen house. In order to obtain homogeneous colonies for both treatments, a colony can be split and requeened with siblings. During the experiment, the following biological parameters should be monitored; firstly, colony development should be measured at least weekly, by counting combs occupied by bees and the degree (%) of brooding and occupation on each single comb (Arpaia, 1996). To assess larval mortality, constant examination of young larvae from all the different combs should be carried out three times per week using a dissecting microscope. At the same time, queen fertility and fecundity should be monitored for assessing potential sublethal effects. Another good indicator is pupal weight, which should be recorded at least weekly. Colony activity should also be monitored by recording (automatically or manually) the number of flights in and/or out of the hive in a specified time span. Studies at colony level can also be performed with bumblebees artificially fed in controlled environments (Morandin and Winston, 2003). The amount of pollen consumption, worker weight, colony size, amount of brood and number of queens and males produced can be chosen as colony health measures. These experiments should be planned at the stage at which the transgenic plant is undergoing transgene event characterization and efficacy testing in the laboratory and enclosed greenhouse, and conducted early during the small-scale field trials.

Attractiveness

ADVERSE-EFFECT HYPOTHESIS. Bt cotton may increase or reduce floral attractiveness. If attractiveness increases, bees may be harmed by increased exposure if they are sensitive to the Cry toxin; if attractiveness decreases, bees may be harmed by reduced effectiveness in finding food. In the latter case, pollination effectiveness will be reduced.

It is necessary to assess both preforaging behaviour and feeding behaviour of honeybees in detail to judge the attractiveness of transgenic plants and their acceptance as a food source, compared to their controls. This experiment can be done in semi-field conditions with foragers that are given the choice between Bt and non-Bt cotton plants. Free-living colonies are put in screen houses with both transgenic and control flowering cotton plants available, recording the number of flowering plants per unit area. Attractiveness is investigated by visual observations and the number of foragers arriving on flowers in a fixed time span recorded. Also, the time spent foraging should be recorded in order to investigate food acceptance upon tasting. Non-parametric statistical tests should be considered for data analyses, as behavioural data are generally not normally distributed. The statistical analysis should also be considered in deciding the number of replications to be done, which should have a minimum of 100 individuals. This level of study can be conducted while transgenic plants are being assessed in small-scale field trials (see Andow *et al.*, Chapter 1, this volume).

Community level

Each of the three potential adverse effects hypothesized above could lead to a cascading decline in pollinator biodiversity in field conditions. To test this possibility, it will be necessary to grow transgenic plants in field conditions; first on a smaller scale and then possibly also on a larger scale. The experimental set-up will basically be the same as for screen house experiments. The main goal for field experiments will be the study of species assemblages and the estimation of biodiversity. In order to assess this, different sampling techniques available should be combined (e.g. direct observations, pollinator collection). The conditions in the experimental area should allow cotton pollen to be the major component of the bees' diet. The estimate of pollinator biodiversity in a given area should be done in a quantitative manner. Analysis of such data requires appropriate statistical methods. It should be remembered, for instance, that alpha diversity indexes are not appropriate for comparing biodiversity in different situations and that multivariate methods are the most appropriate for comparing data that are in matrix format. Specific statistical tests (e.g. Multiple-range permutation procedure, Zimmerman *et al.*, 1985; Indicator-species analysis, Dufrene and Legendre, 1997) are suggested. These experiments will complete the knowledge of the ecology of pollinators in the agroecosystem. As this step will be crucial, we recommend adopting these surveys for both the precommercial field release when small-scale and large-scale field trials are carried out, and possibly to continue them after commercial release, in order to study effects on a more local basis and an ecologically appropriate time span (see Andow *et al.*, Chapter 1, this volume).

Pollen-feeding predators

Feeding bioassays for these otherwise predaceous species are recommended, using pollen from transgenic plants and isogenic controls. A first step should involve bioassays using pollen from the selected transgenic plants. An indication for the quantity of pollen to be used as diet could be obtained by measuring the amount of pollen produced by a single flower, and/or the amount of pollen eaten by a single insect. It is important to consider that several factors may mitigate exposure to transgenic pollen in the field, such as the overlap between flowering time and predators' activity, the availability of different pollen sources, etc. Experiments on these species are discussed more thoroughly in Faria *et al.* (Chapter 8, this volume).

Discussion and Conclusions, and Suggestions for Further Development

Despite knowledge limitations, we have shown that the species-prioritization procedure (evaluating association with the crop plant and habitat *per se*) and the detailed exposure assessment (evaluating exposure to the transgene product and metabolites in the transgenic plant) are useful for selecting important flower-visiting and pollen-feeding species for further risk assessment. All the criteria used for Bt cotton in this case study can be applied to other transgenic plants with few adjustments. The 'worst-case scenario' approach was a helpful tool to overcome uncertainty and lack of knowledge because all possible species were kept in the analysis through the species-selection process (Tables 7.2A, 7.2B, 7.3 and 7.4).

For the specific case of Bt cotton, the process of species prioritization would have been much more accurate if a more comprehensive list of flower-visiting and pollen-feeding species for the cotton agroecosystem in the different regions and cropping systems in Brazil was available. In Brazil, cotton is grown in a great diversity of environments with very different climates and soils, on small- and large-scale holdings with variable levels of technology. Therefore, we need to take into account the possible regional differences in species distribution, abundance and ecological role. This would avoid the error that abundant endemic species potentially exposed to high mortality are ranked low because of their limited distribution (under geographical distribution in Table 7.2A). In addition, specific data on the role of pollinator species on cotton are urgently required. Although *A. mellifera* and *Bombus* spp. were the highest-ranked organisms in the matrix, different species might be selected for each region and cropping system with more accurate knowledge about the fauna of cotton. For example, in one year of sampling in only one location in the Brasília region, 21 different species of native bees that had not been reported for that region before were collected in cotton fields (C.S. Pires and F.A. Silveira, unpublished results). While some knowledge about the wild bee fauna has been collected for natural ecosystems in Brazil (e.g. Martins, 2002; Pinheiro-Machado *et al.*, 2002; Silveira *et al.*, 2002a,b; Viana and Alves dos Santos, 2002), the native bee fauna in Brazilian agroecosystems is still poorly investigated (Pinheiro-Machado *et al.*, 2002).

The highest priority given to honeybees and bumblebees is not an unexpected result, as these taxa are commonly very active in most ecosystems. *A. mellifera*, an exotic species in Brazil, is rather abundant in almost any environment in that country and was already shown to effectively pollinate cotton, other crop plants and even wild plants. For this reason, it will probably appear as a priority species in all risk assessments. However, it should be noted that from the conservation standpoint, more weight should be given to native species, especially those not managed by humans and whose populations will not be easily restored if affected by Bt crops. It will be important to run the process again on a regional scale when more detailed data become available for specific agroecosystems with different cotton varieties and management practices. For a full risk assessment of Bt cotton in Brazil, we therefore recommend collection of further information in order to complete the analysis conducted in this case study. Surveys of cotton flower visitors, including an assessment of relative abundance, are needed for each cotton-producing region of the country because it is known that regional bee fauna can vary greatly across Brazil. This is also true at the field scale within each region, as we can expect to find the most diverse local fauna in fields that are less-intensively managed and therefore less exposed to pesticides or surrounded by tracts of native flora.

References

Akin, D.S., Stewart, S.D. and Knighten, K.S. (2002) Quantification of toxin levels in cottons expressing one and two insecticidal proteins of *Bacillus thuringiensis*. *Proceedings of the 2002 Beltwide Cotton Conferences*. National Cotton Council, Memphis, Tennessee.

Arpaia, S. (1996) Ecological impact of Bt-transgenic plants: 1. Assessing possible effects of CryIIIB toxin on honeybees. *Journal of Genetic Breeding* 50, 315–319.

Birch, A.N.E., Wheatley, R.E., Anyango, B., Arpaia, S., Capalbo, D., Emana Getu, D., Fontes, E., Kalama, P., Lelmen, E., Løvei, G., Melo, I.S., Muyekho, F., Ngi-Song, A., Ochieno, D., Ogwang, J., Pitelli, R., Schuler, T., Sétamou, M., Sithanantham, S., Smith, J., Nguyen Van Son, Songa, J., Sujii, E., Tran Quang Tan, Wan, F.-H. and Hilbeck, A. (2004) Biodiversity and non-target impacts: a case study of Bt maize in Kenya. In: Hilbeck, A. and Andow, D.A. (eds) *Environmental Risk Assessment of Genetically Modified Organisms: A Case Study of Bt Maize in Kenya.* CAB International, Wallingford, UK, pp. 117–185

Brødsgaard, C.J., Ritter, W. and Hansen, H. (1998) Response of in-vitro reared honey bee larvae to various doses of *Paenibacillus larvae* spores. *Apidologie* 29, 569–578.

Buchmann, S.L. and Nabhan, G.P. (1996) *The Forgotten Pollinators.* Island Press, Washington, DC.

Dufrene, M. and Legendre, P. (1997) Species assemblages and indicator species: the need for a flexible asymmetrical approach. *Ecological Monographs* 67(3), 345–366.

Greenplate, J.T. (1997) Response to reports of early damage in 1996 commercial Bt-transgenic cotton (Bollgard™) plantings. *Society for Invertebrate Pathology Newsletter* 29(2), 15–18.

Greenplate, J.T., Head, G.P., Penn, S.R. and Kabuye, V.T. (1998) Factors potentially influencing the survival of *Helicoverpa zea* on Bollgard cotton. *Proceedings of the 1998 Beltwide Cotton Conference* 2, 1030–1033. National Cotton Council, Memphis, Tennessee.

Gretenkord, C. (1997) Laborzucht der dunklen Erdhummel *Bombus terrestris* und toxikologische Untersuchungen unter Labor- und Halbfreiland-Versuchen. Dissertation. Rheinische Friedrich-Wilhelms Universität, Bonn, Germany.

Kearns, C.A. and Inouye, D.W. (1997) Pollinators, flowering plants and conservation biology. *BioScience* 47, 297–307.

Kevan, P.G. and Phillips, T. (2001) The economics of pollinator declines: assessing the consequences. *Conservation Ecology* 5(1), 8. Available at: http://www.consecol.org/vol5/iss1/art8 (accessed July 2004).

Losey, J.E., Rayor, L.S. and Carter, M.E. (1999) Transgenic pollen harms monarch larvae. *Nature* 399, 214.

Malone, L.A. and Pham-Delègue, M. (2001) Effects of transgene products on honey bees (*Apis mellifera*) and bumblebees (*Bombus* sp.). *Apidologie* 32, 287–304.

Martins, C.F. (2002) Diversity of the bee fauna of the Brazilian Caatinga. In: Kevan, P.G. and Imperatriz-Fonseca, V.L. (eds) *Pollinating Bees. The Conservation Link Between Agriculture and Nature.* Ministry of Environment, Brasília, Brazil, pp. 131–134.

McGregor, S.E. (1976) *Insect Pollination of Cultivated Crop Plants.* Agricultural Handbook No. 496. ARS-USDA, Washington, DC.

MMA (2004) Brazilian Ministry of Environment, list of butterflies and bees. Available at: http://www.mma.gov.br/port/sbf/fauna/index.cfm (accessed July 2003).

Monsanto (2002) Safety assessment of Bollgard® cotton event 531. Available at: http://www.essentialbiosafety.com/docroot/decdocs/02-269-001.pdf (accessed July 2003).

Morandin, L.A. and Winston, M.L. (2003) Effects of novel pesticides on bumble bee (Hymenoptera: Apidae) colony health and foraging ability. *Environmental Entomology* 32(3), 555–563.

OGTR (2003) *Risk Assessment and Risk Management Plan Consultation Version: Commercial Release of Insecticidal (INGARD® event 531) Cotton.* Office of the Gene Technology Regulator, Woden, Australia.

Pimentel, D., Acquay, H., Biltonen, M., Rice, P., Silva, M., Nelson, V., Lipner, Y., Giordano, S., Horowitz, A. and D'Amore, M. (1992) Environmental and economic costs of pesticide use. *BioScience* 42, 750–760.

Pinheiro-Machado, C., Alves-dos-Santos, I., Imperatriz-Fonseca, V.L., Kleinert, A. de M.P. and Silveira, F.A. (2002) Brazilian bee surveys. State of knowledge, conservation and sustainable use. In: Kevan, P.G. and Imperatriz-Fonseca, V.L. (eds) *Pollinating Bees: The Conservation Link Between Agriculture and Nature.* Ministry of Environment, Brasília, Brazil, pp. 115–129.

Roubik, D.W. (ed.) (1995) *Pollination of Cultivated Plants in the Tropics.* FAO Agricultural Services Bulletin 118, Rome, Italy.

Sears, M.K., Hellmich, R.L., Stanley-Horn, D.E., Oberhauser, K.S., Pleasants, J.M., Mattila, H.R., Siegfried, B.D. and Dively, G.P. (2001) Impact of Bt corn pollen on monarch butterfly populations: a risk assessment. *Proceedings of the National Academy of Sciences USA* 98(21), 11937–11942.

Silveira, F.A. (2003) As abelhas e o algodão Bt – Uma avaliação preliminar. In: Pires, C.S.S., Fontes, E.M.G. and Sujii, E.R. (eds) *Impacto Ecológico de Plantas Geneticamente Modificadas. O Algodão Resistente a Insetos como Estudo de Caso.* EMBRAPA, Brasília, Brazil, pp. 195–215.

Silveira, F.A., Melo, G.A.R. and Almeida, E.A.B. (eds) (2002a) *Abelhas Brasileiras. Sistemática e Identificação.* Edicion IDMAR, Belo Horizonte, Brazil.

Silveira, F.A., Pinheiro-Machado, C., Alves dos Santos, I., Kleinert, A. de M.P. and Imperatriz-Fonseca, V.L. (2002b) Taxonomic constraints for the conservation and sustainable use of wild pollinators – The Brazilian wild bees. In: Kevan, P.G. and Imperatriz-Fonseca, V.L. (eds) *Pollinating Bees. The Conservation Link Between Agriculture and Nature.* Ministry of Environment, Brasília, Brazil, pp. 41–50.

Simpson, D.M. (1954) Natural cross-pollination in cotton. *USDA Technical Bulletin* 1094, 1–17.

Viana, B.F. and Alves dos Santos, I. (2002) Bee diversity of the coastal sand dunes of Brazil. In: Kevan, P.G. and Imperatriz-Fonseca, V.L. (eds) *Pollinating Bees. The Conservation Link Between Agriculture and Nature.* Ministry of Environment, Brasília, Brazil, pp. 135–153.

Wright, G.A., Skinner, B.D. and Smith, B.H. (2002) Ability of honeybee, *Apis mellifera*, to detect and discriminate odors of varieties of canola (*Brassica rapa* and *Brassica napus*) and snapdragon flowers (*Antirrhinum majus*). *Journal of Chemical Ecology* 28(4), 721–740.

Zimmerman, G.M., Goetz, H. and Mielke, P.W. Jr (1985) Use of an improved statistical method for group comparisons to study effects of prairie fire. *Ecology* 66, 606–611.

8

Assessing the Effects of Bt Cotton on Generalist Arthropod Predators

M.R. de Faria, J.G. Lundgren, E.M.G. Fontes, O.A. Fernandes, F. Schmidt, Nguyen Van Tuat and D.A. Andow

Corresponding author: Eliana M.G. Fontes, Embrapa Cenargen (Genetic Resources and Biotechnology), Caxia Postal 02372, Parque Estação Biológica, Av. W3 Norte – Final, Brasília, DF-707 70-901, Brazil. Fax: +55 61 34484673/3403624, e-mail: eliana@cenargen.embrapa.br

Introduction

We followed a screening methodology described in Hilbeck *et al.* (Chapter 5, this volume) to initiate the assessment of potential risks of Bt cotton to non-target arthropod predators. We first identified arthropod predators as an important functional group for non-target assessment. Arthropod predators are arthropods (insects, spiders, mites) that prey on other species, which are mainly other species of arthropods. For this functional group, we prioritized non-target predators for further screening and evaluation. We identified possible pathways by which the predator may be exposed to transgene products of Bt cotton, by considering trophic webs in the cotton community, and developed hypothetical adverse-effect scenarios for the highest-priority species that include other ecological adverse effects that are not trophically mediated. We then formulated specific testable hypotheses that underlie the exposure pathways and/or adverse-effect scenarios, and described experiments that can test these hypotheses in the laboratory and field.

We identified arthropod predators as an important functional group for non-target assessment (step 1 of Box 5.1). Arthropod predators are an important component of agroecosystems and should they be adversely affected, undesired pest outbreaks could occur. These predators can: (i) regulate and suppress pest populations in most crops (as 'natural enemies'); (ii) alter ecosystem functions via trophic cascades (top-down ecological processes);

(iii) interact with transgenic plants via a broad range of trophic links, as they are often omnivorous on plants and fungi (Hodek and Honek, 1996; McEwen *et al.*, 2001); (iv) link the agroecosystem to the surrounding natural and agricultural ecosystems; and (v) have an intrinsic value as a significant part of the biodiversity in natural systems.

Predators represent a phylogenetically diverse group of species that vary behaviourally and physiologically, which could lead to different levels of susceptibility to transgenic crops. Behaviourally, predators vary in the degree of their feeding specialization; some predators are almost exclusively predatory such as many species of spiders, while others readily accept non-prey food when it is available, for example ladybird beetles and anthocorids (Corey *et al.*, 1998; Harmon and Andow, 2004; Lundgren *et al.*, 2004). Furthermore, the degree of feeding specialization can vary during the life of a predator. Morphologically and physiologically, the structure of the feeding apparatus may restrict the breadth of the diet, and the physiology of the predator digestive system also affects the degree to which predators may be affected by the Cry toxins incorporated into transgenic crops. For these reasons, several kinds of predators should be considered for risk-assessment studies.

This case-example analysis does not constitute a risk assessment of the effects of Bt cotton on non-target arthropod predators in Brazil; a full characterization of risk would need to be carried out after conducting some of the proposed experiments. This chapter illustrates how the screening methodology can be applied to the case of Bt cotton and arthropod predators, specifically illustrating implementation of steps 2–5 of this process (Box 5.1).

Species Selection

Although agricultural research in Brazil has had a strong emphasis on research on cotton, the scientific literature on several issues important for environmental risk assessment is scarce or absent. The predator community associated with cotton is well documented in Brazil, especially in north-east Brazil (Gravena, 1990; Gravena and Cunha, 1991; Ramalho and Wanderley, 1996; Gondim *et al.*, 2001; Silvie *et al.*, 2001; Carvalho and Souza, 2002; Gallo *et al.*, 2002); however, detailed information on the biology, ecology and behaviour of many species is scarce. A lack of literature on key predators and their relative importance creates major problems in assessing the risk of Bt cotton to predators in Brazilian cotton. The initial list of predators potentially at risk (Table 8.1) contained only species that are believed to be effective natural enemies in cotton. This narrowed list of species represented sufficiently the range of physiological and behavioural diversity within generalist predators, and was considered appropriate for conducting the assessment methodology (see Hilbeck *et al.*, Chapter 5, this volume). As will be detailed below, the initial screening process retains only those species that are strongly associated with cotton and are likely to have significant ecological function. Thus, prescreening the list of predators to include only those believed to be important biological control agents avoids wasting time on species that are less

Table 8.1. Predator taxa on cotton in Brazil according to feeding guild, life cycle stage with natural enemy function and main prey.

Feeding guild	Species or species group	Order and family	Life cycle stage with natural enemy function	Main prey
		Description of predator taxa		
Predator of mites	Phytoseiidae (e.g. *Amblesius* spp., *Eusieus* spp.)	Acarina: Phytoseiidae (predatory mites)	All	Mites
Predator of eggs and small larvae	Forficulidae (e.g. *Doru* spp.)	Dermaptera: Forficulidae (earwigs)	Nymphs and adults	Lepidoptera
Predator of eggs, small larvae and pupae	Carcinophoridae (e.g. *Euborellia annulipes* (Lucas))	Dermaptera: Carcinophoridae (earwigs)	Nymphs and adults	Boll weevil, Lepidoptera
Predator of eggs and small larvae	*Geocoris ventralis* (Fieber), *Geocoris* sp.	Heteroptera: Lygaeidae (bigeyed bugs)	Nymphs and adults	Mites, insect eggs, small Lepidopteran larvae, aphids, whiteflies
Predator of eggs and small larvae	*Orius* spp.	Heteroptera: Anthocoridae (minute pirate bugs)	Nymphs and adults	Lepidoptera, aphids, thrips, mites
Predator of eggs and small larvae	*Nabis* spp.	Heteroptera: Nabidae (damsel bugs)	Nymphs and adults	
General predator	Reduviidae (*Zelus laticornis* (Herrich-Schaefer), *Z. leucogrammus* (Perty), *Z. longipes* (L.), *Z. ruficeps* (Stal), *Z. armillatus* (Lep. & Serville), *Apiomerus* sp.), *Cosmoclopius annulosus* (Stal)	Heteroptera: Reduviidae (assassin bugs)	Nymphs and adults	Flying insects Lepidoptera, Coleoptera
General predator	Pentatomidae (e.g. *Podisus nigrispinus* (Dallas), *Supputius cincticeps* (Stal), *Alchaeorhynchus grandis* (Dallas), *Apateticus* sp.)	Heteroptera: Pentatomidae (stink bugs)	Nymphs and adults	Lepidoptera

Continued

Table 8.1. Predator taxa on cotton in Brazil according to feeding guild, life cycle stage with natural enemy function and main prey – cont'd.

Feeding guild	Description of predator taxa		Life cycle stage with natural enemy function	Main prey
	Species or species group	Order and family		
General predator	Thomisidae (e.g. *Misumenops guianensis* (Taczanowski), *Synaemops rubropunctatus* (Mello-Leitão), *Xysticus* spp.)	Araneae: Thomisidae (crab spiders)	All	Boll weevil, bees
General predator	Lycosidae (e.g. *Lycosa* spp.)	Araneae: Lycosidae (wolf spiders)	All	Boll weevil
General predator	*Chrysoperla* species complex; *Chrysoperla externa* (Hagen)	Neuroptera: Chrysopidae (lacewings)	Larvae	Aphids, thrips, mites
General predator	Staphylinidae (e.g. *Paederus* spp.)	Coleoptera: Staphylinidae	Adults	Lepidoptera
General predator	Carabidae (e.g. *Cicindela* spp., *Calosoma* spp.)	Coleoptera: Carabidae	Larvae and adults	Lepidoptera
General predator	Coccinellidae (at least including *Cycloneda sanguinea* (L.), *Scymnus* spp., *Coleomegilla maculata* (De Geer), *Eriopsis connexa* (Germar), *Hippodamia convergens* (Guérin-Méneville), *Olla v-nigrum* (Mulsant), *Hyperaspis festiva* (Mulsant), *Scymnus* sp.	Coleoptera: Coccinellidae (lady beetles)	Larvae and adults	Aphids, mites
Aphid predator	*Pseudodorus clavatus* (Fabricius)	Diptera: Syrphidae	Larvae	Aphids
General predator	*Polistes* spp.	Hymenoptera: Vespidae (wasps)	Adults	Cotton leafworm, boll weevil
General predator	*Camponotus sericeiventris* (Guérin)	Hymenoptera: Formicidae (ants)	Adults	Boll weevil
General predator	*Solenopsis geminata* (Fabricius) and other spp.	Hymenoptera: Formicidae (ants)	Adults	Boll weevil
General predator	*Crematogaster* spp.	Hymenoptera: Formicidae (ants)	Adults	Boll weevil
General predator	*Pheidole* spp.	Hymenoptera: Formicidae (ants)	Adults	Boll weevil
General predator	*Conomyrma* spp.	Hymenoptera: Formicidae (ants)	Adults	Boll weevil
General predator	*Neivamyrmex* spp.	Hymenoptera: Formicidae (ants)	Adults	Boll weevil

significant. Nevertheless, it is possible that some significant predators of minor cotton pests and some predators with significant links to surrounding ecosystems have been overlooked. For example, the authors have identified some further predator taxa that should be considered in a full risk assessment of Bt cotton in Brazil: comb-footed spiders (Theridiidae) such as *Latrodectus* sp., Dipteran larvae in the families Syrphidae (such as *Toxomerus floralis*) and Dolichopodidae (such as *Condylostylus* spp.), praying mantis (Mantidae), dragonflies (Odonata), predatory crickets (such as *Oecanthus* sp.) and the predatory thrips *Franklinothrips* sp.

The selection matrix was used to record the maximum association with cotton and the cotton habitat, and the potential maximum functional significance of each species. Because there are important regional differences in predator assemblages, predators were ranked separately by region based on the published literature and the expertise of the authors and consulted colleagues. The cross-regional importance of certain predators was considered when information was available. In most cases the predators identified as being important in one region were ubiquitous throughout the country.

Potential association with the crop

Ranking the criteria to evaluate the potential association of the species with cotton (Tables 8.2A and 8.2B) revealed significant information gaps. The criterion 'abundance on other plants in cotton habitat' could not be ranked at all due to lack of information and was therefore not used, and there were many information gaps with regard to the strength of association of the predators with prey on cotton versus other habitats such as for the spiders, ants and ground-dwelling beetles. 'Phenology: proportion of cotton-growing season when present' was characterized only as early, mid and late season due to lack of more precise information. Because the results of this method were qualitative rather than quantitative, it is difficult to determine how accurately the rankings reflect actual conditions. However, since the worst-case scenario was considered when relevant information was lacking, the final rankings of species are highly likely to include most of the species that should be considered further. The highest rank for potential association with the crop was given to lacewings, pentatomid bugs and ladybird beetles with some regional variation in Brazil.

Potential functional significance of an environmental effect

For the ranking of the functional significance of a species, and therefore the significance of a potential effect of Bt cotton on a species (Table 8.3), experts among the authors and associates ranked the predators as potentially important biological control agents in Brazilian cotton. This resulted in a list of eight taxa (families or species). From this list, each expert prioritized the four taxa that were most important. Ladybird beetles, lacewings, a lycosid spider

Table 8.2A. Selection matrix – association with the crop and crop habitat.

(A) Non-target predator taxon × crop coincidence
Geographical distribution: degree of overlap of cotton crop and non-target taxon.
Habitat specialization: degree of association between taxon and cotton habitat.
Prevalence: proportion of cotton habitat occupied by taxon.
Abundance on cotton crop and on other plants occurring in the cotton habitat: average or typical densities.
For definitions see Box 5.2 in Hilbeck *et al.* (Chapter 5, this volume).

Selection matrix

			Association with crop – coincidence					
			Prevalence: proportion of suitable cotton habitat occupied			Abundance on cotton crop		
Non-target predator taxon	Geographical distribution	Habitat specialization	Meridian	Midwest	North-east	Meridian	Midwest	North-east
Phytoseiidae	All three regions	Low: on crop area, field margins, wild areas	Most	Most	Most	Medium	Medium	Medium
Forficulidae and Carcino-phoridae	All three regions	Low: on crop area, field margins, wild areas	All	All	All	High	High	High
Geocoris spp.	All three regions	Low: on crop area, field margins, wild areas	All	Patchy	All – most	Medium	Medium	Medium
Orius spp.	All three regions	Low: on crop area, field margins, wild areas	Patchy	Patchy	All – most	Medium	Medium	Medium
Nabis spp.	All three regions	Low: on crop area, field margins, wild areas	Most?	All	Most	Medium	Medium	Medium
Reduviidae	All three regions	Low: on crop area, field margins, wild areas	All	All	?	Low	Low	?
Pentatomidae	All three regions	Low: on crop area, field margins, wild areas	Patchy	Patchy	All	Medium	Low	High
Thomisidae	All three regions	Low: on crop area, field margins, wild areas	All	All	All	Medium to low	Medium to low	Medium to low
Lycosidae	All three regions	Low: on crop area, field margins, wild areas	?	All	?	?	High	?

Chrysoperla complex	All three regions	Low: on crop area, field margins, wild areas	All	Patchy	All	Medium to low	Medium to low	High
Staphylinidae	All three regions	Low: on crop area, field margins, wild areas	?	?	?	?	High	?
Carabidae	All three regions	Low: on crop area, field margins, wild areas	?	All	?	?	High	?
Coccinellidae	All three regions	Low: on crop area, field margins, wild areas	All	All	All	High	High	High
Pseudodorus clavatus	All three regions	Low: on crop area, field margins, wild areas	All	All	?	Medium	High	?
Polistes spp.	All three regions	Low: on crop area, field margins, wild areas	All	All	?	Medium	Medium	?
Camponotus sericeiventris	All three regions	Low: on crop area, field margins, wild areas	?	?	?	?	?	?
Solenopsis spp.	All three regions	Low: on crop area, field margins, wild areas	All	All	All	High to medium	High to medium	High to medium
Crematogaster spp.	All three regions	Low: on crop area, field margins, wild areas	?	?	?	?	?	?
Pheidole spp.	All three regions	Low: on crop area, field margins, wild areas	?	?	?	?	?	?
Conomyrma spp.	All three regions	Low: on crop area, field margins, wild areas	?	?	?	?	?	?
Neivamyrmex spp.	All three regions	Low: on crop area, field margins, wild areas	?	?	?	?	?	?

Table 8.2B. Selection matrix – association with the crop.

(B) Non-target predator taxon × crop trophic relationship
Phenology: temporal overlap (i) from non-target taxon perspective; (ii) from crop perspective.
Linkage: specialization of trophic relationship and dependency on cotton.
For definitions see Box 5.2 in Hilbeck et al. (Chapter 5, this volume).

Selection matrix

	Association with crop–trophic relationship					
	Phenology of predator taxon and cotton crop			Linkage with crop		
				Dependency on prey on cotton		
Non-target predator taxon	Life cycle stage on cotton	Proportion of cotton-growing season when taxon present (predatory life cycle stage)	Degree of feeding specialization	Meridian (MD)	Midwest (MW)	North-east (NE)
Phytoseiidae	All	All	Mites	Medium	Medium	Medium
Forficulidae and Carcinophoridae	All	Mid to late	None	Not strong	Not strong	Not strong
Geocoris spp.	All	Early, mid	None	Medium to not strong	Medium to not strong	Medium to not strong
Orius spp.	All	Early, mid	None	?	?	Medium
Nabis spp.	All	MD, MW: mid to late NE: early, mid	None	Not strong	Not strong	Not strong
Reduviidae	All	Mid to late	None	Weak	Weak	?
Pentatomidae	All	All	None	Not strong	Not strong	?
Thomisidae	All	Early, mid	None	Not strong	Not strong	Not strong
Lycosidae	All	All	None	?	?	?
Chrysoperla complex	All	MD: early, mid MW, NE: all	None	Low	Low	Strong
Staphylinidae	All	All	None	?	?	?
Carabidae	All	All	None	?	?	?
Coccinellidae	All	Early, mid	None	Not strong	Strong	Not strong
Pseudodorus clavatus	All	Early, late	Aphids, whiteflies	Not strong	Not strong	?
Polistes spp.	All	Mid to late	Caterpillars, boll weevil	Not strong	Not strong	?
Camponotus sericeiventris	All	?	?	?	?	?
Solenopsis spp.	All	All	None	Not strong	Not strong	Not strong
Crematogaster spp.	All	?	?	?	?	?
Pheidole spp.	All	?	?	?	?	?
Conomyrma spp.	All	?	?	?	?	?
Neivamyrmex spp.	All	?	?	?	?	?

Table 8.3. Selection matrix – functional significance for cropping system (related to significance of adverse effect).

Selection matrix	Significance as:			
Non-target predator taxon	Biological control in cotton	Biological control in other crops	Food for other natural enemies	Biological control in natural areas
Phytoseiidae	?	?	?	?
Forficulidae and Carcinophoridae	Medium	Medium	?	?
Geocoris spp.	Medium	Medium	?	?
Orius spp.	?	?	?	?
Nabis spp.	?	?	?	?
Reduviidae	Low	Low	?	?
Pentatomidae	?	Medium	Parasitoids	?
Thomisidae	?	?	?	?
Lycosidae	High	High	?	?
Chrysoperla complex	High	High	Parasitoids	?
Staphylinidae	?	?	?	?
Carabidae	?	?	?	?
Coccinellidae	High	High	Parasitoids	?
Pseudodorus clavatus	Low	Low	Parasitoids	?
Polistes spp.	Medium	?	?	?
Camponotus sericeiventris	?	?	?	?
Solenopsis spp.	Medium	Medium	?	?
Crematogaster spp.	?	?	?	?
Pheidole spp.	?	?	?	?
Conomyrma spp.	?	?	?	?
Neivamyrmex spp.	?	?	?	?

species, *Solenopsis* sp., Dermapterans and *Geocoris* received the highest rankings, and pentatomids received the lowest ranking. Next, specialists identified the taxa that were significant to other crops, again by prioritizing the top four. The taxa with highest ranks were lacewings, ladybirds and the lycosid spider. Intermediate rankings were given to *Geocoris*, pentatomids, *Solenopsis* sp., Dermapterans and predatory mites. Lacewings, ladybirds and the unidentified spider were regarded as highest-ranking taxa, based on their known significance as biological control agents for various crops in Brazilian agriculture. Significance as a food for other natural enemies could not be ranked. Information on the significance of these species in Brazil's natural areas is poorly known or non-existent.

From the selection process described above, lacewings (North-east region) and ladybird beetles (all cotton regions) were unanimously prioritized as being the highest priority and most ecologically relevant predator groups for non-target risk assessments (Table 8.4). For Brazilian cotton systems we believe that, given present knowledge, the selected predators are the ones with the greatest likelihood of having a significant role in ecosystem functions.

Table 8.4. Selection matrix – prioritized species[a]

Combination of association with crop (estimates from Tables 8.2 and 8.3) and functional significance in the cropping system (estimate from Table 8.3). See Hilbeck *et al.* (Chapter 5, this volume) for more detail on the procedure followed.

Selection matrix	Ranking of expert group				
	Association with crop			Functional significance for cropping	Prioritized taxon rank[a]
Non-target predator taxon	Meridian (MD)	Midwest (MW)	North-east (NE)	system	
Phytoseiidae	2 (1 if mites as prey alone)			?	2+?
Forficulidae and Carcinophoridae	2	2	2	?	2+?
Geocoris spp.	3	3	3	2	5
Orius spp.	3	3	2/3	3	MD, MW: 6 NE: 5.5
Nabis spp.	3	3	3	3	6
Reduviidae	3	3	?	?	3+?
Pentatomidae	2	2	1	2	MD, MW: 4 NE: 3
Thomisidae	3	3	3	3	6
Lycosidae	?	?	?	1	2?
Chrysoperla complex	2/3	2/3	1	1	MD, MW: 3.5 NE: 2
Staphylinidae	?	?	?	?	?
Carabidae	?	?	?	3	3+?
Coccinellidae	2	2	1+	1	MD, MW: 3 NE: 2+
Pseudodorus clavatus	2/3	2/3	?	?	2/3+?
Polistes spp.	3	3	?	?	3+?
Camponotus sericeiventris	?	?	?	?	?
Solenopsis spp.	2	2	2	2	4
Crematogaster spp.	?	?	?	?	?
Pheidole spp.	?	?	?	?	?
Conomyrma spp.	?	?	?	?	?
Neivamyrmex spp.	?	?	?	?	?

[a]1–2 = highest priority (widely distributed, abundant, mostly present, closely linked, high functional significance; 3–4 = intermediate priority; 5–6 = low priority (distribution, abundance and presence limited, loosely linked, lower functional significance; ? = insufficient information to rank.

Assessment of Potential Exposure Pathways to the Transgenic Crop and Transgene Products

Predators can be exposed to tissues and exudates from transgenic crops through multiple ecological pathways (Andow and Hilbeck, 2004).

Lacewings

Lacewings are important natural control agents of cotton pests in South America (Albuquerque *et al.*, 2001). These lacewings are typically nomadic and are adapted to living in patchy and temporary habitats (Duelli, 2001). Adults are strong fliers; larvae are less mobile and individuals that hatch in cotton fields are dependent on food within this habitat. Nomadic lacewings are generalist predators with only moderately restricted habitat requirements. A number of factors will influence whether lacewings distinguish between prey on crop field weeds and on the crop itself (Szentkiralyi, 2001). Although many species of lacewings are known in Brazilian agroecosystems (Freitas and Penny, 2001), the prominent lacewing species on cotton in each cotton-growing region of Brazil still need to be identified before case-specific risk-assessment experiments can be conducted.

Bitrophic interactions
Lacewing (*Chrysoperla* spp. in particular) larvae and adults differ in their feeding habits; larvae tend to be more predaceous and less phytophagous than adults (McEwen *et al.*, 2001). Green lacewing larvae have piercing-sucking mouthparts, and the food must be a fluid to be ingested, such as the contents of eggs or prey (Canard, 2001). The mouthparts of adults have symmetrical mandibles without any incisor and spoon-like lacini. They are not predaceous and feed on glycophagous diets that include sugar-rich foods such as nectar, plant fluids and honeydew, and will also consume pollen (Bozsik, 1992; Canard, 2001). The larvae do not feed on pollen, except perhaps incidentally in other liquid foods (Patt *et al.*, 2003). Larvae also suck on cotton nectar and plant fluids, such as from damaged tissue (Canard, 2001; Limburg and Rosenheim, 2001; F. Schmidt and A. Hilbeck, Zurich, Switzerland, 2004, personal communication). *Chrysoperla plorabunda* larvae feed on leaf veins of cotton plants in the field, though these larvae do not survive well on a diet of only leaf tissue (Limburg and Rosenheim, 2001). In general, the literature suggests that bitrophic feeding is more important for the adult stage than the immature stage, which probably only feeds on plant liquids in transient periods of prey scarcity (Limburg and Rosenheim, 2001). Actual field observations of lacewing feeding behaviour on cotton in Brazil would aid in making decisions concerning potential bitrophic interactions.

The available information on Bt toxin expression levels in Bt cotton pollen indicates variable levels of expression (see Grossi *et al.*, Chapter 4, this volume; Arpaia *et al.*, Chapter 7, this volume). Information on Bt toxin presence in cotton nectar and exudates is not available, but plant fluids from damaged leaf or stem tissue are very likely to contain Bt toxin. Lacewing larvae and adults may therefore be exposed bitrophically via leaf veins and other plant fluids, and via nectar if the nectar expresses the Bt toxin (Box 8.1). Lacewing adults may also be exposed bitrophically through pollen feeding.

> **Box 8.1.** Hypothesized exposure pathways for lacewings from Bt cotton. Realized exposure depends on the actual presence of Bt toxin in the specific tissue or exudates.
>
> **Bitrophic interactions:**
> - Larvae feed on nectar and small amounts of pollen in the nectar.
> - Larvae feed on leaf veins or other leaf- or stem-tissue fluids of plants.
> - Adults feed on pollen.
> - Adults feed on nectar.
> - Adults feed on leaf- or stem-tissue fluids of plants.
>
> **Multi-trophic interactions:**
> - Larvae feed on herbivores that ingested Bt cotton.
> - Larvae feed on the honeydew of Homoptera feeding on Bt cotton.
> - Adults feed on the honeydew of Homoptera feeding on Bt cotton.

Multi-trophic interactions

All larvae of lacewings found in agricultural habitats feed on herbivores, and thus tritrophic interactions should be considered in risk assessments. *Chrysoperla* spp. larvae and adults can be found generally within the cotton system throughout the growing season (Albuquerque *et al.*, 2001). In other geographic regions, *Chrysoperla* spp. abundance is positively correlated with increasing prey abundance, specifically aphid populations (Pantaleoni, 2001). There is a high likelihood that lacewings encounter some herbivores that are ingesting the transgenic cotton. The dominant prey for many lacewing larvae is phloem-feeding Homoptera, and both larvae and adults will feed on Homopteran honeydew (Canard, 2001). Cry toxins have been found in the cotton aphid *Aphis gossypii* on one kind of Bt cotton in China (Zhang *et al.*, 2004) and in the honeydew of the brown planthopper on rice (Bernal *et al.*, 2002), though not in maize aphids (Head *et al.*, 2001; Raps *et al.*, 2001; Dutton *et al.*, 2002) or in maize phloem sap (Raps *et al.*, 2001). We had no information on the occurrence of Bt toxins in the phloem of the Bt cotton varieties being developed for Brazil, or in the aphids feeding on the phloem.

In summary, lacewing larvae may be exposed tritrophically through feeding on herbivores that have ingested Bt cotton, and both larvae and adults may be exposed tritrophically through feeding on Homopteran honeydew if it contains the Bt toxin (Box 8.1). They are not likely to be exposed tritrophically if the herbivores do not ingest the Bt toxin.

Ladybird beetles

Ladybird beetles are widely regarded as important biological control agents in agricultural habitats around the world. The most abundant ladybirds in cropland are eurytopic, accepting a broad range of habitats and often having broad diets. Moreover, surrounding habitats can affect the diversity and abundance of ladybirds within cropland (Hodek and Honek, 1996). Adult ladybirds are more mobile than immatures, and thus habitat selection occurs during the adult stage. Often communities that occur within agricultural habitat are dominated by two to four species, and these dominant species differ according to a number of factors. The distribution of ladybirds within a field likely depends on food availability, microclimate, host-plant characteristics, competitive interactions and adjacent habitats (Hodek and Honek, 1996). Coccinellid populations need to be sampled throughout the Brazilian cotton-growing region to determine appropriate species for risk analyses.

Bitrophic interactions
Ladybird species are predaceous as larvae and adults, though many species are best described as omnivores rather than predaceous. Larvae and adults of ladybird beetles may feed on pollen, nectar and anthers, all of which may contain Bt toxin. During anthesis in maize, the omnivorous species *Coleomegilla maculata* may become more pollinivorous than predaceous and so bitrophic exposure could become quite important for some species at specific times (Lundgren *et al.*, 2004). Omnivory in certain species can vary among life stages. For instance, larvae of *Harmonia axyridis* can feed on maize pollen, but adults are strictly predaceous (Koch, 2003; Lundgren *et al.*, 2004). Because of this behavioural variation, assays used to evaluate the risk of bitrophic interactions to ladybird predators should be tailored to the biology of the species being examined.

Multi-trophic interactions
The majority of ladybird species are mostly predaceous (exceptions are in the subfamily Epilachninae), and risk assessments should consider potential multi-trophic interactions. Potential prey for ladybirds are aphids (Hemiptera: Aphididae), coccids, pseudococcids, and diaspidids (Hemiptera: Coccoidea), whiteflies (Aleyrodidae), insect eggs and virtually any other soft-bodied insects (Hodek and Honek, 1996). Some feeding specialization occurs within this family, but for the most part ladybirds within agroecosystems feed on phloem-feeding Homopterans. Some ladybirds aggregate to high densities of prey (Schellhorn and Andow, 1999), and it is likely that predaceous species will encounter herbivores feeding on transgenic cotton. Assays designed to evaluate tritrophic interactions should involve ecologically relevant prey species; prey accepted in the laboratory may be unsuitable for development or not fed upon by the selected ladybird species under field conditions. For ladybird beetles, many of the exposure pathways are the same as for lacewings. The primary difference is that the adult stage of ladybirds is usually predaceous (Box 8.2).

Identification of Adverse-Effect Scenarios

From the possible exposure pathways identified above, possible adverse effects on the prioritized species can be suggested. These adverse effects might lead to consequent adverse effects such as a reduction in biological control, but these effects on biological control cannot be evaluated in laboratory or greenhouse

> **Box 8.2.** Hypothesized exposure pathways for ladybirds from Bt cotton. Realized exposure depends on the actual presence of Bt toxin in the specific tissue or exudates.
>
> **Bitrophic interactions:**
> • Larvae feed on pollen.
> • Larvae feed on nectar.
> • Adults feed on pollen.
> • Adults feed on nectar.
>
> **Multi-trophic interactions:**
> • Larvae feed on herbivores that ingest Bt cotton.
> • Adults feed on herbivores that ingest Bt cotton.

settings for cotton. For the two selected prioritized species groups, ladybird beetles and lacewings, the potential identified adverse effects on the species can be estimated by measuring relative generational fitness, i.e. survival and reproduction (Box 8.3). A reduction in generational fitness can lead to lower populations of predators that could result in reduced biological-control efficacy.

The adverse-effect scenarios were formulated as testable hypotheses (Boxes 8.4 and 8.5). These hypotheses serve as the basis for developing experiments in closed environments (laboratory, greenhouse) before field release of the transgenic crop. These experiments are designed to determine if the predator would be exposed to Bt toxin and, if exposed, whether any of a set of potential adverse effects could occur.

Testable Hypotheses and Experimental Designs

We propose a sequence of experiments that could be conducted on lacewings and coccinellids on Bt cotton to confirm or refute the specified hypotheses. Other crops and predators can be evaluated with similar tests that are tailored

> **Box 8.3.** Fitness components considered for evaluation in a prerelease testing programme with natural enemies.
>
> **1.** Traits should be compared between individuals fed on diets with and without Cry toxin.
> **2.** The purified Cry toxin should be verified to be equivalent to that which is expressed in the plant.
>
> • Stage-specific or age-specific mortality of larvae, pupae and adults.
> • Age-specific fecundity (first 7 days after mating, and lifetime-realized fecundity).
> • Size/weight of adult males and females (e.g. dry weights, body, elytral and tibial lengths; these observations should be recorded on individuals of equivalent ages).
> • Adult lifespan.
> • Egg-hatching rate/viability (care must be taken to eliminate egg cannibalism).

Box 8.4. Hypothetical adverse effects on lacewings stemming from a suspected exposure pathway.

• Lacewing larvae that feed on Bt cotton nectar containing Cry toxin may have higher mortality than larvae that feed on nectar of non-Bt cotton.

• Lacewing larvae that feed on leaf veins or other plant fluids of Bt cotton plants containing Cry toxin may have higher mortality than larvae that feed on leaf veins or other plant fluids of non-Bt cotton.

• Lacewing larvae that feed on herbivores that ingested Bt cotton may have higher mortality than larvae feeding on herbivores on non-Bt cotton.

• Lacewing larvae that feed on Homopteran honeydew containing Cry toxin may have higher mortality than larvae feeding on honeydew without Cry toxin.

• Lacewing adults that feed on Bt cotton pollen and nectar containing Cry toxin may have higher mortality and/or lower reproduction than adults that feed on pollen and nectar of non-Bt cotton.

• Lacewing adults that feed on Homopteran honeydew containing Cry toxin may have higher mortality and/or lower reproduction than adults feeding on honeydew without Cry toxin.

• Lacewing adults that emerge from larvae that fed on herbivores that ingested transgenic cotton and feed on Homopteran honeydew containing Cry toxin may have higher mortality and/or lower reproduction than adults from larvae that fed on herbivores from non-Bt cotton and that feed on honeydew without Cry toxin.

• Lacewing larvae and/or adults prefer to feed on Bt cotton tissue. If the predator avoids the Bt cotton tissue, then any adverse effect of bitrophic exposure is likely to be reduced. If it prefers the Bt cotton tissue, then any adverse effect of bitrophic exposure may be increased.

• Lacewing larvae and/or adults prefer to feed on the food containing Cry toxin than the food without Cry toxin. If the predator avoids the Bt food, then any adverse effect of tritrophic exposure is likely to be reduced. If it prefers the Bt food, then any adverse effect of tritrophic exposure may be increased.

---

to the biology of the predator and crop in question. Species should be identified using either morphological or molecular methods.

The experiments are separated into three general categories: (i) evaluations of bitrophic exposure and its effects on predators, including laboratory toxicity testing with purified transgene products; (ii) evaluations of multi-trophic exposure and its effects on predators; and (iii) field-level studies that investigate the relative effects of transgenic crops to predator populations and predation in the field.

The experiments that are recommended in each category consider the toxicity of a specific Bt cotton event (i.e. what proportion of lacewings are killed when they ingest nectar from transgenic cotton?), the degree of exposure that is likely (i.e. what proportion of lacewings are exposed to cotton nectar, and what dose do they likely ingest under natural conditions?), or both of these. Using this framework, scientists will be able to elucidate the mechanisms that underlie the potential adverse effects to predators, rather than simply documenting the effects of transgenic crops on predator populations.

> **Box 8.5.** Hypothetical adverse effects on Coccinellids stemming from a suspected exposure pathway.
>
> • Coccinellid larvae that feed on Bt cotton pollen and nectar containing Cry toxin may have higher mortality than larvae that feed on non-Bt cotton pollen and nectar.
> • Coccinellid larvae that feed on herbivores that ingested Bt cotton may have higher mortality than larvae feeding on herbivores from non-Bt cotton.
> • Coccinellid adults that feed on Bt cotton pollen and nectar containing Cry toxin may have higher mortality and/or lower reproduction than adults that feed on non-Bt cotton pollen and nectar.
> • Coccinellid adults that feed on herbivores that ingested Bt cotton may have higher mortality and/or lower reproduction than adults feeding on herbivores from non-Bt cotton.
> • Coccinellid adults that emerge from larvae that fed on herbivores that ingested Bt cotton, and that feed on herbivores fed on Bt cotton may have higher mortality and/or lower reproduction than adults from larvae that fed on herbivores from non-Bt cotton and feed on herbivores from non-Bt cotton.
> • Coccinellid larvae and/or adults prefer to feed on Bt cotton tissue. If the predator avoids the Bt cotton tissue, then any adverse effect of bitrophic exposure is likely to be reduced. If it prefers the Bt cotton tissue, then any adverse effect of bitrophic exposure may be increased.
> • Coccinellid larvae and/or adults prefer to feed on the food containing Cry toxin than the food without Cry toxin. If the predator avoids the Bt food, then any adverse effect of tritrophic exposure is likely to be reduced. If it prefers the Bt food, then any adverse effect of tritrophic exposure may be increased.

These mechanisms will allow for more focused experimentation and precise estimation of the actual risk to predators in the field.

If either the exposure pathway is unlikely or the adverse-effect scenario is unlikely, then there is likely no significant risk to the predator. Therefore, the initial experiments can focus either on the exposure pathway (Boxes 8.1 and 8.2) or the adverse-effect scenarios (Boxes 8.4 and 8.5). If either part is falsified, then we can infer that any possible risk is below the detection threshold of the experiment, and there will normally be no need to continue testing.

Bitrophic (plant/predator) effects

These tests can be carried out at the stage of development of the transgenic plant when the event is being characterized in the laboratory or greenhouse.

Exposure assays
PLANT TISSUES. Measure the presence or the concentration of Bt toxin in cotton nectar, pollen and leaf veins (see Arpaia *et al.*, Chapter 7, this volume, for more information on measured expression levels). If these tissues contain negligible amount of Bt toxin, then the consequent adverse-effect hypothesis does not need to be tested.

PREDATOR. The predator should be demonstrated to contain Bt toxin after feeding on the plant tissue. Exposure via pollen feeding could be quantified by whole body analysis or, if necessary, through gut content analysis for coccinellids. If the predator does not contain Bt toxin after feeding on the plant tissue then the consequent adverse-effect hypothesis does not need to be tested.

Adverse-effect assays

The specific adverse-effect hypotheses for lacewings and coccinellids are listed in Boxes 8.4 and 8.5. Feeding assays should be conducted using plant tissues or exudates that are acceptable to the predator and contain Bt toxin. Fitness or some fitness components should be measured (Box 8.3). These experiments are only necessary if the predator has been demonstrated to contain Bt toxin after feeding on the plant tissue.

• Larvae and adults reared on each Bt toxin-containing plant tissue should be compared to larvae and adults reared on the same tissue from non-Bt cotton in a no-choice experiment. The duration of the assay should continue throughout the entire larval development period and for at least 15 days for adults. One difficulty that is encountered in this type of assay is that rearing the predators on nutritionally poor foods will result in unacceptably high levels of mortality in the control treatments reducing the likelihood of detecting effects. If this occurs, then the experiment can be designed to estimate mortality rates from a given age onwards. It is likely that these foods would be used in the field when no other foods were available. The experiment would start with well-fed individuals of known age (several ages should be used, e.g. neonates, late instars, newly enclosed adults), switch them to the plant diet and estimate the rate of mortality.
• Recommended sample sizes: four replicated blocks with at least 30 individuals per treatment in each block. The larvae or adults should be reared individually under standardized conditions to minimize variability due to preconditioning, etc. For mortality experiments, it may be necessary to use more individuals per replicate.

Laboratory toxicity testing with purified protein

These experiments are useful for determining worst-case adverse effect, because toxin concentrations much higher than that likely to be experienced in the field are tested. If no adverse effect is observed after exposure to very high concentrations of toxin, it is reasonable to infer that the adverse effect is minimal in the field.

Testable hypothesis: Larvae or adults of lacewings or coccinellids that consume food containing high concentrations of Cry toxin have a lower fitness than the larvae or adults that consume a diet without Bt toxins.

TOXICITY OF PURIFIED CRY PROTEIN TO LACEWINGS AND COCCINELLIDS. These tests can be carried out at the stage of development of the transgenic plant when the event is being characterized in the laboratory or greenhouse (see Andow *et al.*, Chapter 1, this volume). Lethal and sublethal effects of purified truncated Cry proteins to green lacewing larvae and adults can be determined using an artificial diet. Artificial diets are available and are efficient for the rearing of lacewings (Hilbeck *et al.*, 1998b; Nordlund *et al.*, 2001). For larvae, the liquid diet needs to be presented in such a way that the larvae actually ingest the Bt toxin (see bitrophic exposure above). Artificial diets are also available for rearing coccinellids in the laboratory (Atallah and Newsom, 1966).

One should consider conducting single high-dose bioassays with adults and larvae of the selected species, i.e. by incorporating the toxin into artificial diet at a higher dose than the one expressed in the transgenic plant. Alternatively, one may prefer to calculate median lethal dose (LD_{50}) for adults and larvae. This requires conducting a series of feeding trials with increasing Bt toxin concentrations (multiple-dose bioassay).

Important points to consider are:

- Assure that the Cry protein is stable within the diet (e.g. proteases do not degrade the protein).
- Check that the purified Bt toxin is biologically equivalent to the form expressed in the Bt cotton (efficacy, stability, structure, specificity, etc.).
- For a dose-response experiment, use at least eight different concentrations of Bt toxin that relate to the expression levels in the plant. Water should be provided to individuals.
- Continue assay through the entire larval development period; for adults the assays should be run for at least 15 days or until most adults in one of the treatments are dead.
- Sample sizes should be run in at least four blocks (replicates), with at least 30 individuals in each treatment in each replicate. Rear the larvae or adults individually.

Preference test

If any of the bitrophic adverse-effects hypotheses are confirmed, a preference experiment may be useful, complemented with actual field observations of bitrophic feeding behaviour on cotton.

Testable hypothesis: Lacewing or coccinellid larvae and/or adults prefer to feed on transgenic plant tissue (Boxes 8.4 and 8.5).

The appropriate life stage and plant tissue should be used. Larval lacewings can be offered Bt- and non-Bt-containing food sources in an arena in a 10 cm diameter Petri dish or another suitable container (Meier and Hilbeck, 2001). Adult coccinellids can be placed in cages made of hollow PVC tubes (100 mm in diameter × 250 mm) covered with a cotton screen. Adequate water should also be provided. Count the number of visits to each food source in 1 h. If the insect is not very mobile, record the position of the insect at regular time intervals. If possible, measure either the consumption rate of each food type by weight of food over 24 h, controlling for water loss or gain, or duration of the feeding events in the case of liquid or leaf tissues. Sample sizes should be run in several blocks, with at least 50 individuals per treatment for the entire experiment. The larvae or adults should be tested individually under standardized conditions.

Multi-trophic exposure

These tests can be carried out at the stage of development of the transgenic plant when the event is being characterized in the laboratory or greenhouse.

Exposure assays
Tritrophic exposure requires that the Cry toxin is present in prey. *A. gossypii* and first-instar *Spodoptera* sp. (most likely *Spodoptera frugiperda* or another *Spodoptera* species that is prevalent in a particular region) are expected to be relevant prey species for chrysopids under Brazilian conditions. The consequent adverse-effect hypotheses do not need to be tested if the prey and their products contain no or negligible amounts of Bt toxin. The predator should also be demonstrated to contain Bt toxin after feeding on the prey. This can be tested using methods similar to those suggested under bitrophic exposure.

Adverse-effect assays
Specific tritrophic adverse-effect hypotheses for lacewings and coccinellids are listed in Boxes 8.4 and 8.5. If the larvae or adults feed on prey or prey products that contain Bt toxin, then hazard experiments should be conducted, using only those prey or prey products known to contain Bt toxin. Lacewing larvae are provided either *A. gossypii* or *Spodoptera* larvae as their sole food. The prey are harvested from either Bt cotton or non-Bt cotton plants grown in greenhouses. Fitness parameters should be measured, as outlined in Box 8.3. The lacewing larvae may be exposed via consumption of the aphids or their honeydew. The aphid honeydew is not likely to be a nutritionally complete food, so mortality experiments that are similar to those described in the bitrophic tests above for poor-quality foods could be conducted. Other experimental conditions follow the bitrophic adverse-effects assays above. Coccinellid larvae and adults should be tested on their preferred prey, provided that this prey contains Bt toxin.

Preference tests
If any of the tritrophic adverse-effects hypotheses are confirmed, a paired preference experiment may be useful.

Testable hypothesis: Lacewing or coccinellid adults/larvae prefer to feed on the food containing Cry toxin than the food without Cry toxin (Boxes 8.4 and 8.5).

To design this experiment, information on the relative abundance and species composition of the prey that is available in the field to lacewing larvae or coccinellid larvae and adults is required. The appropriate life stage and prey types should be used. Use methods similar to that described in the bitrophic preference test above. If prey are marked (either externally with a pen or paint or internally with stable isotopes) prey consumption can be estimated.

Preference tests for aphid honeydew will be difficult and are not recommended, because this potential food will be difficult to control. If multiple foods adversely affect lacewings, these experiments can become quite complicated. Rather than attempt a comprehensive assessment, one can assess if there is evidence for any interaction among food types using a multiple preference test. All Bt foods and possible non-Bt foods can be placed in a single arena to estimate a preference hierarchy. Multiple preference tests are poor at distinguishing among many preferences, so the results should be interpreted with caution.

Field tests

These can be conducted at the small-scale field test stage and large-scale field test stage prior to the commercial release of transgenic plants. The aims are to determine if any of the exposure pathways and adverse-effect scenarios supported by the laboratory and greenhouse experiments are likely to occur in the field, and if so, to quantify the effect and estimate the consequent effects on the predator population size and predation rates on important prey. Data from population-dynamics studies can provide a firm basis for making a quantitative estimate of the risk to non-target predators posed by large-scale commercial production of Bt cotton.

Hypotheses
• A possible risk (for a definition, see Hilbeck *et al.*, Chapter 5, this volume) identified in the prerelease testing stage (above) actually occurs in the environment.
• In the environment, predator populations are lower in Bt cotton than in non-Bt cotton.
• Predators will provide lower biological control in Bt cotton than in non-Bt cotton.

An experimental test of the first hypothesis will depend on the specific possible risk being examined. For example, if *A. gossypii* containing Bt toxin increased mortality of lacewing larvae in the laboratory, the field experiment should be designed to estimate larval mortality associated with Bt cotton versus non-Bt cotton when *A. gossypii* is the main prey. Such an experiment could be conducted in small plots by augmenting aphid populations or spraying with an appropriate selective insecticide that releases the aphid populations from natural enemies, and then releasing a known density of neonate larvae (number per plant). Plants can be carefully inspected or harvested at frequent intervals to determine the number of surviving larvae. The intervals and total duration of the experiment would be determined by the results from the laboratory experiment. In this way, the mechanistic details revealed by the laboratory tests can be used to focus the field experiments. In general, the experiments would be designed to have abundant target food, either by adding more, manipulating the system so more appeared or choosing a time to conduct the experiment when more was available. Then, either natural populations of the appropriate predator stage could be observed or the predator could be augmented in the field to facilitate estimation of the specific parameter associated with potential risk. In addition, it will be important to determine that the predators actually eat the food in the field. For example, nectar consumption can be determined by the anthrone test (Olson *et al.*, 2000). It is also possible to use high performance liquid chromatography (HPLC) that could distinguish nectar from aphid honeydew. It will be difficult to determine all of the necessary details for the experiment until the specific hypothesis is clarified. In some cases, a semi-field experiment involving open greenhouses or field cages might be more appropriate than a field experiment.

These mechanistic field experiments will need to consider the importance of intraguild predation on lacewing larvae (Rosenheim *et al.*, 1993). Intraguild predation by predaceous Hemiptera on lacewing larvae can be high and obscure any effect of Bt toxin. One possible method would be to conduct a field experiment that excludes intraguild predators, using a method similar to Rosenheim *et al.* (1993), and then conduct the experiment with intraguild predators.

An experimental test of the second risk hypothesis will require appropriate field sizes and sampling methods. It is unclear what would be an ideal plot size and, in many instances, this has to be defined on a case-by-case basis. However, experimental plots much smaller than 0.5 ha are not likely to provide biologically meaningful information unless additional care is taken in the experimental design and interpretation of results. Field sampling of eggs, larvae and adults to monitor population sizes should be carried out once or twice per week throughout the growing season. Sampling should be conducted in a statistically sound manner. Hatched versus unhatched lacewing eggs can be distinguished in the field by the colour of the egg (unhatched young eggs are green, eggs shortly before hatching are grey and empty hatched eggs are white) or other methods. If adults are abundant enough, direct counts on plants in the field are preferable. Yellow sticky impact traps (Udayagiri *et al.*, 1997) or direct aspiration from plants also can be used for capturing coccinellid adults. Lacewing species should be identified using both morphological (collect eggs and rear larvae) and molecular methods. Rearing the lacewing eggs to maturity will provide additional information about parasitism rates of this life stage, one type of intraguild interaction.

The third hypothesis focuses on the biological control function of the predators. This can be evaluated by estimating predation rates on prey. Quantification of predation rates under field conditions is complicated by variability in prey size and distinguishing between living and dead prey. It is possible to do experiments with small field cages, in which a cohort of predators are released into a number of the cages; in the control cages none are released. Quantitative serological methods might be adapted for use with lacewings. These can be used to estimate consumption rates of predators. A frequently used method is to place sentinel prey out into the field for fixed periods of time. Disappearance of sentinels is often related to predation rates. If the prey are attached firmly to a substrate, their exoskeleton with characteristic puncture holes will remain attached to the substrate after predation by lacewings.

Discussion and Conclusions, Suggestions for Further Development

The focus of this chapter was to describe and illustrate a methodology for identifying and evaluating possible adverse effects of Bt cotton on biocontrol by predators. The methodology was elaborated to evaluate these possible adverse effects prior to field release of Bt cotton, and an outline for continuing the evaluation after field release was provided. In using the methodology, the authors identified and prioritized significant gaps of knowledge that would need to be filled before a full risk assessment could be carried out. Many of the predators could only be considered as whole families in this chapter, and we recommend repeating the methodology at the species level once additional baseline knowledge about the cotton agroecosystem in each region, season and crop-production system is available. For instance, information on species presence and prey preference on cotton for the Chrysopidae, Coccinellidae, Lycosidae and Formicidae would be particularly useful. We recommend involving taxonomic experts before selecting a focal species in the Chrysopidae and Coccinellidae, as cryptic species complexes occur in both of these groups.

We carried out an analysis of possible adverse effects on the lacewings (Chrysopidae) and ladybird beetles (Coccinellidae), two groups that were highly associated with cotton and probable high biological-control activity. This analysis revealed knowledge gaps such as the status of Bt protein expression in nectar and phloem and whether Homoptera can take it up (thereby possibly exposing the predator life cycle stages that feed on Homoptera and their honeydew), but these gaps can be filled with relatively simple experiments. We generally agreed that the analysis was useful and helpful in making decisions to prioritize questions for research, and can be used in an iterative way as information accrues.

Acknowledgements

We thank Dr Francisco Ramalho, Dr Angelo Pallini and Dr Pierre Silvie for providing valuable information on general predators on cotton in Brazil, and Evelyn Underwood for assistance with the tables.

References

Albuquerque, G.S., Tauber, C.A. and Tauber, M.J. (2001) *Chrysoperla externa* and *Ceraeochrysa* spp.: potential for biological control in the New World tropics and subtropics. In: McEwen, P., New, T.R. and Whittington, A.E. (eds) *Lacewings in the Crop Environment.* Cambridge University Press, Cambridge, UK, pp. 408–423.

Andow, D.A. and Hilbeck, A. (2004) Science-based risk assessment for non-target effects of transgenic crops. *BioScience* 54, 637–649.

Atallah, Y.H. and Newsom, L.D. (1966) Ecological and nutritional studies on *Coleomegilla maculata* DeGeer (Coleoptera: Coccinellidae). I. The development of an artificial diet and a laboratory rearing technique. *Journal of Economic Entomology* 59, 1173–1179.

Bernal, C.C., Aguda, R.M. and Cohen, M.B. (2002) Effect of rice lines transformed with *Bacillus thuringiensis* toxin genes on the brown planthopper and its predator *Cyrtorhinus lividipennis*. *Entomologia Experimentalis et Applicata* 102, 21–28.

Bozsik, A. (1992) Natural adult food of some important *Chrysopa* species (Planipennia: Chrysopidae). *Acta Phytopathologica et Entomologica Hungarica* 27, 141–146.

Canard, M. (2001) Natural food and feeding habits of lacewings. In: McEwen, P., New, T.R. and Whittington, A.E. (eds) *Lacewings in the Crop Environment.* Cambridge University Press, Cambridge, UK, pp. 116–129.

Carvalho, C.F. and Souza, B. (2002) Potencial de insetos predadores no controle biológico aplicado. In: Parra, J.R., Botelho, P.S.M., Corrêa-Ferreira, B.S. and Bento, J.M.S. (eds) *Controle biológico no Brasil.* Manole, São Paulo, Brazil, pp. 191–208.

Corey, D., Kambhampati, S. and Wilde, G. (1998) Electrophoretic analysis of *Orius insidiosus* (Hemiptera: Anthocoridae) feeding habits in field corn. *Journal of the Kansas Entomological Society* 71, 11–17.

Duelli, P. (2001) Lacewings in field crops. In: McEwen, P., New, T.R. and Whittington, A.E. (eds) *Lacewings in the Crop Environment.* Cambridge University Press, Cambridge, UK, pp. 158–171.

Dutton, A., Klein, H., Romeis, J. and Bigler, F. (2002) Uptake of Bt-toxin by herbivores feeding on transgenic maize and consequences for the predator *Chrysoperla carnea. Ecological Entomology* 27, 441–447.

Freitas, S. and Penny, N.D. (2001) The green lacewings (Neuroptera: Chrysopidae) of Brazilian agro-ecosystems. *Proceedings of the California Academy of Sciences* 52, 245–395.

Gallo, D., Nakano, O., Carvalho, R.P.L., Baptista, G.C. de, Berti Filho, E., Parra, J.R.P., Zucchi, R.A., Alves, S.B., Vendramin, J.D., Marchini, L.C., Lopes, J.R.S. and Omoto, C. (2002) *Entomologia Agrícola.* Fealq, Piracicaba, Brazil.

Gondim, D.M.C., Belot, J.L., Silvie, P. and Petit, N. (2001) *Manual de identificação das pragas, doenças, deficiências minerais e injúrias do algodoeiro do Brasil.* Boletim Técnico, 33, 3ª edn. Codetec/CIRAD, Cascavel, Brazil.

Gravena, S. (1990) Estratégias e táticas do MIP algodoeiro no Brasil. In: Fernandes, O.A., Correia, A. do C.B. and Bortoli, S.A. de (eds) *Manejo integrado de pragas e nematóides.* FUNEP/UNESP, Jaboticabal, Brazil, Vol. 1, pp. 1–14.

Gravena, S. and Cunha, H.F. da (1991) *Artrópodos predadores da cultura algodoeira: atividade sobre* Alabama argillacea *(Hub.) com breves referências a* Heliothis sp. *(Lepidóptera: Noctuidae).* Boletim Técnico, 1. FUNEP, Jaboticabal, Brazil.

Harmon, J.P. and Andow, D.A. (2004) Indirect effects between shared prey: predictions for biological control. *BioControl* 49(6), 605–626.

Head, G., Brown, C.R., Groth, M.E. and Duan, J.J. (2001) Cry1Ab protein levels in phytophagous insects feeding on transgenic corn: implications for secondary exposure risk assessment. *Entomologia Experimentalis et Applicata* 99, 37–45.

Hilbeck, A., Moar, W.J., Pusztai-Carey, M., Filipini, A. and Bigler, F. (1998b) Toxicity of the *Bacillus thuringiensis* Cry1Ab toxin on the predator *Chrysoperla carnea* (Neuroptera: Chrysopidae) using diet incorporated bioassays. *Environmental Entomology* 27, 1255–1263.

Hodek, I. and Honek, A. (1996) *Ecology of Coccinellidae.* Kluwer Academic Publishers, Dordrecht, The Netherlands.

Koch, R.L. (2003) The multicolored Asian lady beetle, *Harmonia axyridis*: a review of its biology, uses in biological control, and non-target impacts. *Journal of Insect Science* 3, 32.

Limburg, D.D. and Rosenheim, J.A. (2001) Extrafloral nectar consumption and its influence on survival and development of an omnivorous predator, larval *Chrysoperla plorabunda* (Neuroptera: Chrysopidae). *Environmental Entomology* 30, 595–603.

Lundgren, J.G., Razzak, A.A. and Wiedenmann, R.N. (2004) Population responses and feeding behaviors of the predators *Coleomegilla maculata* and *Harmonia axyridis* (Coleoptera: Coccinellidae) during anthesis in an Illinois cornfield. *Environmental Entomology* 33, 958–963.

McEwen, P., New, T.R. and Whittington, A.E. (eds) (2001) *Lacewings in the Crop Environment.* Cambridge University Press, Cambridge, UK.

Meier, M.S. and Hilbeck, A. (2001) Influence of transgenic *Bacillus thuringiensis* corn-fed prey on prey preference of immature *Chrysoperla carnea* (Neuroptera: Chrysopidae). *Basic and Applied Entomology* 2, 35–44.

Nordlund, D.A., Cohen, A.C. and Smith, R.A. (2001) Mass rearing, release techniques, and augmentation. In: McEwen, P., New, T.R. and Whittington, A.E. (eds) *Lacewings in the Crop Environment.* Cambridge University Press, Cambridge, UK, pp. 303–319.

Olson, D.M., Fadamiro, H., Lundgren, J.G. and Heimpel, G.E. (2000) Effects of sugar feeding on carbohydrate and lipid metabolism in a parasitoid wasp. *Physiological Entomology* 25, 17–26.

Pantaleoni, R.A. (2001) Lacewing occurrence in the agricultural landscape of *Pianura Padana.* In: McEwen, P., New, T.R. and Whittington, A.E. (eds) *Lacewings in the Crop Environment.* Cambridge University Press, Cambridge, UK, pp. 447–470.

Patt, J.M., Wainright, S.C., Hamilton, G.C., Whittinghill, D., Bosley, K., Dietrick, J. and Lashomb, J.H. (2003) Assimilation of carbon and nitrogen from pollen and nectar by a predaceous larva and its effects on growth and development. *Ecological Entomology* 28, 717–728.

Ramalho, F.S. and Wanderley, P.A. (1996) Ecology and management of the boll weevil in South American cotton. *American Entomologist* 42, 41–47.

Raps, A., Kehr, J., Gugerli, P., Moar, W.J., Bigler, F. and Hilbeck, A. (2001) Detection of Cry1Ab in phloem sap of *Bacillus thuringiensis* corn and in the selected herbivores *Rhopalosiphum padi* (Homoptera: Aphididae) and *Spodoptera littoralis* (Lepidoptera: Nocutidae). *Molecular Ecology* 10, 525–533.

Rosenheim, J.A., Wilhoit, L.R. and Armer, C.A. (1993) Influence of intraguild predation among generalist insect predators on the suppression of an herbivore population. *Oecologia* 96, 439–449.

Schellhorn, N.A. and Andow, D.A. (1999) Cannibalism and interspecific predation: role of oviposition behavior. *Ecological Applications* 9, 418–428.

Silvie, P., Leroy, T., Michel, B. and Bournier, J. (2001) *Manual de identificação dos inimigos naturais no cultivo de algodão.* Boletim Técnico, 35, CODETEC/CIRAD-CA, Cascavel, Brazil.

Szentkiralyi, F. (2001) Lacewings in fruit and nut crops. In: McEwen, P., New, T.R. and Whittington, A.E. (eds) *Lacewings in the Crop Environment*. Cambridge University Press, Cambridge, UK, pp. 172–238.

Udayagiri, S., Mason, C.E. and Pesek, J.D. Jr (1997) *Coleomegilla maculata, Coccinella septempunctata* (Coleoptera: Coccinellidae), *Chrysoperla carnea* (Neuroptera: Chrysopidae), and *Macrocentrus grandii* (Hymenoptera: Braconidae) trapped on colored sticky traps in corn habitats. *Environmental Entomology* 26, 983–988.

Zhang, G.-F., Wan, F.-H., Guo, J.-Y. and Hou, M.-L. (2004) Expression of Bt toxin in transgenic Bt cotton and its transmission through pests *Helicoverpa armigera* and *Aphis gossypii* to natural enemy *Propylea japonica* in cotton plots. *Acta Entomologica Sinica* 47, 334–341.

9 Non-Target and Biodiversity Impacts on Parasitoids

A. Pallini, P. Silvie, R.G. Monnerat, F. de S. Ramalho, J.M. Songa and A.N.E. Birch

Corresponding author: Dr Angelo Pallini, Department of Animal Biology, Federal University of Viçosa, Campus Universitário s/n, 36570-000 Viçosa, Brazil. e-mail: pallini@ufv.br

Introduction

We followed a screening methodology described in Hilbeck *et al.* (Chapter 5, this volume) to initiate the assessment of potential risks of *Bacillus thuringiensis* (Bt) cotton to parasitoids. We first identified parasitoids as an important functional group for non-target assessment (step 1 of Box 5.1). Parasitoids are wasps or flies that parasitize and kill their insect hosts, through the development of the parasitoid larva inside (endoparasitoids) or on the body (ectoparasitoids) of the egg, larva or adult of the host insect. Detailed definitions of the different kinds of parasitoids are given in Parra *et al.* (2002). For this functional group, we prioritized parasitoids for further screening and evaluation. We identified possible exposure pathways by which the parasitoid may be exposed to transgene products of Bt cotton, by considering trophic food webs in the cotton community, and developed hypothetical adverse-effect scenarios for the highest-priority species. We then formulated specific testable hypotheses that underlie the exposure pathways and adverse-effect scenarios, and describe experiments that can test these hypotheses in the laboratory or field.

Parasitoids are considered the main natural enemies of one of the most important pests in Brazilian cotton fields, the cotton boll weevil *Anthonomus grandis*. A complex community of these parasitic wasps suppresses the boll weevil population in areas where pesticide use is low (north-eastern region). The most important species of these parasitic wasps are *Catolaccus grandis* (Burks), *Bracon vulgaris* (Ashmead), *Bracon mellitor* (Say), *Urosigalphus rubicorpus*, *Eurytoma* sp. and *Eupelmus* sp. (Ramalho *et al.*, 1993, 2000; Ramalho, 1994; Ramalho and Wanderley, 1996). Aphids can be heavily parasitized by species in the Braconidae, Aphidiinae (*Lysiphlebus testaceipes* (Cresson)) or the Aphelinidae (*Aphelinus gossypii* (Timberlake)) frequently at the end of the cotton production cycle (Araújo and Moraes, 1998; Fernandes *et al.*, 2000).

Mass rearing procedures have been developed in Brazil for some parasitoid species as part of boll weevil biocontrol and integrated pest management (IPM) programmes for Lepidopteran pests (Ramalho and Gonzaga, 1991; De Almeida, 1996; Parra and Zucchi, 1997; De Almeida *et al.*, 1998; Aquino *et al.*, 2000).

These beneficial insects belong to a complex food web of arthropods that interact in direct and indirect ways in interlinked ecosystems (Montoya *et al.*, 2003). On cotton, as well as in other agroecosystems, parasitoids interact directly with plants (crops and weeds) when they feed on pollen, nectar and exudates, and in an indirect way when they parasitize the eggs or larvae of herbivorous hosts feeding on cotton. Parasitoids interact with other organisms over at least four trophic levels (weed and crop plants, hosts, parasitoids, predators and hyperparasitoids). Hyperparasitoids, such as the ones belonging to the common genus *Conura* or the species *Syrphophagus aphidivorus* (Mayr) found in Brazil, are not included in this chapter because of the lack of published knowledge about these fourth trophic-level organisms. Parasitoids are well known for their communication via semiochemicals, e.g. using volatile odours produced by their hosts and by injured plants as cues during their foraging (De Moraes *et al.*, 1998; Schuler *et al.*, 1999). A full risk assessment of Bt cotton in Brazil should include them based on the current knowledge at that time.

The cultivation of Bt cotton plants could change the way the arthropod communities interact, due to the possible effects on the cotton food web of Bt toxins and/or other alterations to the Bt cotton. Some examples of the effects of other Bt plants on parasitoids are mentioned in a recent review (Pires *et al.*, 2003). The impact of Bt cotton can be evaluated through preplanned experiments in the laboratory, glasshouse or semi-field and field conditions. These experiments should also include investigations on the feeding, mating and oviposition behaviour, rate of parasitism, longevity, fecundity and population dynamics of the parasitoid community.

This case-example analysis does not constitute a risk assessment of the effects of Bt cotton on parasitoids in Brazil; a full characterization of risk would need to be carried out after conducting some of the proposed experiments. This chapter illustrates how the screening methodology can be applied to the case of Bt cotton and parasitoids, specifically illustrating implementation of steps 2–5 of this process (Box 5.1).

Species Selection

The approach used considered publications, reports from the research stations and personal communications of researchers of the Brazilian Agricultural Research Corporation (Embrapa) and other institutions working in each of the major cotton-producing areas. A list of 14 parasitoid species or species groups, such as families previously known to be associated with Brazilian cotton ecosystems, was compiled from Ramalho (1994) and Ramalho and Wanderley (1996).

The list contains parasitoid species reported to be associated with the main pests occurring on cotton throughout the country, i.e. Lepidopteran species, the cotton boll weevil and the cotton aphid (see Fontes *et al.*, Chapter 2, this volume). In one case, the parasitoid species has not been identified to species level, so was defined as an unknown Bethylidae. It is likely that the list needs to be revisited as knowledge of the parasitoids of these pests is available, particularly for aphids and Lepidoptera. For better knowledge about the parasitoids of the Noctuid *Spodoptera frugiperda*, the main maize pest that is increasing as a cotton pest in some regions, the cotton leafworm *Alabama argillacea* or the bollworm *Heliothis virescens*, the reader should consult publications of the Brazilian researchers Cruz (1995), Valicente and Barreto (1999), Fernandes *et al.* (1999), Gravena and Pazetto (1987), De Almeida *et al.* (1998) and Menezzes Jr *et al.* (1994).

The species, their function (host-stage attacked), taxonomic information and main host species are listed in Table 9.1. By limiting the list to species associated with main pests, we ensure that we are starting with species that could potentially be involved in the greatest agricultural risks. For a full risk assessment, other species could be included that also have a role beyond pest control and the agroecosystem.

We used the selection matrix, as described by Hilbeck *et al.* (Chapter 5, this volume), to rank the parasitoids for association with the cotton plant and the cotton habitat in Brazil, and their functional significance in cotton. This process did not require any specific information about Bt cotton, but did require good information on cotton agroecosystems and cotton production in Brazil.

Association with the crop

The first step was to decide if the 14 candidate species were widely distributed in cotton fields. Since production systems, scales of fields and distributions differ widely across Brazil, we scored species for geographical distribution, habitat specialization on cotton, relative abundance on cotton and relative abundance on other plants and prevalence on cotton (the proportion of suitable cotton habitat occupied) in each of the three main regions considered (Midwest, North-east and Meridian; see Fontes *et al.*, Chapter 2, this volume), as shown in Tables 9.2A and 9.2B.

In order to rank the linkage of the parasitoid species with cotton, we evaluated the proportion of the cotton plant growth cycle when the parasitoid is present, the life cycle stage of the pest on cotton and the linkage of the parasitoid with cotton pests (= the degree of feeding specialization, the strength of the association with its prey on cotton and the specificity of host finding) (Table 9.2C). Host finding is a key stage of the parasitoids' foraging process, due to their ability to use herbivore-induced plant volatiles to find their hosts (De Moraes *et al.*, 1998). Based on distribution, abundance, prevalence and linkage, we assigned the species a rank for potential association with cotton by region (Table 9.4).

Table 9.1. List of parasitoid species and species groups on cotton in Brazil – taxonomy, natural enemy function and host type.

Parasitoid function	Species or species group	Order and family	Main host
Egg endoparasitoid	*Trichogramma pretiosum* (Riley)	Hymenoptera: Trichogrammatidae	Lepidoptera
Egg endoparasitoid	*Chelonus insularis* (Cresson)	Hymenoptera: Braconidae	*Spodoptera frugiperda*
Larval ectoparasitoid	*Euplectrus comstockii* (Howard)	Hymenoptera: Eulophidae	Lepidoptera
Larval endoparasitoid	*Netelia* spp.	Hymenoptera: Ichneumonidae	Lepidoptera
Larval/pupal endoparasitoid	*Brachymeria* spp.	Hymenoptera: Chalcididae	Lepidoptera
Larval ectoparasitoid	*Bracon vulgaris* (Ashmead)	Hymenoptera: Braconidae	Boll weevil/pink bollworm
Larval ectoparasitoid	*Bracon mellitor* (Say)	Hymenoptera: Braconidae	Boll weevil
Larval ectoparasitoid	*Urosigalphus rubicorpus* (Gibson)	Hymenoptera: Braconidae	Boll weevil
Larval endoparasitoid	*Hyalomyodes brasiliensis* (Townsend)	Diptera: Tachinidae	Boll weevil
Larval ectoparasitoid	*Eurytoma* spp.	Hymenoptera: Eurytomidae	Boll weevil
Larval ectoparasitoid	*Catolaccus grandis* (Burks)	Hymenoptera: Pteromalidae	Boll weevil
Larval ectoparasitoid	*Catolaccus hunteri* (Crawford)	Hymenoptera: Pteromalidae	Boll weevil
Larval ectoparasitoid	*Eupelmus cushmani* (Crawford)	Hymenoptera: Eupelmidae	Boll weevil
Larval ectoparasitoid	Unknown sp.	Hymenoptera: Bethylidae	Boll weevil
Aphid endoparasitoid	*Lysiphlebus testaceipes* (Cresson)	Hymenoptera: Braconidae, Aphidiinae	Aphids

Functional significance of environmental effect

Table 9.3 shows the rankings for the potential functional significance and therefore the significance of an impact on the parasitoid species according to the main environmental effect, i.e. a loss or gain of the biological-control function of the parasitoid, but also considering the following criteria: significance of biological control on other crops and in natural areas, and the significance of the species as food for other natural enemies. The rank took into account all Brazilian cotton-producing regions together, because region-specific

Table 9.2A. Selection matrix – association with the crop and crop habitat.

Non-target parasitoid taxon × crop coincidence
Geographical distribution: degree of overlap of cotton crop and non-target taxon.
Habitat specialization: degree of association between taxon and cotton habitat or agroecosystem.
(For definitions see Box 5.2, Hilbeck *et al.*, Chapter 5, this volume.)

Selection matrix	Association with crop – coincidence					
	Geographical distribution			Habitat specialization		
Non-target parasitoid taxon	Midwest	North-east	Meridian	Midwest	North-east	Meridian
Trichogramma pretiosum	Yes	Yes	Yes	Low	High	Low
Chelonus insularis	?	?	Yes	?	?	?
Euplectrus comstockii	?	Yes	?	?	Low	?
Netelia spp.	?	Yes	?	?	Low	?
Brachymeria spp.	?	Yes	Yes	High	Low	?
Bracon vulgaris	Yes	Yes	Yes	?	High	High
Bracon mellitor	?	?	Yes	?	?	Low
Urosigalphus rubicorpus	?	Yes	?	?	Low	?
Hyalomyodes brasiliensis	?	?	Yes	?	?	Low
Eurytoma spp.	?	Yes	?	?	Low	?
Catolaccus grandis	?	Yes	Yes		High	High
Catolaccus hunteri	?	Yes	?	?	Low	?
Eupelmus cushmani	?	?	Yes	?	Low	?
Unknown (Bethylidae)	?	Yes	?	?	Low	?
Lysiphlebus testaceipes	Yes	Yes	Yes	?	High	?

data are scarce. There are generally no untreated plots to measure or characterize natural biological control. All the significance criteria are considered important for the study of the impact of Bt cotton on parasitoid ecology because of: (i) the importance of this group of natural enemies for controlling the key pests on cotton, the boll weevil and caterpillars; (ii) their importance for controlling several other pests in many other agroecosystems (sugar cane, tomato, soybean, maize, etc.); and (iii) their role in food webs in both natural and agroecosystems. It is also reasonable to consider the importance of parasitoids as food items for other species. There are many predators that use parasitoids as part of their diet, such as lacewings, coccinellids, earwigs, ants and vespids. All these predators are key natural enemies on several crops that coexist or alternate with cotton (maize, sugar cane) or are used as intercrops with cotton (sorghum, beans, etc.).

Prioritization

Table 9.4 shows the overall rank for each species of each main production region that combines the ranks for likelihood of association with crop and sig-

Table 9.2B. Selection matrix – association with the crop.

Non-target parasitoid taxon × crop coincidence
Prevalence: proportion of cotton habitat occupied by taxon.
Abundance on cotton crop and on other plants occurring in the cotton habitat:
average or typical densities.
(For definitions see Box 5.2, Hilbeck *et al.*, Chapter 5, this volume.)

Selection matrix	Association with crop – coincidence				
Non-target parasitoid taxon	Prevalence: proportion of suitable cotton habitat occupied	Abundance on cotton crop			Abundance on other plants in cotton agro-ecosystem
		Midwest	North-east	Meridian	
Trichogramma pretiosum	No data	High	High	High	Low
Chelonus insularis	No data	Low	Low	Low	?
Euplectrus comstockii	No data	?	High	?	Low
Netelia spp.	No data	?	Low	?	Low
Brachymeria spp.	No data	?	High	Low	Low
Bracon vulgaris	No data	?	High	High	Low
Bracon mellitor	No data	?	?	Low	Low
Urosigalphus rubicorpus	No data	?	Low	?	Low
Hyalomyodes brasiliensis	No data	Low	Low	Low	Low
Eurytoma spp.	No data	?	High	?	Low
Catolaccus grandis	No data	?	High	Low	Low
Catolaccus hunteri	No data	?	Low	?	Low
Eupelmus cushmani	No data	?	?	?	Low
Unknown (Bethylidae)	No data	?	Low	?	Low
Lysiphlebus testaceipes	No data	?	High	?	Low

nificance. The highest-ranked species are listed below. These species received the highest rank because they are widely distributed and abundant in the cotton-growing regions of Brazil, are specialized on cotton in at least some of the regions, are closely linked to the cotton habitat and cotton pests (and specialized on one or a few species), and because they are considered to have a high biological-control potential in cotton and sometimes in other crops associated with cotton.

- *Trichogramma pretiosum*, an egg endoparasitoid of Lepidopteran species, mainly *Alabama argillacea* and *Pectinophora gossypiella*.
- *C. grandis*, a larval ectoparasitoid of the cotton boll weevil in Brazil.
- *B. vulgaris* (Ashmead) was reported to be a larval ectoparasitoid of the pink bollworm and the boll weevil. The species was ranked no. 1 in the North-east (Ramalho *et al.*, 1993) and Meridian regions (Pierozzi and Habib, 1992a, 1993b); the rank is unknown in the Midwest due to lack of information

Table 9.2C. Selection matrix – association with the crop.

Non-target parasitoid taxon × crop–trophic relationship
Phenology: temporal overlap (i) from non-target taxon perspective;
(ii) from crop perspective.
Linkage: specialization of trophic relationship and dependency on cotton.
(For definitions see Box 5.2, Hilbeck *et al.*, Chapter 5, this volume.)

Selection matrix	Association with crop–trophic relationship				
	Phenology of parasitoid taxon and cotton crop		Linkage with crop		
Non-target parasitoid taxon	Degree of overlap of species life cycle with cotton[a] (%)	Proportion of cotton-growing season when taxon is present (%)	Degree of feeding specializa-tion	Strength of association with host on cotton	Host finding
Trichogramma pretiosum	70	70	High	High	High
Chelonus insularis	?	?	?	?	?
Euplectrus comstockii	70	70	Low	Low	?
Netelia spp.	100	100	High	High	?
Brachymeria spp.	100	100	Medium	Medium	?
Bracon vulgaris	70	70	High	High	High
Bracon mellitor	70	70	Low	Low	?
Urosigalphus rubicorpus	70	70	Low	Low	?
Hyalomyodes brasiliensis	70	70	Low	Low	?
Eurytoma spp.	70	70	Low	Low	?
Catolaccus grandis	100	100	High	High	High
Catolaccus hunteri	70	70	Low	Low	?
Eupelmus cushmani	100	100	Low	Low	?
Unknown (Bethylidae)	70	70	Low	Low	?
Lysiphlebus testaceipes	100	100	High	High	High

[a]Assumes that the phenology of Bt cotton is the same as conventional cotton in Brazil.

on habitat specialization and abundance. *B. vulgaris* is a primary parasitoid of the pink bollworm according to Sauer (1938). After the introduction of the boll weevil in Brazil, the species was also found parasitizing boll weevil larvae in the field (Araujo *et al.*, 1993; Ramalho *et al.*, 1993, 1996). No genetic analysis has been carried out so far to determine whether the *B. vulgaris* is

Table 9.3. Selection matrix – functional significance in cropping system (related to significance of possible adverse effect).

Selection matrix	Significance as			
Non-target parasitoid taxon	Biological control in cotton	Food for other natural enemies	Biological control on other crops	Biological control in natural areas
Trichogramma pretiosum	High	Lacewings, coccinellids	High (sugarcane, maize, tomatoes)	No data
Chelonus insularis	Low	No data	High (maize)	No data
Euplectrus comstockii	Medium	Ants and earwigs	Medium (maize, tomato, bean)	No data
Netelia spp.	High	No data	Medium	No data
Brachymeria spp.	Medium	No data	Medium	No data
Bracon vulgaris	High	Ants and earwigs	High	No data
Bracon mellitor	Low	No data	Medium	No data
Urosigalphus rubicorpus	High	Ants and earwigs	Medium	No data
Hyalomyodes brasiliensis	Low	No data	Low	No data
Eurytoma spp.	Low	Ants and earwigs	Low	No data
Catolaccus grandis	High	Ants and earwigs	Low	No data
Catolaccus hunteri	Low	Ants and earwigs	Medium	No data
Eupelmus cushmani	High	No data	Low	No data
Unknown (Bethylidae)	Low	No data	Low	No data
Lysiphlebus testaceipes	High	No data	Medium (horti-cultural crops)	No data

indeed the same species on both hosts or if the parasitoids on each host belong to different subspecies or biotypes.

• *Netelia* spp., larval endoparasitoids of Lepidopteran larvae. The species was ranked no. 1 in the North-east region; the rank is unknown in the Meridian and Midwest regions due to lack of information on the geographical distribution, habitat specialization and abundance.

• *L. testaceipes* (Cresson) was ranked no. 1 in the North-east; the rank is unknown in the Meridian and the Midwest regions due to lack of information on the geographical distribution, habitat specialization and abundance.

The analysis has limitations when data are unavailable for some production regions, especially the Midwest region, where the most extensive cotton fields in the country are found with thousands of hectares of continuous cotton

Table 9.4. Selection matrix – prioritized species.[a]

Combination of association with crop (average estimates from Tables 9.2A–C) and functional significance in the cropping system (estimate for significance from Table 9.3).

Selection matrix	Ranking of expert group						
	Association with crop			Functional significance for cropping system	Prioritized rank for taxon		
Non-target parasitoid taxon	Midwest	North-east	Meridian		Midwest	North-east	Meridian
Trichogramma pretiosum	1	1	1	1	2	2	2
Chelonus insularis	3	3	3	3	6	6	6
Euplectrus comstockii	3	3	3	2	5	5	5
Netelia spp.	?	1	?	2	?	3	?
Brachymeria spp.	?	2	2	2	?	4	4
Bracon vulgaris	?	1	1	1	?	2	2
Bracon mellitor	?	3	?	2	?	5	?
Urosigalphus rubicorpus	?	3	?	1	?	4	?
Hyalomyodes brasiliensis	3	3	3	3	6	6	6
Eurytoma spp.	3	3	3	3	6	6	6
Catolaccus grandis	1	1	1	1	2	2	2
Catolaccus hunteri	?	3	?	3	?	6	?
Eupelmus cushmani	3	3	3	2	5	5	5
Unknown (Bethylidae)	?	3	?	?	?	?	?
Lysiphlebus testaceipes	?	1	?	2	?	3	?

[a]1–2 = highest priority (widely distributed, abundant, mostly present, closely linked, significant possible adverse effect); 3–4 = intermediate priority; 5–6 = low priority (distribution, abundance and presence limited, loosely linked, less significant possible adverse effect).

fields sprayed with insecticides. The heavy use of pesticides may be keeping parasitoid populations low, but additional research is needed to characterize parasitoid communities in this region. New data from the Midwest region will probably not change the list of species but could change their ranking. It will be necessary to survey the parasitoid species in the cotton fields, surrounding crops, weeds and the more natural areas near cotton in the Midwest to have enough information to rank the species with confidence.

Assessment of Potential Exposure Pathways to Transgenic Crop and Transgene Products

Parasitoids interact directly with cotton plants when they feed on pollen, nectar and plant exudates, such as fluid from damaged tissue, and in an indirect way when they parasitize eggs or larvae of their hosts, pest insects that have fed on the cotton plant or feed on cotton pest products such as honeydew or insect frass. Two of the highest-priority parasitoid species, *B. vulgaris* (Ashmead) and *C. grandis* (Burks), both natural enemies of the cotton boll weevil, were selected in order to carry out the assessment of direct or trophically mediated exposure pathways to the Bt protein or metabolites, or to unintended changes caused by the Bt protein or the transformation process in the cotton plant, according to the procedure described in Hilbeck *et al.* (Chapter 5, this volume). The authors recommend that the other parasitoid species assigned the highest priority should also be analysed in the same way for a full risk assessment of Bt cotton in Brazil. For instance, aphid parasitoids may be exposed to Bt via their hosts, as aphids feeding on some Bt cotton events may contain Cry1A toxins (Zhang *et al.*, 2004).

B. *vulgaris* is reported to parasitize the larvae of both pink bollworm, *P. gossypiella* (Saunders), and the cotton boll weevil, *A. grandis*. The pink bollworm is a target species of Bt cotton (see Sujii *et al.*, Chapter 6, this volume; Fitt *et al.*, Chapter 12, this volume), and so will probably be present in very low numbers on Bt cotton. However, the cotton boll weevil is probably not affected (Sujii *et al.*, Chapter 6, this volume), so *B. vulgaris* could parasitize cotton boll weevil larvae that have fed on Bt cotton and therefore the *B. vulgaris* larva could be exposed to the Bt toxin inside this host. Prey-mediated tritrophic exposure of both *B. vulgaris* and *C. grandis* is therefore possible via the cotton boll weevil larvae on Bt cotton.

We questioned whether Bt cotton contains Bt toxin in the nectar, pollen or plant exudates. The available information on expression levels in Bt-cotton pollen indicates variable levels of expression (see Arpaia *et al.*, Chapter 7, this volume). Bt toxin presence in cotton nectar and exudates has never been assessed, but plant fluids from damaged leaf or stem tissue are very likely to contain Bt toxin. As the highest-priority parasitoid species are all present on cotton during most of the cotton growth cycle (Table 9.2C), the worst-case scenario can be assumed (Hilbeck *et al.*, Chapter 5, this volume) and the highest expression level used.

Parasitoids are known to feed on the honeydew produced by aphids on cotton plants (Stapel *et al.*, 1997; Stapel and Cortesero, 2004). Although Bt has not been estimated in the vascular tissue of the Bollgard cottons, it may be present in some Bt cotton events. Because aphids feeding on some Bt cotton events contain Cry1A toxins (Zhang *et al.*, 2004), it is possible that their honeydew also contains Cry1A toxins that could be consumed by parasitoids. It has been confirmed that the honeydew of Homoptera feeding on other transgenic plants contains the transgene product (Shi *et al.*, 1994; Couty and Poppy, 2001; Bernal *et al.*, 2002).

There is considerable evidence from non-transgenic crops (reviewed in Groot and Dicke, 2002) that natural plant-defence compounds are taken up by herbivorous insects (the hosts for parasitoids) and then passed on to parasitoids via tritrophic interactions. Uptake of plant toxins by the herbivore can adversely affect parasitoid fitness parameters (e.g. prolonged larval periods, reduced adult weights, reduced adult longevity). Similar experiments using Bt toxin-fed host insects and parasitoids, have shown a range of trophic interactions on parasitoids, ranging from positive effects, no effects or negative effects (reviews in Schuler *et al.*, 2001, 2003, 2004; Bernal *et al.*, 2002; Groot and Dicke, 2002; Cowgill *et al.*, 2004; Prütz and Dettner, 2004; O'Callaghan *et al.*, 2005). On Bt cotton, potential negative tritrophic impacts have been found in laboratory studies with the Braconidae *Cotesia marginiventris* (Cresson) and the Encyrtidae *Copidosoma floridanum* (Dalman) parasitizing soybean looper *Pseudoplusia includens* (Walker) larvae (Baur and Boethel, 2003). The ecological fitness parameters quantified in these various experiments on Bt-toxin interactions with parasitoids were not standardized, but have included attraction to host, host feeding behaviour, host size and suitability for parasitism, exposure time of hosts to the parasitoid, parasitism rate, parasitoid emergence rate, parasitoid development rate and mortality. It is very likely that experimental design (e.g. choice versus no-choice situations) and scale of testing (e.g. laboratory, semi-field, field, according to Hassan, 1998) will have large effects on the outcomes of such experiments. Field studies of parasitoids and natural enemies in general on Bt cotton in China have found decreases (Yang *et al.*, 2001), increases (Sun *et al.*, 2002) and no change (Ning *et al.*, 2001).

Other changes in the Bt cotton plant (some of which could be 'unintended' or 'unexpected') could affect the behaviour of parasitoids, such as changes in the cotton-plant odours following insect herbivore feeding that the parasitoid uses to locate its host. Other changes in secondary plant chemistry could also affect host location and choice (Gouinguené and Turlings, 2002). Host-species odours are also used by parasitoids for host location, and any change in host physiology may alter the parasitoids' host-location behaviour (Hérard *et al.*, 1988; Takasu and Lewis, 2003). Unexpected changes associated with transformation can increase or decrease the crops' susceptibility to non-target pests (e.g. aphids on Bt maize; Lumbierres *et al.*, 2004). This could lead to altered densities of host insects and thus alter interactions with parasitoid species that will respond to host density at different stages of crop development.

Cotton species or cultivars other than the approved Bt cotton varieties that are sold on the seed market could receive the Bt gene via gene flow (see Johnston *et al.*, Chapter 11, this volume), and on these cotton plants the same exposure pathways as above would apply.

Identification of Adverse-Effect Scenarios and Formulation of Testable Hypotheses

B. vulgaris and *C. grandis* are used here to identify adverse-effect scenarios. For a full risk assessment we recommend that the other high-priority species be analysed in the same way. From the possible exposure pathways identified above, possible adverse effects on the prioritized species can be suggested. In the subsequent section, appropriate experimental designs for testing the resulting hypotheses in laboratory or greenhouse settings are described. The possible consequence of the scenario is a reduction in biological control, but these effects on biological control cannot be evaluated before commercial release in laboratory or greenhouse settings. The proposed experiments therefore measure relative generational fitness, i.e. survival and reproduction (see Box 8.3, Faria *et al.*, Chapter 8, this volume). A reduction in generational fitness can lead to lower populations of parasitoids, which could result in reduced biological-control efficacy.

The main host of *B. vulgaris*, the pink bollworm, is expected to be present in low numbers on Bt cotton. On Bt cotton, *B. vulgaris* could parasitize an early pink bollworm instar and die along with its host, or it may be indirectly affected through lack of host larvae. Both scenarios would likely lead to decrease or disappearance of *B. vulgaris* from the cotton field. If boll weevil is present as an alternate host, *B. vulgaris* populations might be maintained on the boll weevil, but this will not necessarily imply that the parasitization potential on the pink bollworm is maintained. If Bt-resistant pink bollworm larvae were to develop, an important adverse-effect scenario would be that parasitism of resistant larvae is reduced.

However, the main adverse-effect scenarios considered here focus on the role of *B. vulgaris* and *C. grandis* as natural enemies of boll weevil, a pest that is likely to be unaffected by Lepidopteran-active Bt cotton (Sujii *et al.*, Chapter 6, this volume). Among the parasitoids of the boll weevil, *C. grandis* and *B. vulgaris* are considered to be the most efficient biocontrol agents in the North-east region in Brazil (Ramalho *et al.*, 1993, 2000; Ramalho, 1994), and *Bracon* species are considered the most important mortality factor of *A. grandis* in the region of Campinas, São Paulo (Pierozzi and Habib, 1992b, 1993a). Therefore, both species may be generally important in suppressing boll weevil populations.

Between cotton-growing seasons, the parasitoids survive in alternative hosts such as *Euscepes postfasciatus* (Fairmaire) (Coleoptera: Curculionidae), and – for *B. vulgaris* – Lepidopteran species such as *Plodia interpunctella*, with high reproductive success as shown by ongoing research at the Biological Control Unit of Embrapa Cotton by Ramalho *et al.* (1998) in the North-east and by Pierozzi and Habib (1993b) in the Meridian region. *C. grandis* preferably parasitizes third instar larvae of the boll weevil, which are mainly in the fallen buds on the ground, while *B. vulgaris* prefers larvae of boll weevil in the buds and bolls on the plant (Silva, 2002), so the species have different ecological niches on cotton.

It is possible that *B. vulgaris* and *C. grandis* are exposed to the Bt toxin or metabolites through their parasitism of boll weevil larvae that have fed on Bt cotton – this is a possible tritrophic exposure pathway to Bt cotton. The adverse effect associated with this exposure pathway is that this exposure may have a negative effect on the parasitoid, which will result in a lower parasitism rate on the cotton boll weevil in Bt cotton, leading to increased cotton boll weevil damage. The first part of this adverse-effect scenario could be tested by the following hypothesis:

Testable hypothesis 1: Bt cotton-fed host larvae of the cotton boll weevil adversely affect the survival and development time of parasitoid larvae of *B. vulgaris* and *C. grandis*.

The second part of this adverse-effect scenario could be tested by the following hypothesis:

Testable hypothesis 2: The longevity, fecundity and rate of parasitism of *B. vulgaris* or *C. grandis* adults are reduced when they developed in cotton boll weevil larvae raised on Bt cotton compared to non-Bt cotton.

Feeding on pollen, nectar and plant exudates, such as fluid from damaged plant tissues, can be significant for the survival and fitness of the adults of parasitoid species, which means that they are more likely to survive until they find a suitable host larva and be able to produce enough eggs to successfully parasitize it. For example, extrafloral nectaries were found to be essential to the survival of *T. pretiosum* adults on cotton, and resulted in a greater parasitism of *Helicoverpa zea* eggs (Treacy *et al.*, 1987). Does this food source (Bt cotton versus conventional cotton) affect the fitness of *B. vulgaris* or *C. grandis*? If so, is this due to Bt content in the food source or due to some other components in the food source that could have been altered unexpectedly? The adverse effect associated with the exposure pathway is that this bitrophic exposure may have a negative fitness impact on the parasitoid, which could result in a lower parasitism rate on cotton boll weevil in Bt cotton, potentially in turn resulting in increased cotton boll weevil damage. The exposure pathway could be verified or disproved by experiments to test the following hypotheses:

Testable hypothesis 3: The Bt toxin is present in the parasitoid adult after it feeds on nectar, pollen and plant exudates of the Bt cotton event(s) being considered in Brazil.

Testable hypothesis 4: Adult feeding on nectar, pollen or plant exudates from Bt cotton adversely alters the oviposition behaviour, mating behaviour and/or fitness of *B. vulgaris* and *C. grandis*.

Parasitoids are strongly influenced by host-plant odours for locating their host larvae (Turlings *et al.*, 1990). Any changes in the Bt cotton-plant odours may therefore influence the host finding and oviposition behaviour of *B. vulgaris* or *C. grandis*. The cotton boll weevil larvae raised on Bt cotton may also have altered physiology that results in a changed chemical odour, and parasitoids also use host-larvae odours as host-finding cues (Vet *et al.*, 1995; Takasu and Lewis, 2003). The adverse-effect scenario is therefore that Bt cotton has altered plant odours and/or affects host-larvae odours, which affects the host finding and oviposition behaviour of the parasitoid, resulting in a lower parasitism rate on the boll weevil in Bt cotton potentially leading to increased cotton boll weevil damage. The first part of the adverse-effect scenario can be tested by the following two hypotheses:

Testable hypothesis 5: Bt cotton adversely alters the oviposition behaviour of *B. vulgaris* and *C. grandis* due to behavioural changes in reaction to odour patterns from the plant.

Testable hypothesis 6: The oviposition behaviour of *B. vulgaris* and *C. grandis* is adversely altered on cotton boll weevil larvae raised on Bt cotton compared to non-Bt cotton due to behavioural changes in reaction to odour patterns from the plant and/or host insect and its excretion products.

Mating behaviour is also mediated by odours. Parasitoids use chemical cues (normally pheromones) to find their mate (Godfray, 1994). So, it is possible that any of the previous exposure pathways could change their ability to use pheromone odours to locate their partner and consequently impact their population growth.

Testable hypothesis 7: Bt cotton adversely alters the mating behaviour of *B. vulgaris* and *C. grandis*.

The third part of all of the adverse-effect scenarios can be tested in the field by the following hypothesis:

Testable hypothesis 8: Population parameters of *B. vulgaris* and *C. grandis* are adversely altered in Bt cotton fields compared to non-Bt cotton fields.

To test the final part of all of the adverse-effect scenarios, it is necessary to determine whether there is any evidence that a decrease in parasitism by *B. vulgaris* or *C. grandis* on the boll weevil could result in increased boll weevil survival. These data are not available, but there is evidence from the North-east cotton production region that the two species are suppressing boll weevil populations (Ramalho *et al.*, 1993, 2000; Ramalho, 1994).

Experimental Designs

In this chapter, we present a series of experiments that address the above hypotheses for *B. vulgaris* and *C. grandis* parasitizing boll weevil larvae on Lepidopteran-active Bt cotton in Brazil. Similar tests can be developed for the other high-priority species. They are based on standard protocols that are commonly used by researchers for testing similar hypotheses relating to parasitoid behaviour and fitness: choice/field cage experiments (Pallini *et al.*, 1997, 1999; Monnerat and Bordat, 1998), olfactometer experiments (Turlings *et al.*, 1990; Janssen *et al.*, 1995), population dynamics (Moran *et al.*, 1996; Schmitz *et al.*, 1997) and protein detection (Bravo *et al.*, 1992a,b). Both *B. vulgaris* and *C. grandis* have been successfully raised on insect and artificial diets in the laboratory in Brazil (Guerra *et al.*, 1993; Pierozzi and Habib, 1993b). If both resistant and non-resistant strains of a susceptible host are available, comparative experimental designs can be used (Schuler *et al.*, 2003, 2004). With some modification, the protocols can also be used for other parasitoid species. The experiments are designed for the stage of development of the transgenic plant when the transgene event is characterized in the laboratory, before field release (see Andow *et al.*, Chapter 1, this volume). Further experiments must also be carried out at the stage of development of the transgenic plant at which small-scale field testing is done.

The experimental designs may be adjusted using pilot experiments to assess the optimal conditions for allowing test species (herbivores and their parasitoids) to express their full repertoire of behaviour in ways closely mimicking behaviour under field conditions. Each species has peculiarities associated with flight behaviour and responses to odour sources, either in a more open (here represented by experimental field cage units) or in closed environments (e.g. small-scale olfactometer experiments). Therefore, species-specific behavioural particularities expressed at different scales have to be taken into account before choosing the best design for testing the hypotheses.

The series of experiments covers a range of possible adverse-effect scenarios for the parasitoids when the Bt cotton plant is introduced in the environment. From the experiments listed, we do not imply that all need to be performed to assess the impact of Bt cotton exposure on parasitoids. The number and type of tests required at any stage in the risk-assessment process depends on the possible risk being evaluated, the available methods and on the time and other resources needed to provide scientifically robust answers. Nevertheless, we suggest that at least experiments measuring ecological fitness (rates of parasitism, longevity and fecundity), foraging/searching behaviour (odour-based choices) and population dynamics (preferably at field scale) should always be undertaken to assess the effects of exposure of Bt cotton to key parasitoids in regionally-specific agroecosystems. The proposed experiments do not completely test the hypotheses, but they are proposed as the initial experiments for testing the hypotheses. In most cases, the experiments are designed to reject the hypothesis so that if the hypothesis is rejected, no further testing of that particular hypothesis is necessary.

Testable hypothesis 1: Bt cotton-fed host larvae of the cotton boll weevil adversely affect the survival and development time of parasitoid larvae of *B. vulgaris* and *C. grandis*.

Experiment 1: Survival and development time of parasitoid larvae on Bt cotton-fed host boll weevils
Laboratory experiment to be conducted prior to field release.
Treatment: Boll weevils fed on Bt cotton.
Control: Boll weevils fed on an isoline and/or standard cotton variety from the region.
Replications: Four assays with 30 host larvae feeding on Bt and non-Bt cotton.
Conditions: Standardize the number of host insects per plant and their feeding time on test plants prior to exposure to parasitoids (see Sujii *et al.*, Chapter 6, this volume). Hosts are presented to parasitoids in a way that ensures successful parasitism.
Parameters measured: Emergence rate, development time from egg to emergence of adult parasitoids, and adult size and sex.

Testable hypothesis 2: The longevity, fecundity and rate of parasitism of *B. vulgaris* or *C. grandis* adults are reduced when they developed in cotton boll weevil larvae raised on Bt cotton compared to non-Bt cotton.

Experiment 2: Rates of parasitism, parasitoid longevity and fecundity (measures of ecological fitness) on boll weevil on Bt cotton
Greenhouse experiment to be conducted prior to field release.
Treatment: Parasitoids reared on boll weevil hosts on Bt cotton and as adults provided boll weevil hosts on Bt cotton.
Control: Parasitoids reared on boll weevil hosts on an isoline and/or standard cotton variety from the region, and as adults provided with boll weevil hosts on the same variety.
Replications: Four assays with 20 male–female parasitoid pairs.

Conditions: Standardize the number of host insects per plant and their feeding time on test plants prior to the release of parasitoids (see Sujii *et al.*, Chapter 6, this volume). Because *C. grandis* preferentially parasitizes third instar boll weevil larvae in fallen cotton buds and bolls on the ground, for this species, hosts should be presented in the appropriate plant material (e.g. cut-off buds).

Parameters measured: Number of hosts parasitized per day, longevity of females and number of offspring produced.

Parameters calculated: Net reproductive rate.

Testable hypothesis 3: The Bt toxin is present in the parasitoid adult after it feeds on nectar, pollen and plant exudates of the Bt cotton event(s) being considered in Brazil.

Experiment 3: No-choice feeding experiment with nectar, pollen or plant exudates from Bt cotton
Laboratory experiment to be conducted prior to field release.
Treatment: Nectar, pollen or plant exudates from Bt cotton provided to adults.
Control: Nectar, pollen or plant exudates from an isoline and/or standard cotton variety from the region.
Conditions: Parasitoids reared uniformly and separated by sex. Adult parasitoids starved for between 6 h and 24 h. Nectar, pollen and plant exudates (fluids from damaged tissues collected from various tissues at various stages of cotton growth) collected from cotton plants and provided to starved adults.
Replications: Thirty individuals of each sex for each food type.
Detection method: Protein detection by QuikStiks (presence/absence) or Western-blot (quantitative), immunocytochemical detection of Bt toxin in the parasitoid gut (Bravo *et al.*, 1992a,b). Measured immediately after feeding and up to 24 h later.
Parameters measured: Presence of protein in adult, concentration of protein in adult, weight and sex of adult, presence of protein in parasitoid gut.
Parameters calculated: Rate of change of protein concentration in adult parasitoids.

Testable hypothesis 4: Adult feeding on nectar, pollen or plant exudates from Bt cotton adversely alters the oviposition behaviour, mating behaviour and/or fitness of *B. vulgaris* and *C. grandis*.

Experiment 4: Rates of parasitism, parasitoid longevity and fecundity (measures of ecological fitness) on boll weevil when feeding on nectar, pollen or plant exudates of Bt cotton
Greenhouse experiment to be conducted prior to field release.
Treatment: Nectar, pollen or plant exudates from Bt cotton fed to adults.
Control: Nectar, pollen or plant exudates from an isoline and/or standard cotton variety from the region.
Replications: Four assays with 20 male–female pairs in each replicate.
Conditions: If hypothesis 3 is rejected, then hypothesis 4 does not need to be investigated. Cotton plants are used to provide the necessary nectar, pollen or plant exudates. Boll weevil hosts reared on artificial diet are provided daily

separate from cotton plants. Boll weevils are not presented on cotton plants but presented in a way that parasitism can occur.

Parameters measured: Number of hosts parasitized per day, longevity of females and number of offspring produced.

Parameters calculated: Net reproductive rate.

Testable hypothesis 5: Bt cotton adversely alters the oviposition behaviour of *B. vulgaris* and *C. grandis* due to behavioural changes in reaction to odour patterns from the plant.

Experiment 5.1: Choice of plant-odour study in the laboratory
Laboratory experiment to be conducted prior to field release.
Treatment: Bt cotton variety.
Control: Isoline and/or standard cotton variety from the region.
Replications: Four assays with 20 female parasitoids per treatment.
Conditions: Insects are tested individually using a Petterson olfactometer, with standardized airflow rate (optimized from pilot experiments) and activated carbon-filtered air supply. A new set of odour sources should be used for each individual tested. Tests should be completed with intact plants (not damaged by herbivores) to evaluate response to possible habitat location cues by the parasitoids. Tests should be repeated on plant parts that have been damaged by boll weevil at a stage in plant development appropriate for the parasitoid.
Parameter measured: Choice of female for an odour.

Experiment 5.2: Adult feeding choice experiment in field cages
Semi-field experiment to be conducted early during small-scale field testing.
Treatment: Potted Bt cotton plant at flowering stage.
Control: Potted isoline or standard cotton variety from the region.
Replications: Repeat at least four times (replicated in time) – one experiment per day or until 40 females have been observed.
Conditions: If hypothesis 3 is rejected, then this experiment is not essential. Array six potted plants (alternated Bt and non-Bt cotton plants) in a circular pattern. Release a small number of mated females of known age and dietary history at the centre of the array of pots. The number should be small enough that the behaviours of individuals can be tracked and recorded. Experiment can last for 2–8 h. It will be essential to determine if the parasitoids prefer to feed during a limited part of the day in pilot observations prior to the experiment, so that the experiment can be conducted during the appropriate time of day. In the repeat experiments, use new groups of potted plants and insects each time. The experiments can also be repeated with aphid honeydew from aphids feeding on Bt or control cotton as the food source for the parasitoid.
Parameters measured: Record foraging behaviour parameters (e.g. number of visits to each food source, duration of feeding bout, frequency and duration of other behaviours associated with feeding). Analyse the number of parasitoids that have Bt toxin in them, by the use of polymerase chain reaction (PCR) or other techniques like enzyme-linked immunosorbent assay (ELISA) at the end of the experiment.

Parameters calculated: Time budget or behavioural ethogram.

Testable hypothesis 6: The oviposition behaviour of *B. vulgaris* and *C. grandis* is adversely altered on cotton boll weevil larvae raised on Bt cotton compared to non-Bt cotton due to behavioural changes in reaction to odour patterns from the plant and/or host insect and its excretion products.

Experiment 6.1: Choice of odour study in the laboratory
Use the same experimental procedure as experiment 5.1 with the following changes.
Treatment: Boll weevil larvae feeding on Bt cotton.
Control: Boll weevil larvae feeding on isoline and/or standard cotton variety from the region.

Experiment 6.2: Oviposition behaviour potentially affected by host odour
Laboratory or semi-field experiment to be conducted early during small-scale field trials.
Treatment: Potted Bt cotton plant at flowering stage infested with boll weevil for *B. vulgaris*, dropped buds and bolls infested with boll weevil for *C. grandis*.
Control: Similarly treated isoline or standard cotton variety from the region.
Replications: Repeat at least four times (replicated in time) – one experiment per day or until 40 females have been observed.
Conditions: If no preference is observed in experiment 6.1, then this experiment is not essential. Array six potted plants or six groups of buds/bolls (alternated Bt and non-Bt cotton) in a circular pattern in the experimental arena (semi-field cages for *B. vulgaris* and laboratory cages for *C. grandis*). Release a small number of mated females of known age and dietary history at the centre of the array. The number should be small enough that the behaviours of individuals can be tracked and recorded. Experiment can last for 2–8 h (superparasitism should be infrequent). In the repeat experiments, use new groups of potted plants, buds/bolls and insects each time to avoid pseudoreplication.
Parameters measured: Record foraging behaviour parameters (e.g. number of visits to each host, duration of foraging bout, occurrence of stinging behaviour, frequency and duration of other behaviours associated with foraging). Parasitism rates by rearing hosts.
Parameters calculated: Time budget or behavioural ethogram. Female oviposition preference.

Testable hypothesis 7: Bt cotton adversely alters the mating behaviour of *B. vulgaris* and *C. grandis*.

Experiment 7.1: Female mating behaviour and success
Laboratory experiment to be conducted prior to field release.
Treatment: Female parasitoids reared on boll weevil hosts on Bt cotton.
Control: Female parasitoids reared on boll weevil hosts on isoline and/or standard cotton variety from the region.

Replications: Four assays with 20 females each.

Conditions: Each virgin female is provided with three unrelated virgin males in an experimental arena (more than one male is needed to average out variation among males, but the number should not be too high). All parasitoids should be provided honey or nectar sources *ad libitum* prior to and during the experiment. Pilot experiments are needed to determine when females will accept mates after emergence, and to delineate behaviourally male mating attempts from successful male matings.

Parameters measured: Time to each mating attempt, duration of each mating attempt, time to successful mating, duration of successful mating, number of attempts prior to successful mating.

Experiment 7.2: Male mating behaviour and success
Laboratory experiment to be conducted prior to field release.

Treatment: Male parasitoids reared on boll weevil hosts on Bt cotton.

Control: Male parasitoids reared on boll weevil hosts on isoline and/or standard cotton variety from the region.

Replications: Four assays with 20 males each.

Conditions: Each virgin male is provided with three unrelated, receptive virgin females in an experimental arena. All parasitoids should be provided honey or nectar sources *ad libitum* prior to and during the experiment. Pilot experiments are needed to determine when after emergence males become sexually active, and to delineate behaviourally male mating attempts from successful male matings.

Parameters measured: Time to each mating attempt, duration of each mating attempt, time to each successful mating, duration of each successful mating, number of attempts prior to each successful mating.

Experiment 7.3: Female mating choice and male mate competition
Laboratory experiment to be conducted prior to field release.

Compare the mating behaviour of *B. vulgaris* and *C. grandis* on Bt and non-Bt cotton plants.

Treatment: Female parasitoids reared on boll weevil hosts on Bt cotton.

Control: Female parasitoids reared on boll weevil hosts on isoline and/or standard cotton variety from the region.

Replications: Four assays each with 20 females per treatment.

Conditions: This experiment is needed only if female or male mating behaviours or success are affected by Bt cotton. Rear male parasitoids on boll weevils on either Bt cotton or isoline and/or standard cotton variety from the region. Mark males so that the cotton source can be readily identified visually. Provide each virgin female with two to three equal-aged virgin males from each plant source. All parasitoids should be provided honey or nectar sources *ad libitum* prior to and during the experiment.

Parameters measured: Time to and male identity of first mating attempt, duration and male identity of each mating attempt, time to and male identity of first successful mating, duration and male identity of first successful mating, number of attempts by each male type prior to first successful mating.

Discussion and Suggestions for Further Development

Parasitoids are key natural enemies for cotton crops in Brazil and other locations, due to their ability to control or help in controlling pests such as cotton boll weevil *A. grandis* (Boheman), cotton leaf worm *A. argillacea* (Hübner) and pink bollworm *P. gossypiella* (Saunders). This can be linked to other benefits, such as the reduction of pesticide use, and therefore the preservation of biodiversity, the reduction of environmental contamination, an improvement in the farmers' quality of life and a reduction of production costs.

They have evolved sophisticated sensory systems to detect their hosts and to avoid competition, using semiochemicals. This means that there are numerous, complex ways in which the introduction of a modified crop plant could influence the behaviour, physiology and ecological fitness of parasitoids, either directly (via ingestion of the toxin from the plant) or indirectly (via its hosts or via unintended changes in the crops' physiology and metabolism).

To predict which of these many potential interactions of parasitoids with Bt cotton are most likely to be affected, procedures to prioritize species associated with the highest potential risks were applied to Brazilian cotton parasitoids in this chapter, based on Hilbeck *et al.* (Chapter 5, this volume).

Several important gaps in knowledge became apparent that need to be filled by research in Brazil. These range from taxonomic uncertainties about the true identity of key parasitoid species (*B. vulgaris*, Table 9.1, and all the species of the *Bracon* group in fact), insufficient regional data on geographical distributions and habitat associations in each of the three main cotton-production areas (Table 9.2A), information needs on prevalence, abundance, strength of association with hosts on cotton (Tables 9.2B, 9.2C) and lack of data on the role and importance of key species in natural habitats associated with cotton (Table 9.3). Despite these important knowledge gaps that should be filled, we were able to outline experiments to test adverse-effect hypotheses resulting from the prioritized parasitoid taxa and most likely pathways of exposure. These experimental outlines require further development to specify conditions that are ecologically realistic, manageable and which produce results useful to the precautionary risk-assessment approach that we adopt. By using transparent criteria in the selection matrix, we aim to open up discussion on the 'value judgements' used by scientists in different ways to frame risk-assessment questions and to produce selection criteria. We realize that in most cases the resources available to carry out such risk assessments are very restricted, particularly in developing countries. Our proposed system at least allows researchers to prioritize key species/ecological functions and exposure pathways for testing, so that critical experiments can be performed with a scientifically argued priority order according to key biological questions rather than just regulators' needs, funding availability and any imposed time constraints.

As pointed out by Levidow (2003), uncertainty and certainty about scientific evidence on complex issues like transgenic crop interactions across multiple trophic levels depends on several subjective factors: what is regarded as the threshold level for 'unacceptable harm' to the test species or ecosystem, how the scientific questions are posed about cause–effect pathways for this level of harm and how the experimental methods are designed to answer the specific scientific questions posed.

Acknowledgements

We thank Dr Walkymario de Paulo Lemos for valuable comments on the early version of the chapter. AP was supported by CNPq-Brazil.

References

Aquino, I.S., Ramalho, F.S., Payton, M.R. and Eikenbary, R.D. (2000) Waxfilm (Pat. pend.): an alternative film for rearing parasitoids *Anthonomus grandis* (Col., Curculionidae). *Journal of Applied Entomology* 124(9–10), 387–390.
Araújo, L.H.A. and Moraes, A.C.S. (1998) Parasitismo do pulgão do algodoeiro por *Lysiphlebus testaceipes* Cresson (Hymenoptera: Aphidiidae) em casa de vegetação. Embrapa/Algodão, *Pesquisa em andamento* 92, 3.
Araújo, L.H.A., Sobrinho, R.B., Almeida, R.P. and Mesquita, C.K. (1993) Biological control of the boll weevil. *Pesquisa Agropecuária Brasileira* 28(2), 257–261.
Baur, M.E. and Boethel, D.J. (2003) Effect of Bt-cotton expressing Cry1A(c) on the survival and fecundity of two hymenopteran parasitoids (Braconidae, Encyrtidae) in the laboratory. *Biological Control* 26, 325–332.
Bernal, C.C., Aguda, R.M. and Cohen, M.B. (2002) Effect of rice lines transformed with *Bacillus thuringiensis* toxin genes on brown planthopper and its predator *Cyrtorhinus lividipennis*. *Entomologia Experimentalis et Applicata* 102, 21–28.
Bernal, J.S., Griset, J.G. and Gillogly, P.O. (2002) Impacts of developing on Bt maize-intoxicated hosts on fitness parameters of a stemborer parasitoid. *Journal of Entomological Science* 37, 27–40.
Bravo, A., Jancens, S. and Peferoen, M. (1992a) Immunocytochemical localization of *Bacillus thuringiensis* insecticidal crystal proteins in intoxicated insects. *Journal of Invertebrate Pathology* 60, 237–246.
Bravo, A., Hendricks, K., Jancens, S. and Peferoen, M. (1992b) Immunocytochemical analysis of specific binding of *Bacillus thuringiensis* crystal proteins to lepidopteran and coleopteran midgut membranes. *Journal of Invertebrate Pathology* 60, 247–253.
Couty, A. and Poppy, G.M. (2001) Does host-feeding on GNA intoxicated aphids by *Aphelinus abdominalis* affect their longevity/fecundity? *Physiological Entomology* 26, 287–293.
Cowgill, S.E., Danks, C. and Atkinson, H.J. (2004) Multitrophic interactions involving genetically modified potatoes, nontarget aphids, natural enemies and hyperparasitoids. *Molecular Ecology* 13, 639–647.
Cruz, I. (1995) Manejo integrado de pragas de milho com ênfase para o controle biológico. In: *Anais IV Ciclo de palestras sobre controle biológico de pragas*, Campinas, Instituto Biológico de São Paulo/Sociedade Entomológica do Brasil, Brazil, pp. 48–92.
De Almeida, R.F. (1996) Biotecnologia de produção massal de *Trichogramma* spp. a través do hospedeiro alternativo *Sitotroga cerealella*. EMBRAPA-CNPA, Circular técnica No. 19, 36 pp.

de Almeida, R.P., da Silva, C.A.D. and de Medeiros, M.B. (1998) Biotecnologia de produção massal e manejo de *Trichogramma* para o controle biológico de pragas. Embrapa-CNPA, Campina Grande, *Documentos* 60, 61.

de Almeida, R.P., Braga Sobrinho, R., Araújo, L.H.A., de Souza, J.E.G. and Dias, J.M. (1998) Parasitismo de *Trichogramma* nativo sobre *Alabama argillacea*, em áreas de algodoeiro arbóreo. Embrapa-Algodão, Comunicado técnico, 6.

De Moraes, C.M., Lewis, W.J., Pare, P.W., Alborn, H.T. and Tumlinson, J.H. (1998) Herbivore-infested plants selectively attract parasitoids. *Nature* 393(6685), 570–573.

Fernandes, A.M.V., Farias, A.M.I., de Faria, C.A. and Tavares, M.T. (2000) Occorência de *Aphelinus gossypii* Timberlake (Hymenoptera: Aphelinidae) parasitando *Aphis gossypii* (Hemiptera: Aphididae) em algodão no estado de Pernambuco. *Anais da Sociedade Entomologica do Brasil* 29, 831–833.

Fernandes, M.G., Busoli, A.C. and Degrande, P. E. (1999) Parasitismo natural de ovos de *Alabama argillacea* Hüb. e *Heliothis virescens* Fab. (Lep.: Noctuidae) por *Trichogramma pretiosum* Riley (Hym.: Tichogrammatidae) em algodoeiros no Mato Grosso do Sul. *Anais da Sociedade Entomologica do Brasil* 28, 695—701.

Godfray, H.C.J. (1994) *Parasitoids: Behavioural and Evolutionary Ecology*. Princeton University Press, Princeton, New Jersey.

Gouinguené, S.P. and Turlings, T.C.J. (2002) The effects of abiotic factors on induced volatile emissions in corn plants. *Plant Physiology* 129, 1296–1307.

Gravena, S. and Pazetto, J.A. (1987) Predation and parasitism of cotton leafworm eggs, *Alabama argillacea* (Lep.: Noctuidae). *Entomophaga* 32, 241–248.

Groot, A.T. and Dicke, M. (2002) Insect-resistant transgenic plants in a multi-trophic context. *The Plant Journal* 31, 387–406.

Guerra, A.A., Robacker, K.M. and Martinez, S. (1993) *In vitro* rearing of *Bracon mellitor* and *Catolaccus grandis* with artificial diets devoid of insect components. *Entomologia Experimentalis et Applicata* 68(3), 303–307.

Hassan, S.A. (1998) Introduction to standard characteristics of test methods. In: Haskell, P.T. and McEwen, P. (eds), *Ecotoxicology: Pesticides and Beneficial Organisms*. Kluwer Academic Publishers, Dordrecht, The Netherlands, pp. 55–68.

Hérard, F., Keller, M.A., Lewis, W.J. and Tumlinson, J.H. (1988) Beneficial arthropod behavior mediated by airborne semiochemicals. IV – Influence of host diet on host oriented flight chamber responses of *Microplitis demolitor* Wilkinson. *Journal of Chemical Ecology* 14, 1597–1606.

Janssen, A., van Alphen, J.J.M., Sabelis, M.W. and Bakker, K. (1995) Odour-mediated avoidance of competition in *Drosophila* parasitoids: the ghost of competition. *Oikos* 73, 356–366.

Levidow, L. (2003) Precautionary risk of Bt maize: what uncertainties? *Journal of Invertebrate Pathology* 83, 113–117.

Lumbierres, B., Albajes, R. and Pons, X. (2004) Transgenic Bt maize and *Rhopalosiphum padi* performance. *Ecological Entomology* 29, 309–317.

Menezzes, A.O. Jr, Miguel, M. and Berbel, B.M. (1994) Biologia de *Euplectrus puttleri* Gordh, 1980 (Hymenoptera: Eulophidae) sobre lagartas do curuquerê-do-algodoeiro *Alabama argillacea* Hubner, 1818 (Lepidoptera: Noctuidae). *Semina Ciências Agrárias, Londrina* 15(1), 23–28.

Monnerat, R.G. and Bordat, D. (1998) Influence of HD-1 *Bacillus thuringiensis* spp. *kurstaki* on the developmental stages of *Diadegma* sp. (Hym.: Ichneumonidae) parasitoid of diamondback moth (Lep.: Yponomeutidae). *Journal of Applied Entomology* 122, 49–51.

Montoya, J.M., Rodriguez, M.A. and Hawkins, B.A. (2003) Food web complexity and higher-level ecosystem services. *Ecology Letters* 6(7), 587–593.

Moran, M.D., Rooney, T.P. and Hurd, L.E. (1996) Top down cascade from a bitrophic predator in an old field community. *Ecology* 77, 2219–2227.

Ning, X.Z., Song, Q.P., Kong, X.H., Chen, H. and Meng, J.W. (2001) A preliminary research on the regularity of population fluctuations of major insects and natural enemies in the field of Bt transgenic cotton in the Xinjiang region. *China Cotton* 28, 12–13.

O'Callaghan, M., Glare, T.R., Burgess, E.P.J. and Malone, L.A. (2005) Effects of plants genetically modified for insect resistance on nontarget organisms. *Annual Review of Entomology* 50, 271–292.

Pallini, A., Janssen, A. and Sabelis, M.W. (1997) Odour-mediated responses of phytophagous mites to conspecific and heterospecific competitors. *Oecologia* 110, 179–185.

Pallini, A., Janssen, A. and Sabelis, M.W. (1999) Spider mites avoid plants with predators. *Experimental and Applied Acarology* 23, 803–815.

Parra, J.E.P. and Zucchi, R.A. (eds) (1997) *Trichogramma e o controle biológico aplicado*. Piracicaba, FEALQ, 324.

Parra, J.R.P., Botelho, P.S.M., Corrêa-Ferreira, B.S. and Bento, J.M.S. (2002) Controle biológico: terminologia. In: Parra, J.R.P. *et al.* (eds) *Controle biológico no Brasil: parasitóides e predadores*. São Paulo, Manole, Brazil, pp. 1–16.

Pierozzi, I. Jr and Habib, M.E.M. (1992a) Levantamento e análise da eficiência de fatores bióticos de mortalidade natural de *Anthonomus grandis* Boheman, 1843 (Coleoptera, Curculionidae), na região de Campinas, SP. *Revista de l'Academia Colombiana de Ciencias*, Bogotá, Colombia.

Pierozzi, I. Jr and Habib, M.E.M. (1992b) Proposta e análise de components básicos para um programa de M.I.P. para algodoais infestados por *Anthonomus grandis* Boheman, 1843 (Coleoptera, Curculionidae), na região de Campinas-SP. *Revista de Agricultura* 67(3), 253–269.

Pierozzi, I. Jr and Habib, M.E.M. (1993a) Aspectos biológicos e de comportamento dos principais parasitos de *Anthonomus grandis* Boheman, 1843 (Coleoptera: Curculionidae), na região de Campinas, SP. *Anais da Sociedade Entomologica do Brasil* 22(2), 317–323.

Pierozzi, I. Jr and Habib, M.E.M. (1993b) Identificação de fatores de mortalidade natural dos estágios imaturos de *Anthonomus grandis* Boheman, 1843 (Coleoptera: Curculionidae), na região de Campinas, SP. *Anais da Sociedade Entomologica do Brasil* 22, 325–332.

Pires, C.S.S., Sujii, E.R. and Fontes, E.M.G. (2003) Avaliação ecologica de risco de plantas geneticamente modificadas resistentes a insetos sobre inimigos naturais. In: Pires, C.S.S. *et al.* (eds) *Impacto ecológico de plantas geneticamente modificadas. O algodão resistente a insetos como estudo de caso*. Embrapa Recursos Genéticos e Biotecnologia, Brasília, Brazil, pp. 85–115.

Prütz, G. and Dettner, K. (2004) Effect of Bt corn leaf suspension on food consumption by *Chilo flavipes* and life history parameters of its parasitoid *Cotesia flavipes* under laboratory conditions. *Entomologia Experimentalis et Applicata* 111, 179–187.

Ramalho, F.S. (1994) Cotton pest management. Part 4. A Brazilian perspective. *Annual Review of Entomology* 39, 563–578.

Ramalho, F.S. and Gonzaga, J.V. (1991) Methodology of the application of pyrethroids against cotton boll weevil and pink bollworm. *Tropical Pest Management* 37, 324–328.

Ramalho, F.S. and Wanderley, P.A. (1996) Ecology and management of the boll weevil in South American cotton. *American Entomologist* 42, 41–47.

Ramalho, F.S., Gonzaga, J.V. and Silva, J.R.B. (1993) Métodos para determinação das causas de mortalidade natural do bicudo-do-algodoeiro. *Pesquisa Agropecuária Brasileira* 28, 877–887.

Ramalho, F.S., Wanderley, P.A. and Santos, T.M. (1996) Natural enemies and programs of biological control of cotton boll weevil in Brazil. In: Stadler, T. (ed.) *Workshop Proceedings 'Integrated Pest Management of the Cotton Boll Weevil in Argentina, Brazil and Paraguay'.* Buenos Aires, Argentina, pp. 142–148.

Ramalho, F.S., Wanderley, P.A. and Mezzomo, J.A. (1998) Influência da temperatura na fecundidade e ataque de *Catolaccus grandis* (Burks) (Hymenoptera: Pteromalidae), parasitóide do bicudo-do-algodoeiro, *Anthonomus grandis* Boheman (Coleoptera: Curculionidae). *Revista Brasileira de Entomologia* 42, 71–78.

Ramalho, F.S., Medeiros, R.S., Lemos, W.P., Wanderley, P.A., Dias, J. and Zanuncio, J.C. (2000) Evaluation of *Catolaccus grandis* (Burks) (Hym.: Pteromalidae) as biological control agent against cotton boll weevil. *Journal of Applied Entomology* 124, 359–364.

Sauer, N.F.G. (1938) Inimigos naturais de *Platyedra gossypiella* (Saunders) no estado de São Paulo: vespas depredadoras e espécies de parasitas com notas sobre a sua biologia. *Arquivos do Instituto Biológico* 9, 187–200.

Schmitz, O.J., Beckerman, A.P. and O'Brien, K.M. (1997) Behaviorally mediated trophic cascades: effects of predation risk on food web interactions. *Ecology* 78, 1388–1399.

Schuler, T.H., Potting, R.P.J., Denholm, I. and Poppy, G. (1999) Parasitoid behaviour and Bt plants. *Nature* 40, 825–826.

Schuler, T.H., Denholm, I., Jouanin, L., Clark, A.J., Clark, S.J. and Poppy, G.M. (2001) Population-scale laboratory studies of the effect of transgenic plants on nontarget insects. *Molecular Ecology* 10, 1845–1853.

Schuler, T.H., Potting, R.P.J., Denholm, I., Clark, S.J., Clark, A.J., Stewart, C.N. and Poppy, G.M. (2003) Tritrophic choice experiments with Bt plants, the diamondback moth (*Plutella xylostella*) and the parasitoid *Cotesia plutellae*. *Transgenic Research* 12, 351–361.

Schuler, T.H., Denholm, I., Clark, S.J., Stewart, C.N. and Poppy, G.M. (2004) Effects of Bt plants on the development and survival of the parasitoid *Cotesia plutella* (Hymenoptera: Braconidae) in susceptible and Bt-resistant larvae of the diamondback moth, *Plutella xylostella* (Lepidoptera: Plutellidae). *Journal of Insect Physiology* 50, 435–443.

Shi, Y., Wang, M.B., Powell, K.S., Vandamme, E., Hilder, V.A., Gatehouse, A.M.R., Boulter, D. and Gatehouse, J.A. (1994) Use of the rice sucrose synthase-1 promoter to direct phloem-specific expression of beta-glucuronidase and snowdrop lectin genes in transgenic tobacco plants. *Journal of Experimental Botany* 45, 623–631.

Silva, A.M.C. (2002) *Competição entre populações dos parasitóides* Catolaccus grandis *(Burks) (Hymenoptera: Pteromalidae) e* Bracon vulgaris Ashmead *(Hymenoptera: Braconidae).* Trabalho de conclusão de curso (Graduação em Ciências Biológicas), Universidade Estadual da Paraíba, Campina Grande, PB, Brazil.

Stapel, J.O. and Cortesero, A.M. (2004) Importance of nectar sources for adult parasitoids in biological control programs. In: Midwest Biological Control News Online IV (5). Available at: http://www.entomology.wisc.edu/mbcn/misc405.html (accessed July 2004).

Stapel, J.O., Cortesero, A.M., DeMoraes, C.M., Tumlinson, J.H. and Lewis, W.J. (1997) Effects of extrafloral nectar, honeydew and sucrose on searching behavior and efficiency of *Microplitis croceipes* (Hymenoptera: Braconidae) in cotton. *Environmental Entomology* 26, 617–623.

Sun, C.G., Xu, J., Zhang, Q.W., Feng, H.B., Wang, F. and Song, R. (2002) Effect of transgenic Bt cotton on population of cotton pests and their natural enemies in Xinjiang. *Chinese Journal of Biological Control* 18, 106–110.

Takasu, K. and Lewis, W.J. (2003) Learning of host searching cues by the larval parasitoid *Microplitis croceipes*. *Entomologia Experimentalis et Applicata* 108, 77–84.

Treacy, M.F., Benedict, J.H., Walmsley, M.H., Lopez, J.D. and Morrison, R.K. (1987) Parasitism of bollworm (Lepidoptera: Noctuidae) eggs on nectaried and nectariless cotton. *Environmental Entomology* 16, 420–423.

Turlings, T.C.J., Tumlinson, J.H. and Lewis, W.J. (1990) Exploitation of herbivore-induced plant odors by host-seeking parasitic wasps. *Science* 250, 1251–1253.

Valicente, F.H. and Barreto, M.R. (1999) Levantamento dos inimigos naturais da lagarta do cartucho do milho, *Spodoptera frugiperda* (J.E. Smith) (Lepidoptera: Noctuidae), na região de Cascavel, PR. *Anais da Sociedade Entomologica do Brasil* 28, 333–337.

Vet, L.E.M., Lewis, W.J. and Carde, R.T. (1995) Parasitoid foraging and learning. In: Carde, R.T. and Bell, W.J. (eds) *Chemical Ecology of Insects 2*. Chapman, New York, pp. 65–101.

Yang, Y.Z., Yu, Y.S., Ren, L., Shao, Y.D. and Qian, K. (2001) Effect of Bt transgenic cotton on parasitism of cotton bollworm. *Entomological Knowledge* 38, 435–437.

Zhang, G.-F., Wan, F.-H., Guo, J.-Y. and Hou, M.-L. (2004) Expression of Bt toxin in transgenic Bt cotton and its transmission through pests *Helicoverpa armigera* and *Aphis gossypii* to natural enemy *Propylea japonica* in cotton plots. *Acta Entomologica Sinica* 47, 334–341.

10 Non-Target and Biodiversity Impacts in Soil

L.C. Mendonça Hagler, I.S. de Melo, M.C. Valadares-Inglis, B.M. Anyango, J.O. Siqueira, Pham Van Toan and R.E. Wheatley

Corresponding author: R.E. Wheatley, Scottish Crop Research Institute, Invergowrie, Dundee DD2 5DA, Scotland, UK. Fax: +44 1382 568502, e-mail: R.Wheatley@scri.sari.ac.uk

Introduction

Possibly the most diverse of ecosystems can be found in soil. Soil contains many thousands of different species of bacteria, fungi, micro-, meso- and macrofauna, the numbers and activities of which are very temporally and spatially variable (see Box 10.1 for definitions of the different categories of soil organisms used in this chapter). The numerically dominant groups, the bacterial and fungal communities, can perform many functions and transformations such as the release and transformation of mineral nitrogen for plant growth, nitrogen fixation, plant-growth promotion, pathogen inhibition and phosphorus mobilization (van Elsas *et al.*, 1997). Soil meso- and macrofauna, such as earthworms, nematodes and arthropods, feed on living and dead plant tissues and play a vital part in soil nutrient cycling. For instance, they break the plant material into smaller pieces and redistribute it within the soil, making it better available for microbial activity and enhancing its incorporation into the soil organic matter. These biologically mediated processes are essential for providing the resources for sustainable plant growth and crop production and for the maintenance of all terrestrial ecosystems. Plants use the nutrients released during these transactions, and the soil organic matter and soil biota profoundly affect the soil structure and so water retention and drainage and the soil atmosphere.

Changes in the physiology, nutrient content or morphology of the crop plant caused by changes in the soil can strongly influence herbivore–plant interactions above ground, and therefore pest damage and/or the efficacy of the biocontrol community (see Chapters 6–9, this volume; Scheu, 2001).

> **Box 10.1.** Definitions of categories of soil flora and fauna used in this chapter.
>
> Soil microorganisms: 0.0002–0.002 mm in length and diameter.
> Bacteria; fungi (yeasts).
>
> Soil microfauna: length 0.004–0.2 mm, diameter < 0.1 mm.
> e.g. slime moulds in the cellular phase (Acrasiomycetes); slime nets (Labyrinthulomycota); fungi; algae; amoebae; flagellates and ciliates (Protozoa); flatworms (Platyhelminthes).
>
> Soil mesofauna: length 0.2–4 mm, diameter 0.1–2 mm.
> e.g. plasmodial slime moulds (Myxomycetes); slime moulds in the pseudoplasmodia phase (Acrasiomycetes); mites (Acarina: Gamasina, Actineda and Oribatida); springtails (Collembola); nematodes (Nematoda); water bears (Tardigrada); rotifers (Rotatoria); enchytraeid worms (Oligochaeta); some flatworms (Platyhelminthes).
>
> Soil macrofauna: length 4–80 mm, diameter 2–20 mm.
> e.g. earthworms (Oligochaeta); pauropods (Pauropoda); symphylans (Symphyla); slugs and snails (Mollusca); millipedes (Diplopoda); centipedes (Chilopoda); woodlice (Isopoda); spiders (Arachnida); insect larvae and adults – some sucking bugs (Hemiptera: Cydnidae, Aphididae), some caterpillars or pupae (Lepidoptera), some fly larvae (Diptera), some beetle larvae and adults (Coleoptera: e.g. Carabidae, Scarabaeidae, Staphylinidae, Elateridae, Curculionidae, Silphidae, Cantharidae, Cerambycidae), termites (Isoptera), web-spinners (Embioptera), ants, etc. (Hymenoptera).
>
> Soil megafauna: > 20 mm in diameter and length.
> e.g. vertebrates, some large earthworms, some slugs and snails, some Carabid beetles.

There might also be an impact on the weed community in cotton that influences yield and insect populations, although control is generally intensive in most cotton systems (see Fontes *et al.*, Chapter 2, this volume). Impacts on soil functions may also negatively affect the sustainability of soil fertility and health in the longer term, even if they do not affect cotton production in the short term, due to compensatory effects of fertilizer and other farming inputs and operations.

The diversity of the soil biota can result in extremely complex food webs that are subject to a wide range of interactive influences. There are many different soil types, each essentially defined by texture. Texture is dependent on the relative proportions of mineral particles, sand, silt and clay, present. Soil also contains air spaces, water and organic matter. In such highly heterogeneous environments, physical conditions and chemical gradients change spatially and temporally, thus permitting a wide variety of biogeochemical processes to occur. Species and functional process diversity in plant–soil systems is immense and a vast range of compounds is produced, many of which are then further transformed in other processes. These spatial and temporal variations over small to large scales make system predictions and modelling extremely difficult.

However, activity in the soil ecosystem is usually limited by the availability of nitrogen and energy from fixed carbon compounds derived from plant primary production, mainly plant-derived inputs from the roots, such as root exudates, cellular remains and root debris, and plant residues that fall on the soil surface (leaves, stems, flowers and fruit). Although microbial activity can occur in the bulk soil, it is much greater in the zone closest to the plant roots, the rhizosphere (Fig. 10.1). Inputs to the soil from the growing plant are greatest in the rhizosphere, and as these change with both the type and stage of growth of the plant, so do microbial functional dynamics. Microbial populations can increase by a hundred or a thousand times after plant material is added to soil. Plant species and genotype and soil physical conditions determine the amount and types of compounds entering the soil, and therefore indirectly drive the rates and types of biotic functioning. There can be large differences in the relative proportions of individual bacterial genera on different plant species and cultivars and in different soil types, and the influence of field site interacts with the influence of plant variety on the microbial community (Grayston *et al.*, 1998; Siciliano *et al.*, 1998; Dunfield and Germida, 2001). Soil ecosystem functional dynamics are also affected by other factors such as weather and cultivation techniques, but soil fertility depends primarily on microbial activity that is responsive to plant inputs, particularly in the plant-rooting zone.

As well as providing a basic energy source, these carbon inputs can also affect microbial function in more subtle ways, as they play a role in microbial interactions and signalling. Temporally variable interactions between general soil heterotrophs and specific groups of microorganisms, such as the autotrophic nitrifiers, occur in arable soils in response to additions of carbon and nitrogen. Transgenic insecticidal plants such as *Bacillus thuringiensis* (Bt) cotton produce and release relatively large amounts of a variety of novel proteins, including the active toxins. Microbial processes have been shown to be particularly responsive to protein substrates (Wheatley *et al.*, 1997, 2001). Changes in the root exudates of a transgenic plant may significantly alter the rhizosphere community associated with it (Brusetti *et al.*, 2004).

The authors looked at the three commercial Brazilian cotton-growing regions, the Midwest, the Meridian and the North-east. The three regions vary widely in both soil types and climate (see Fontes *et al.*, Chapter 2, this volume). In the North-east, cotton is grown at low altitudes in a xerophylic climate with relatively low irregular rainfall on relatively fertile soils. In contrast, in the Midwest, which is the major cotton-producing region, cotton is grown at altitudes from 400 m to 1000 m during humid summers with an average seasonal precipitation of 1100 mm, on relatively poor soils. In the Meridian region, cotton is grown at altitudes from 400 m to 700 m in a climate of hot and humid summers and cold and dry winters with average seasonal precipitation of between 900 mm and 1000 mm. Differences in crop-management

Fig. 10.1. Microcolonies of *Pantoea ananatis* strain A5M2 on the cotton root epidermis, 7 days after inoculation. Root colonization is often the limiting step in the use of rhizobacteria capable of producing indole-3-acetic acid in soil and promoting plant growth. Image taken in a field emission scanning electron microscope by Itamar Soares de Melo.

regimes between regions and between farmers will also influence soil dynamics and so affect the risk assessment. For instance, farmers in the Midwest use partial minimum tillage and plant millet after the cotton harvest. The next crop of cotton is then directly drilled into the millet residues, after desiccating herbicides have been applied. As a pest-control measure, they are obliged to chop up cotton crop residues before the following crop of millet. Disturbance of the soil has a tremendous effect on the functional dynamics of the soil organisms, mainly because of the redistribution and so enhanced availability of nutrients and organic matter that was previously spatially unavailable in soil aggregates, etc. Minimum-tillage systems disturb only the surface layers and so result in lower subsurface decomposition rates (Feng *et al.*, 2003). Therefore the crop-management system examined in the soil risk assessment is crucial to the outcome. We recommend that representative examples of all the alternative cropping systems are compared as multiple controls during the assessment of Bt cotton in Brazil.

Because of the vast complexity of the soil ecosystem, it would be both impractical and unreliable to study the system with rationales based on species lists. Although numerically there are far fewer functional properties, it is still impractical to measure all of them in the soil ecosystem, so choices of reliable parameters have to be made. In this chapter, we will consider the functional group 'soil ecosystem processes', and prioritize them for case-specific testing, based on the exposure pathway (association with plant residues and Bt toxin) and the significance or possibility for an adverse effect (importance as an indicator of soil health and effect on cotton development). We will then identify adverse-effect scenarios for the highest-priority processes, formulate testable hypotheses and describe experiments that can test these hypotheses in the laboratory and field. See Box 5.1 in Hilbeck *et al.* (Chapter 5, this volume) for further detail.

This analysis does not constitute a complete risk assessment of the effects of Bt cotton on all soil ecosystems in Brazil. For example, as no significant commercial cropping occurs using 'organic farming' practices, that system is not considered. So consideration here is focused on determining a range of options that can be customized to a specific commercial scenario. A full characterization of risk would need to consider case-specific data from each of the Brazilian cotton-growing regions and agromanagement systems. Horizontal gene transfer between plant cells, bacteria and viruses was not considered in this chapter, but it should also be addressed in a full risk assessment of Bt cotton in Brazil.

Assessment of Input Routes and Possible Exposure Pathways in Soil

Bt proteins are present as active toxins in most of the cells of a Bt-transformed plant, and so are present in all plant residues, and may be released from root tissues (Fig. 10.2) into the surrounding soil. There are several reports of the long-term persistence of active Bt proteins in some soil types after the cultivation of Bt-transformed crops. As most of the activity in the plant–soil system occurs in the rhizosphere and is driven by plant inputs, the possibility that these Bt proteins may have a direct effect on all the biotic dynamics associated with plant production requires investigation before commercial release.

Input routes of Bt plant material and Bt toxin into soil

Bt plant material and Bt proteins can enter the soil through various potential routes, depending on the crop and environment:

1. The direct input of Bt proteins via root exudates and leachates from root injuries. This is a continual input of free, active Bt protein during the whole growing season, increasing in quantity and spatial extent as the root system grows.
2. Bt protein in the plant matrix of sloughed off root debris, e.g. root cap cells and root hairs. This is a continual input during the whole growing season, increasing in quantity and spatial extent as the root system grows.

The Cry1Ac protein in cotton with event 531 (INGARD® cotton) is expressed in the roots (OGTR, 2003). From 4 to 9 weeks after germination, Cry1Ac concentrations of between 1 µg/g and 43 µg/g dry weight of roots

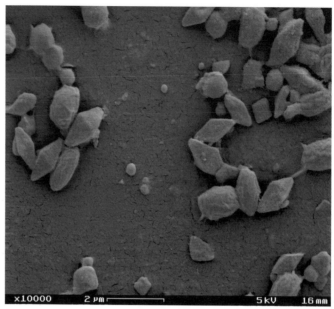

Fig. 10.2. Scanning electron micrograph of Bt toxin crystals and bacterial cells. Crystal produced by *Bacillus thuringiensis*. Image taken in a field emission scanning electron microscope LEO (Zeiss + Leica) at Embrapa Environment by Itamar Soares de Melo.

were found in Australia (Gupta *et al.*, 2002); the roots released Cry1Ac into soil during growth, although this was not quantified. Saxena *et al.* (2004) reported that Bt proteins are not exuded from Bt-cotton roots, though Bt maize, potato and rice roots do exude Bt. However, a certain amount of Bt protein will enter the soil throughout the growth period from decaying root tissues and leachates from root breakage. Bollgard cottons contain the same transgene and promoter as INGARD® cotton, so can be expected to express the protein in the roots similarly.

3. Input of Bt-expressing plant parts (leaves, squares, bolls, pollen) falling to the ground under the plants. This is a continual input during the whole growing season, increasing in quantity as the plants develop.
4. The input of plant residues (dead or alive, e.g. as seeds) remaining in the field after harvest. This is an annual or biannual input of relatively large amounts (depending on crop-management practices).

Sources of Bt-protein inputs and the pathways and processes by which they may have effects on soil ecosystems are summarized in Table 10.1. In this table, any item in the source column can be linked via any of the properties in the pathways and process columns to realize any of the outcomes in the effects column: e.g. pollen (source) may be ingested by fauna (pathway), degraded (process) and so decayed (effects).

Table 10.1. Effects pathways matrix.

Source	Pathway	Processes	Effects
Plant residues	Decomposition	Adsorption	Decay rate
Pollen	Protein release	Denaturation	Persistence
Root exudates	Faunal ingestion	Degradation	Bioactivity
Damaged tissues	DNA transfer	Plant uptake	Accumulation
Bt protein in soil		Elimination	
Transgene		Leaching	
		Run off	

These interaction routes between the Bt plant material or Bt proteins and the soil biota are strongly influenced by the tillage system. Under conventional tillage the plant litter will be mechanically incorporated into the soil, diluting the concentration of the protein but increasing the number of organisms exposed. Under zero tillage, crop residues are left concentrated on the soil surface and the only incorporation that occurs is through the action of soil macroinvertebrates.

Fate of the Bt plant material and the Bt proteins in soil

The fate of the plant materials containing Bt protein and the free Bt proteins in soil ecosystems depends on the rate of input and the rate of dissipation. For exposure and impact analysis of the incorporated and released Bt proteins on the soil ecosystem, it is essential to know how long they persist in the system.

Persistence – microbial or chemical degradation
Laboratory studies on the persistence of ground up lyophilized Bt cotton plant material in soil (in which the conditions for degradation were usually near optimal) have reported that the toxins were still detectable and retained their insecticidal activity when the experiments terminated after 28 days (Palm *et al.*, 1994; Donegan *et al.*, 1995), 120 days (Sims and Ream, 1997) and 140 days, at which point 25–30% of the Cry1Ac proteins produced by Bt-cotton leaf residues were still bound to the soil particles at pH 6.2 (Palm *et al.*, 1996) (see adsorption below). Even longer time periods have been reported using Bt maize residues in the laboratory (Saxena and Stotzky, 2000, 2001b; Hopkins and Gregorich, 2003) and in the field (Zwahlen *et al.*, 2003a,b). The general pattern is similar to that for other proteins and organic compounds in soil. Initially, a rapid decline of the soluble toxin concentration is observed, followed by a more gradual decline to low concentrations of the toxin that may then remain almost unchanged for several weeks and probably months (Palm *et al.*, 1996; Zwahlen *et al.*, 2003a). The persistence of Bt proteins in soil could therefore be long enough to be of concern through the growth cycle of the following crop. In the Midwest and North-east regions in particular, cotton is grown repeatedly on the same land year after year. It is

also possible that other Bt-transformed crops such as Bt maize will be grown in rotation with Bt cotton in future.

Decomposition rates are very dependant on soil pH, temperature and water availability, and vary with soil structure and texture as indicated, for instance, in Zwahlen *et al.* (2003a) and Donegan *et al.* (1995). Finer, clayey soils show, on average, slower decomposition rates and higher retention of organic matter than coarse, sandy soils. The recovery of the Bt protein from soil is also temperature and pH dependant (see also under adsorption below), and procedures used to extract Bt toxins from plant residues need modification to the experimental set-up before use (Palm *et al.*, 1994).

A field study on six fields, where Bt cotton had been grown for 3–6 years and the crop residues incorporated into the soil after harvest, found no detectable Cry1Ac toxin (by enzyme-linked immunosorbent assay (ELISA) and bioassays with *Heliothis virescens* on ground homogenized soil samples) in soil samples taken at 3 months after postharvest tillage (Head *et al.*, 2002). Three fields were sandy loam and three silt loam, and the experiment did not indicate differences between soil types. The authors take this as a demonstration that Bt protein from Bt cotton will not accumulate in the soil under temperate field conditions. Another study compared conventional-till and no-till cotton systems with Bt/glyphosate-tolerant cotton and glyphosate-tolerant cotton without Bt (Lachnicht *et al.*, 2004). They found no significant differences in decomposition between the two transgenic cotton cultivars, measured by change of weight and C:N content of buried litterbags. There was a significant difference between the tillage treatments.

Uptake by soil organisms or plants
Roots are known to reabsorb previously exuded organic compounds, a mechanism used in the uptake of insoluble cations. The uptake of Bt proteins from the soil into growing plants is therefore a possibility, but has not been investigated for cotton. One study investigated the uptake of Bt toxins from soils in which Bt maize had been grown (so containing Bt from root exudates, etc.), or soils in which Bt maize residue or purified Bt toxin was added. None of the test plants grown on these soils (maize, carrot, radish and turnip) took up any Bt proteins from the different soil types over 120 or 180 days. Also, isogenic maize did not take up the Bt toxin when grown in a hydroponic solution in which its Bt counterpart had been grown previously (Saxena and Stotzky, 2001a).

Soil fauna can contact Bt plant roots, residues and the Bt proteins still within the plant cells through their feeding activity as well as by physical contact with any Bt proteins bound to organic matter and clays in the soil. In addition, soil herbivores and detritivores may utilize the Bt protein as a novel food source, or they may sequester it and pass it on up the food chain. Trophic relationships in soil are very complex and require precise experimental design. Researchers found no detectable effects of Bt-cotton leaf tissue on two detritivores, the springtail *Folsomia candida* and the mite *Oppia nitens* (Yu *et al.*, 1997). However, it is not clear whether the organisms actually consumed the Bt material during the experiment, because they are both species that preferably consume fungi growing on decaying plant material (Seniczak and Stefaniak, 1981; Fountain and Hopkin, 2005). It has not yet been demonstrated whether or not fungi growing on decaying Bt plant material also contain the Bt toxins.

Adsorption of the Bt protein

The mobile Bt protein will be rapidly degraded by microbial degradation if it is not leached first (see below; Palm *et al.*, 1996; Saxena *et al.*, 1999; Saxena and Stotzky, 2000). However, laboratory experiments with microbially produced Bt toxin showed persistence and insecticidal activity at 234 days (Tapp and Stotzky, 1998). Bt toxins have been shown to bind to clay minerals and humic acids within a few hours of entering the soil and remain bound for long periods of time (Venkateswerlu and Stotzky, 1992; Tapp *et al.*, 1994; Tapp and Stotzky, 1995; Crecchio and Stotzky, 1998). Correspondingly, most (about 10% to 30%) of the Bt toxin can be recovered (washed out) from soils with low organic matter and high sand content, and least from soils high in clay content and organic matter (Palm *et al.*, 1994; Saxena and Stotzky, 2000). Adsorption is lower at pH much below or above 6. The Bt proteins only partially intercalate the clay mineral montmorillonite and do not intercalate kaolinite, so it can be assumed that they bind primarily on the external surface of the minerals (Tapp *et al.*, 1994; Saxena and Stotzky, 2000). The adsorbed Bt toxins retain their insecticidal activity (Koskella and Stotzky, 1997; Crecchio and Stotzky, 1998; Tapp and Stotzky, 1998). Over 80% of the microorganisms in soil are adsorbed onto organic matter and clay minerals (Bruinsma *et al.*, 2002). The majority of microorganisms could therefore be in close proximity to the adsorbed Bt proteins.

Adsorption of the proteins to clays and humic acids in the soil will increase the greater the concentration of the protein entering the soil, but at high concentrations binding will level off at an equilibrium and more of the protein will remain mobile, indicating a saturation of the adsorption sites (Stotzky, 2000). When the proteins are bound to clay minerals or humic acids in the soil they are resistant to microbial degradation, therefore overall degradation rates are related to the rate of mobilization of the Bt protein (Koskella and Stotzky, 1997; Crecchio and Stotzky, 1998).

Removal from soil (e.g. through leaching)

What is the persistence of the transgene product in the soil water and what is the likelihood of transgenic crop residues or the transgene product being transported in runoff from the transgenic field to adjacent fields or water bodies? A laboratory experiment was carried out using purified Bt toxin, using growing Bt maize exuding Bt from the roots and using Bt maize residues in soil columns containing differing quantities of clay minerals (montmorillonite and kaolinite; Saxena *et al.*, 2002). The Bt (Cry1Ab) protein was found in the leachates of all soil columns and not in the leachates of the respective isolines. The protein exhibited stronger binding and higher persistence, as well as remaining nearer the soil surface, in soil that contained higher clay concentrations (i.e. had a higher cation-exchange capacity and specific surface area). This experiment indicates that there is a possibility that Bt proteins can be transported in water, either in surface waters by runoff and erosion or downwards.

Conclusions: input routes and possible exposure pathways

The Bt proteins will enter the soil inside both fresh and decaying cotton plant material and also as free protein in leachates from plant material. The soil fauna can therefore be simultaneously exposed to a complex combination of living and degrading plant material, free Bt proteins and Bt proteins bound to soil particles. Concentrations of Bt toxin in soil can fluctuate greatly with the highest concentrations occurring around decaying organic matter and roots. There is evidence that Bt toxins are present as long as the incorporated Bt plant residues are present, a 'time release' mechanism (Zwahlen et al., 2003a). It is not clear whether this was tested in a Bt-cotton field study (Head et al., 2002), as the cotton plant residue content of the samples was not stated, and so it is unclear if the bioassay H. virescens was actually exposed to any Bt plant material. It is also not clear whether or not the Bt toxins from Bt cotton will persist in the field outside of the plant material, bound to clays or humic acids, as has been shown for purified Bt in the laboratory and for Bt maize in the field (Saxena and Stotzky, 2000).

Soils in the Midwest, the major cotton-producing area, are mainly latosols or argisols (Fontes et al., Chapter 2, this volume). As the clay component in argisols is moderate and moderate to high in latosols, it is possible that considerable amounts of Bt proteins may become bound in these soils.

Prioritization and Selection of Soil Processes

Soil ecological processes were considered in five main categories (Table 10.2):

- plant residues decomposition;
- biogeochemical nutrient cycling;
- plant/microorganism/meso- and macrofaunal interactions;
- cotton pests and diseases;
- role of biological activity in soil chemical and physical properties.

The significance of the ecological process, and therefore the significance of an adverse effect, was ranked by considering the importance of the process as an indicator of soil health and by considering how directly a variation in the process affects crop development or whether other ecosystem processes intervene to mitigate the effect of this variation. Direct effects involve a change to the process that directly affects crop development, and indirect/mediated effects are changes in the process that are mitigated by other ecosystem processes.

The exposure of the soil biota responsible for the process to cotton plant material (and proteins released from it) was considered by identifying the

Table 10.2. Ranking of soil processes in cotton-production areas in Brazil, main parameters.

Soil process	Associated biotic groups in cotton in Brazil	Importance as an indicator of soil health in cotton	How directly does the process affect cotton development?	Location of process in soil	Association with cotton plant material or released proteins	Expert ranking
(a) Plant residue decomposition						
Plant residue incorporation (biological)	Springtails, woodlice, ants, termites, epigeic and anecic earthworms, millipedes, beetles, molluscs, mites	High	Direct	Soil surface, upper soil layer (depth is specific to soil type and tillage system)	Direct (leaves, stalks, flowers, pollen, bracts, lint, seeds on soil surface)	1
Plant residue diminution (biological)	Springtails, woodlice, millipedes, ants, termites, anecic earthworms, mites, enchytraeid worms, symphylans, slugs and snails	High	Direct	Soil surface, upper soil layer (depth is specific to soil type and tillage system)	Direct (plant residues in the soil)	1
Organic matter decomposition (microbial)	Bacteria, fungi, slime moulds	High	Direct	Upper soil layer, particularly rhizosphere (depth is specific to soil type and tillage system)	Direct (plant residues in the soil)	1
Organic matter decomposition: cellulose breakdown (microbial)	Cellulytic bacteria and fungi	Medium	Semi-direct	Upper soil layer, particularly rhizosphere (depth is specific to soil type and tillage system)	Direct (plant residues in the soil)	2

Table 10.2. Ranking of soil processes in cotton-production areas in Brazil, main parameters – *cont'd.*

Soil process	Associated biotic groups in cotton in Brazil	Importance as an indicator of soil health in cotton	How directly does the process affect cotton development?	Location of process in soil	Association with cotton plant material or released proteins	Expert ranking
Organic matter decomposition: cellulose breakdown (meso- and macrofaunal)	Termites, oribatid mites, earthworms, slugs and snails	Medium	Semi-direct	Upper soil layer, particularly rhizosphere (depth is specific to soil type and tillage system)	Direct (plant residues in the soil)	2
(b) Biogeochemical nutrient cycling						
Ammonification (microbial)	Ammonifying bacteria	High	Semi-direct	Upper soil layer, particularly rhizosphere	Direct (plant residues in the soil)	1
Nitrification (microbial)	Nitrifying bacteria	High	Semi-direct	Aerobic soil layer	Direct (free Bt toxin in the soil) Indirect (NH_4^+ concentration)	1
Denitrification (microbial)	Denitrifying bacteria	Low	Semi-direct	Anaerobic soil layer/conditions (e.g. waterlogging)	Direct (free Bt toxin in the soil) Indirect (NO_3^- concentration)	3
Nitrogen fixation (microbial)	Nitrogen-fixing bacteria and algae (e.g. *Rhizobia*)	Low	Indirect/mediated	In aerobic, neutral pH soil layer – free living in soil, associated with grass roots, symbiotic on legume roots	Direct (free Bt toxin in the soil)	3

Phosphorus mobilization (fungal)	Mycorrhizal fungi on plant roots, mycorrhizal propagules	High	Indirect/mediated	On cotton roots – rate is very dependent on fertilizer use – principally in poor soil Fungal ecology	Direct (association with plant roots)	2
Mobilization of micronutrients (fungal)	Mycorrhizae on plant roots	Medium	Semi-direct	On cotton roots Depends on plant nutritional status	Direct (association with plant roots)	2
(c) Plant/microbial/meso- and macrofaunal interactions						
Parasitism or commensalism in plant roots (microorganisms)	Endophytic bacteria, endophytic fungi and endophytic nematodes	Medium	Semi-direct	In root plant tissue in between cells (phyllosphere)	Direct (root cells)	2
Parasitism or commensalism on plant roots and in rhizosphere (microorganisms)	Rhizosphere bacteria, parasitic or symbiotic fungi (Mycorrhizae)	High	Indirect/mediated	On root surface, and in rhizosphere (consumption of root exudates)	Direct (root cells)	2
Bitrophic feeding on plant roots (crop pests)	Nematodes, beetle larvae, termites, some symphylids, some endogeic earthworms, some springtails, some Dipteran larvae, root aphids, sucking bugs (Cydnidae)	High	Direct and mediated (through disease attack)	On roots, in rhizosphere	Direct (plant roots)	1

Continued

Table 10.2. Ranking of soil processes in cotton-production areas in Brazil, main parameters – *cont'd.*

Soil process	Associated biotic groups in cotton in Brazil	Importance as an indicator of soil health in cotton	How directly does the process affect cotton development?	Location of process in soil	Association with cotton plant material or released proteins	Expert ranking
Saprophytic feeding on plant residues (overlap with decomposition function above)	Bacteria, fungi, slime moulds, water bears, enchytraeid worms, pauropods, symphylans, mites, slugs and snails, woodlice, millipedes	Medium	Indirect/ mediated	Soil surface, upper soil layer (depth is specific to soil type and tillage system)	Direct (plant residues)	3
Feeding on soil fungi (funct-ions: keeping colony in young state, dispersal of fungi in soil, etc.)	Slime moulds, springtails, fungiphagal bacteria, symphylans, pauropods, nematodes, enchytraeids, woodlice, millipedes, Dipteran larvae	Medium/low	Semi-direct	Rhizosphere and detritus	Possible tritrophic (via residue and roots/root exudate-consuming fungi)	3
Feeding on other micro-bial flora and fauna (functions: keeping colonies in young growing state, dispersal of bacteria in soil, etc.)	Protozoa, slime moulds, symphylids, pauropods, springtails, nematodes, some water bears, rotifers, endogeic earthworms	Medium	Semi-direct	On plant roots, in rhizosphere, in soil free-living, in soil bound (bacteria)	Possible tritrophic (via residue and root/root exudate-consuming fungi and bacteria)	2

Predation on soil meso- and macro-fauna	Carabid beetles, Staphylinid beetles, spiders, ants, centipedes some Dipteran larvae, some slugs, some gamasid and actinid mites and some springtails (eating nematodes)	High	Semi-direct (trophically mediated)	Soil surface, upper soil layer	Possible tritrophic or multitrophic (via residue and root-consuming meso- and macro-fauna)	2
(d) Cotton pests and diseases						
Disease transmission	• Fusarium wilt (*Fusarium oxysporium* f. sp. *vasinfectum*) spores and hyphae • *Verticillium dahliae* spores • Other fungal hyphae and spores e.g. *Rhizoctonia solani* • Bacterial blight	High	Semi-direct	• On roots and plant residues in soil • Hyphae and spores in plant residues on soil surface • In residues on soil surface • In residues in soil	Direct (roots) Direct (plant residues on soil surface) Direct (plant residues in soil)	1
Disease suppression (by grazing or antagonism) (e.g. Lartey *et al.*, 1994)	Bacteria e.g. *Bacillus thuringiensis, Pseudomonas fluorescens* Antibacterial and parasitic fungi e.g. *Trichoderma viride* Fungal feeders (see above)	Low	Indirect/ mediated	Rhizosphere and roots	Direct (free Bt protein) Direct (root exudates on plant roots) Indirect (in soil)	2
Predation on crop pests in soil	Predators of root feeders, see predation above	Medium/ high?	Semi-direct	On cotton plant roots, rhizosphere, in upper soil layer	Possible tritrophic – via prey exposed to roots, plant residues	2

Continued

Table 10.2. Ranking of soil processes in cotton-production areas in Brazil, main parameters – cont'd.

Soil process	Associated biotic groups in cotton in Brazil	Importance as an indicator of soil health in cotton	How directly does the process affect cotton development?	Location of process in soil	Association with cotton plant material or released proteins	Expert ranking
(e) Influence of biological activity on soil chemical and physical properties						
Binding/storage of nutrients (production of humic substances)	Calcitrant (humic) soil organic matter (producers: bacteria, fungi, enchytraeid worms, slugs and snail mucus)	High	Indirect/mediated	Strongly dependent on soil type, pH, crop management (tillage, crop rotation, etc.)	Indirect via rate of production of calcitrant soil organic matter	2
Soil aggregate formation (soil aeration, structure and stability, resistance to erosion)	Soil organic matter and carbon compounds (polysaccharides) from decomposition: fungal hyphae (including Mycorrhizae), bacterial colonies, enchytraeid worms, slugs, snail and earthworm mucus, etc.	High	Indirect/mediated	Strongly dependent on soil type, pH, crop management (tillage, crop rotation, etc.)	Indirect via impacts on soil organic matter production, impacts on fungi, bacteria, earthworms, etc.	2
Water movement in the soil (drainage)	Pores formed from earthworm burrows, fungal hyphae, etc. Soil aggregate structure	High	Indirect/mediated	Strongly dependent on soil type, pH, crop management (tillage, crop rotation, etc.)	Indirect via impacts on micro and macro-organisms and impacts on soil aggregate structure	2
Water retention in the soil (soil moisture content)	Soil aggregates and spaces formed from soil organic matter	High	Indirect/mediated	Strongly dependent on soil type, pH, crop management (tillage, crop rotation, etc.)	Indirect via impacts on soil organic matter production and soil aggregate structure	2

location of the functional activity in the soil, i.e. possible direct-exposure routes to cotton plant material and proteins or whether an effect might occur through an indirect impact. Considering both significance and exposure in cotton systems in Brazil, the soil processes were ranked on a 1, 2, 3 basis for each process, 1 being of the greatest priority and 3 the lowest. The highest-priority processes have direct importance as indicators of soil health, as the functional biota are closely associated with living or dead crop plant material or exudates and an adverse effect on the process would likely have a direct impact on crop development and yield stability.

One concern could be the possibility of an alteration in the rate, timing or magnitude of soil organic matter decomposition that could have consequences for soil health and fertility, including a change in energy and nutrient supply for other microbial processes, and for plant growth. Similarly, the stability of soil aggregate structure may be affected with consequences for root development and water release and holding capacities. The consequences of any deleterious effects on residue decomposition are so great that this is ranked 1. As nitrogen-cycling dynamics in agricultural environments are known to be entirely dependent on the quality and quantity of plant inputs, these processes might be directly affected by input changes, particularly with accompanying potential toxicity effects, and so the two important steps in nitrogen cycling in high to medium input cotton production (ammonification and nitrification) are ranked 1. A possible impact on disease transmission might have significant consequences because of the importance of soil-transmitted fungal diseases in limiting cotton production in Brazil (see Fontes *et al.*, Chapter 2, this volume), and, although it was considered to be likely to appear over a relatively longer time scale, was ranked 1.

The selection process could be further developed by considering criteria such as the vulnerability of the process to change in the soil, the amount of redundancy in the soil biota responsible for the process and the experimental accessibility (Kowalchuk *et al.*, 2003).

Selection matrix for soil macrofauna

A functional assemblage approach was also applied to the meso- and macro-fauna involved in decomposition using a selection matrix (Table 10.3). The advantage of this approach is that it enables a transparent selection process to be defined, despite the paucity of species-specific information available on soil macroorganisms in Brazilian cotton cultivation. This approach can be modified and refined as information becomes available. We did not consider vertebrates in cotton fields in Brazil.

The macrofauna was divided into three functional groups:

- decomposers;
- root feeders;
- disseminators of residues.

For reasons stated before, the priority concern is that any impact on soil macroorganism functions may slow down the rate of residue breakdown and

Table 10.3. Macrofaunal rankings for decomposers.

Functional group	Taxa	Other functions	Possible indicator groups	State of knowledge[a]	Priority[b]
Decomposers (litter breakdown)	Collembola, mites, Isopoda, Diplopoda, Symphyla, beetles, ants, etc.	• Many are also disseminators of residues	Millipedes	1	1
Root feeders	Nematodes (pathogenic), insect larvae	• Damage to crop plant • Facilitators of entry for disease pathogens	Nematodes	1	2
Disseminators of residues	Earthworms, other soil-burrowing species; Collembola, ants; beetles, (many species are multifunctional)		Earthworms	1	1 and 2 (earthworms rank 1)

[a]1, fairly good; 2, partial; 3, not satisfactory.
[b]1, high priority; 2, intermediate priority; 3, low priority.

incorporation into soil organic matter, and as this drives the rate of organic matter decomposition by the microbial community in soil, may have consequences for soil health and plant growth. The macroinvertebrate disseminator group was therefore ranked 1.

The soil meso- and macrofauna fulfil many other important functions in soil such as dispersers of beneficial microbial populations in the rhizosphere, as dispersers of saprophytic fungi (Rantalainen *et al.*, 2004), as seed dispersers and as predators of crop pests. In addition, epigeic macroinvertebrates can be an essential food source for generalist predators above ground (e.g. for Carabid beetles, Staphylinid beetles and spiders (Ekschmidt *et al.*, 1997), as well as for vertebrates such as birds or shrews). This can be particularly important in between crops or at the start of the crop-growing season (Settle *et al.*, 1996).

The analysis of the soil meso- and macrofauna in this chapter is incomplete as little is known now about the diversity present in Brazilian cotton fields. To encourage knowledge gathering in this area, we propose a starting model for this important grouping that can be refined and developed as further information becomes available. In this chapter, earthworms are used as an example of the disseminator group. Earthworms perform several functions, such as residue distributors and incorporators, assisting in decomposition, and are abundant and diverse. Site- and region-specific information on their relative significance and species diversity is required for a full analysis.

Identification of Adverse-Effect Scenarios and Testing Hypotheses

In this section, analysis of input routes and exposure pathways, and the results of the soil ecological-process prioritization, are used to develop adverse-effect scenarios for the soil ecosystem, and testing hypotheses to test the scenarios. In the final section of the chapter, specific experimental protocols are described to provide definitive scientific evidence that any potential effect or effect pathway is likely to occur or not.

Persistence in the soil

The first requirement for the verification of exposure routes is the assessment of the amounts of Bt proteins introduced into the soil ecosystem during and after the cultivation of Bt cotton, and their persistence. As the Bt proteins are in the active form there may be an enlarged target range that might adversely affect both micro- and macrofaunal organisms, with subsequent detrimental effects on functional dynamics in the soil ecosystem.

Verification of exposure pathway: how long will Bt proteins in plant residues and Bt proteins released into the soil from such plant residues and from the roots persist in the soil, and associated groundwater? Do they persist to the next cropping season?

Microbial (bacterial and fungal) decomposition of organic matter, and the coincident nitrogen-transformation processes

A vast array of microorganisms and higher trophic groups of organisms such as the micro- and mesofauna interact in the incorporation and breakdown of plant material in soil ecosystems. A study of the impact of leaf material from three Bt-cotton lines found no differences in the numbers of protozoa, but did find changes in culturable bacterial diversity, and significantly greater increases in culturable bacterial and fungal population levels with the transgenic material compared to the parental line in the 2 weeks after the start of the experiment (Donegan *et al.*, 1995). At the end of the experiments (28 or 56 days), these changes were no longer observed. This suggests that the transgenic plants may have decomposed faster than the parental plants and thus more rapidly provided nutrients for microbial growth, though decomposition was not measured. On two sampling occasions there was significantly greater utilization of asparagine, aspartic acid and glutamic acid in soil with material from the two transgenic lines compared to the parental line. These substrates are important intermediates in nitrogen-assimilation reactions. Because the changes were only observed for two of the transgenic lines and not the third, and not for the purified toxin, the authors conclude that they might be due to unintended changes in the physiology of these transgenic lines rather than directly due to the Bt toxins. Further investigations are required to determine what changes might occur with repeated incorporations. Studies on other transgenic plants have also found differences in microbial communities associated with the plants at the senescence growth stage, indicating an association with the decomposer community (Lottmann *et al.*, 1999, 2000; Lukow *et al.*, 2000).

Transforming a plant can influence its physiology in other than the specifically designed effect, either through pleiotropic effects of the transgene (Grossi *et al.*, Chapter 4, this volume), or metabolic effects of the transgenic protein, or changes arising from the transformation process (e.g. tissue culture) and subsequent breeding (Andow *et al.*, Chapter 1, this volume). Studies with other transgenic plant residues in soil have shown differences in the nutritive quality of the transgenic plant for soil microbial and macrofaunal communities, compared to the control or other varieties (Donegan *et al.*, 1997; Escher *et al.*, 2000; Saxena and Stotzky, 2001b,c). Such differences may have significant impacts on plant residue-degradation rates, and the non-transformed parental or near-isoline may not provide a completely satisfactory control for this reason (Andow and Hilbeck, 2004). Therefore, any comparative experiments may not only be assessing effects of the toxins but also changes in the organic constituents resulting from these effects on the plants' physiology. We recommend examination and comparison of the nutritive qualities of the transgenic and control plants (e.g. cellulose and lignin content, C–N ratio) prior to carrying out degradation experiments. See also earlier discussion (section on persistence) for the influence of environmental conditions (pH, temperature, water) and soil type on decomposition.

Adverse-effect scenario: The functioning of soil ecosystems is dependent on the breakdown of plant residues to provide energy. Microbial functional dynamics are dependent on the quality and constituents of this input. During decomposition, the nitrogen-containing compounds required for many other microbial functions are released. So these plant–microbe relations have consequences for soil fertility and crop nutrition. As the patterns of such inputs are specific to the crop variety, it can be anticipated that microbial dynamics under Bt cotton plants will be directly affected by these plants. A change in degradation patterns in the soil may result in a disruption or asynchrony of energy and nutrient supply in relation to the demands of the crop, and so requires quantification.

Testable hypothesis 1: The rate, timing or magnitude of the decomposition of Bt-cotton plant residues will be altered compared to decomposition rates of conventional cotton, due to impacts of the residues on microbial activity.

Nitrogen cycling: ammonification

The recycling of inorganic nitrogen from plant residues for further crop production is a vital function of the soil ecosystem. These transformations require the interaction of a vast array of microorganisms and higher trophic groups, especially in the first step of ammonification. Ammonification, the conversion of organic-nitrogen to ammonium-nitrogen, is closely allied to decomposition.

Adverse-effect scenario: Ammonifying bacteria will come into direct contact with Bt toxins in the rhizosphere and around decomposing cotton plant residues. These toxins may affect their activity and therefore affect the rate of ammonification in the rhizosphere and soil near to the cotton plant, with further consequences for nitrification rates, decreasing the availability of nitrogen for the cotton and subsequent crops grown in the field.

Testable hypothesis 2: Ammonification rates will be reduced in soils containing Bt-cotton residues, with consequential effects on soil fertility and crop production.

Nitrogen cycling: nitrification

Nitrification is the next step in the N-cycle, in which the immobile ammonium form is converted to the mobile nitrate form that is preferentially taken up by plants. The process involves two bacterially mediated steps: first the oxidization of ammonia to nitrite and then the oxidization of this nitrite to nitrate. Nitrification is often a rate-limiting step in the nitrogen cycle and influences nitrogen availability to plants. It can decrease soil pH and lead to leaching of nitrogen from soil ecosystems, and is also the first step in the loss of fixed nitrogen from soil via denitrification. Denitrification can also lead to a significant loss of nitrogen in systems receiving nitrogen fertilizer (Bruinsma *et al.*, 2002), as does much of the cotton production in Brazil. Due to their relatively low redundancy, their sensitivity and their influence on an important nutrient cycle, nitrifying bacteria could be suitable indicators of soil health (Kowalchuk and Stephen, 2001; Bruinsma *et al.*, 2002).

To date, all known ammonia-oxidizing bacteria (AOB) in soil belong to a narrow clade within the β-subclass of the protoebacteria consisting of two genera, *Nitrosospira* (containing former genus designations *Nitrosolobus* and *Nitrosovibrio*) and *Nitromonas*. Although *Nitromonas europae* is the best-characterized species of AOB, probably due to the relative ease with which it can be recovered in pure culture, numerous studies have suggested that *Nitrosospira* species are dominant in terrestrial ecosystems (Kowalchuk and Stephen, 2001; Wheatley *et al.*, 2003). Some AOB groups may be indicative for specific environmental conditions.

Nitrite oxidization, the second step towards nitrate formation, is spread across several bacterial groups, making comprehensive molecular studies of this group more difficult. The step may be more susceptible to stress conditions, as nitrite can accumulate under various stress conditions (Bollag and Kurek, 1980), but ammonia oxidation is thought to be the rate-limiting step in nitrogen turnover in most terrestrial ecosystems (Prosser, 1989).

Adverse-effect scenario: Nitrifying bacteria need an aerobic environment, and are frequently found as colonies on the surface of soil aggregates. Bt toxins released from plant residues and roots will be adsorbed onto these same particles. This proximity could have consequences for the rates of nitrate formation, specifically ammonia oxidization, in soils cultivating Bt plants. Previous works (Wheatley *et al.*, 2001; Mendum and Hirsch, 2002) have shown that nitrification rates are particularly susceptible to changes in carbon inputs, particularly proteins.

Testable hypothesis 3: Inorganic nitrogen-transformation rates will be reduced in soils with Bt-cotton residues, with consequent effects on soil fertility and subsequent crop production.

Role of macroorganisms in decomposition of organic matter

Although ultimately organic matter decomposition is effected by microorganisms, the dynamics of this process are greatly increased by the activities of many macrofauna in many ecological niches. Although little information is available on the macrofaunal diversity of the cotton-cultivation areas in Brazil, it can be assumed that they also play an equally important role there. We propose initial basal studies to validate an indicator model to confirm this. This is based on examining one of the macrofaunal functional groups described previously: the disseminators. In a worst-case scenario, other groups can be added to complement the picture.

We suggest the use of earthworms as a model example for the examination of any possible effects of Bt-cotton cultivation on macrofaunal activity. Zwahlen *et al.* (2003b) reported no lethal effects of Bt maize litter on immature and mature *Lumbricus terrestris*, but a significant weight loss compared to worms eating litter of the non-transformed isoline in temperate maize production. The earthworms excrete the Bt toxin in a concentrated form in their casts. Casts from Bt maize-fed earthworms were found to be toxic to the Lepidopteran tobacco hornworm (*Manduca sexta*) (Saxena and Stotzky, 2001b).

Adverse-effect scenario: Earthworms may be sublethally affected by the Bt protein in the residues and soil they consume, and so their roles as incorporators and disseminators of plant material in the soil might be affected, which may have a slowing effect on the rate of decomposition of organic matter in the soil of Bt-cotton fields. They may also cause tritrophic effects on their predators from Bt in their guts, and on detritivores via the Bt toxin in their casts.

Impacts on residue-eating macroorganisms in soil will be dependent on the nutritional characteristics of the transgenic plant compared to the control, as mentioned above; therefore this should be taken into account. As well as any immediate lethal effects on macrofaunal members of the soil community, long-term sublethal effects possibly limiting functioning need to be considered. We recommend that other components of the soil macro- and mesofauna are also assessed.

Testable hypothesis 4: Macrofaunal activity and interactions (e.g. earthworms) will be adversely affected with a resultant decrease in plant-residue diminution and incorporation rates.

Cotton pest damage and disease transmission in soil

Most of the economically significant pests of cotton in Brazil are not soil dwellers at any life stage (see Sujii *et al.*, Chapter 6, this volume). Some weevil larvae are root borers (e.g. *Chalcodermus niger* (Hustache) and *Eutinobothrus brasiliensis* (Hambleton) [Coleoptera: Curculionidae]) and some Hemipteran bugs are root suckers (e.g. *Atarsocoris brachiariae* (Becker) and *Scaptocoris castanea* (Perty) [Hemiptera: Cydnidae]). A few Lepidopteran pests pupate in the soil, but as this is a non-feeding stage it is expected that there will be no susceptibility to any Bt proteins present in the soil. Contrastingly, there are several significant soil-borne fungal diseases of cotton (see Fontes *et al.*, Chapter 2, this volume). For example, *Fusarium oxysporum* that enters cotton plants through damage caused by the root-knot nematode *Meloidogyne incognita*, and is most severe in sandy soils, and *Verticillium dahliae*, which is most common in neutral to alkaline silt and clay soils of arid regions, particularly in irrigated cotton. Most other fungal diseases are transmitted to the growing cotton plant from infected plant residues via wind, water or insects, e.g. *Rhizoctonia solani*. Bacterial blight, caused by the bacterium *Xanthomonas axonopodis* pv. *malvacearum,* which causes damage from the seedling stage onwards, is carried over in infested seed and in crop residues on the soil surface.

Adverse-effect scenario: Disease-causing fungi or bacteria may be directly affected by the Bt toxin inside plant residues on which they are actively growing or resting as a dormant phase, or they may be indirectly affected by Bt-cotton cultivation due to changes in rates of input and degradation of plant residues. They may be able to increase their survival rate, which could increase infection rates of Bt cotton and negatively impact yields, or they may have reduced survival rates, which could positively influence Bt-cotton cultivation by decreasing the incidence of fungal and bacterial disease.

Testable hypothesis 5: Inputs from Bt cotton will alter the survival and efficacy of soil-borne fungal and bacterial diseases of cotton, compared to the influence of the non-transformed isoline.

Microbial biodiversity in soil

Studies on a range of different transgenic crops and traits have found differences in microbial and fungal community structure in the transgenic plant rhizosphere, compared to the non-transgenic control (for a review see Dunfield and Germida, 2004). These studies have used recently developed methods for the direct assessment of the degree of structural diversity in microbial communities. These molecular techniques can describe microbial populations as a whole, or in constituent parts (e.g. fungi), or functional groups (e.g. nitrifiers), depending on the choice of primers. Variable regions of 16S ribosomal genes are amplified from environmental DNA extracts by polymerase chain reaction (PCR), using primers specifically targeted at conserved regions in the bacterial genome; the primer set used determines the specificity. It is recommendable that the primers are tested and, if necessary, new primers developed for each transgenic trait and cultivar to be tested. The products can then be separated into different DNA bands on a gel by differential gradient gel electrophoresis (DGGE) or temperature gradient gel electrophoresis (TGGE) (Muyzer and Smalla, 1998). A similar technique is terminal restriction fragment length polymorphism (T-RFLP) (Liu *et al.*, 1997; Lukow *et al.*, 2000; Osborn *et al.*, 2000; Blackwood *et al.*, 2003).

Such methods overcome some of the limitations inherent in using culture techniques for identification and population descriptions, namely that only a small fraction of the microbial community can be successfully cultured in the laboratory and these may not necessarily be the most common components in that population (Hugenholtz *et al.*, 1998). Investigations of whole microbial population DNA by DGGE have shown changes in the whole soil population profile during the crop-growing season and under different cultivated crops (Smalla *et al.*, 2001; Pennanen *et al.*, 2003). This is shown by changes in the DGGE profiles of bands relating to taxonomic or functional groups of microbes in the population (Fig. 10.3). The methods are not quantitative, but the presence or absence of bands shows an appearance or disappearance of a group, and differences in the intensity of bands indicate that bacterial numbers are changing even if the group is still present. The banding patterns can be compared between the transgenic plant and non-transgenic controls, taking care to compare the same field sites and stages of crop growth. If comparing different soils, field sites and stages of crop growth, due allowance must be given to the amount of variation this causes compared to what the magnitude of a possible effect might be. It is to be anticipated that such protocols will show differences between different plants and also between the same plant when cultivated in different soils or under a different management system. However, data from experimentation with non-transformed plants show that these are transient changes that are dynamically responsive to the conditions prevailing at the time of sampling (Gomes *et al.*, 2001, 2003).

Additional information can be obtained from community-level physiological profiling (CLPP) using Biolog plates (Griffiths *et al.*, 2000; Buyer *et al.*, 2001). The Biolog method is based on the growth on a variety of substrates, so gives some information on both species and functional diversity. However, as the functions detected in the plates can be performed by one of several species, any loss of species biodiversity will not be seen. Phospholipids fatty acid profiles (PLFA) (Blackwood and Buyer, 2004) can also be used to examine the phenotypic composition of communities. As different subsets of the microbial community have differing PLFA profiles, changes in the ratios between them can be determined. Preferably, such methods should be targeted to specific known functional groups or organisms, and not used for general community analysis.

Changes in microbial community structure may have adverse effects on functional dynamics, especially those associated with decomposition and nutrient cycling. Many soil macrofauna feed on the microbial community,

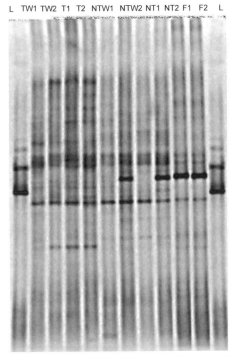

Fig. 10.3. DGGE banding pattern of *rpo*B PCR amplification of soil samples (0–5 cm depth). L = marker (from top to bottom *Staphylococcus aureus* MB, *Bacillus subtilis* IS 75, *Escherichia coli* HB101); TW = tillage with winter cover crop; T = tillage without winter cover crop; NTW = no-tillage with winter cover crop; NT = no-tillage without winter cover crop; F = native forest; 1 = first sampling; 2 = second sampling. Image from Raquel Peixoto, MSc dissertation, Federal University of Rio de Janeiro.

and so soil macroinvertebrates may also be adversely affected if microbial diversity is reduced. It is important to consider that microbial communities with similar structures as determined by these methods may still have ecologically significant differences in species composition, as the methods are not generally sensitive to changes in community structure at the individual strain or species level. The methods may only assess any changes in the numerically dominant portions of the microbial populations in soil. Minority microbial populations may not be represented as the template DNAs from these populations will form only a small fraction of that from the total community, and so are not adequately amplified to levels that can be detected above the background (Liu *et al.*, 1997). Therefore, these methods may not provide definitive answers to specific adverse-effect hypotheses, but indicate changes in community structure that may or may not have consequences on soil functional dynamics.

Changes in microbial communities are not permanent, as further changes in the bacterial community structure will occur if another type of crop replaces that grown before, or if the soil remains fallow. However, exposure of the microbial community will continue whilst the transgenic crop residues remain in the soil, and this may be the case over a relatively long time period after the crop is harvested (see persistence above), particularly as cotton is grown repeatedly year after year in many of the cotton-producing areas in Brazil.

Prerelease Experimental Designs for Potential Effects or Effect Pathways

Experimental protocols were designed to provide scientific evidence of the magnitude and duration of any effects of the Bt toxin on the soil ecosystem. These encompass appropriate spatial and temporal scales, sufficient replication, samples and relevant controls so that statistical power is not an issue for interpretation of the results. Because such a wide range of soil ecosystem processes might be affected in some way by changes in any inputs to the system, these experimental processes should be begun early in the development of any transgenic plant at the stage where the plant is being characterized in the laboratory, before field release (Andow *et al.*, Chapter 1, this volume). Some experiments can be carried out using microbially produced toxin, before the transgenic plant material is available. However, caution is urged when comparing the results, as the addition of plant material to soil has a strongly stimulatory effect on microbial activity, which the purified protein will not have (Donegan *et al.*, 1995; Palm *et al.*, 1996), and results may not be comparable for other reasons. It is important to note that on some occasions significant and contradictory differences have been found between experiments with the purified Bt toxin, even if this is identical to the Bt toxin that is produced in the transformed plant, and experiments using Bt plant material. Further experiments can be carried out using transgenic plants and soils in the greenhouse. Protocols to address the hypotheses in this chapter are described below (see Birch *et al.*, 2004, for further experimental protocols for laboratory and greenhouse).

We recommend that representative examples of all the alternate systems (chemical control, biological control and transgenic) are compared as multiple controls for the assessment of Bt cotton in Brazil. We also recommend that several local non-transgenic cotton cultivars are used as controls, as well as the non-transgenic isoline, provided it is available. The Bt-cotton variety to be tested should be locally adapted, with sufficient non-transgenic host plant resistance to the most damaging diseases and pests in each region. If a non-locally adapted variety is the only one available, it is necessary to consider that the plants may be disproportionately affected by disease and pest damage in comparison with the local non-transgenic varieties and this may influence the results.

It is important that investigations of the impact of transgenic plants sample at various points in the crop-growing season, so as to cover the variation in the microbial community as the crop develops (Smalla *et al.*, 2001; Heuer *et al.*, 2002; Dunfield and Germida, 2003). The spatial heterogeneity of the soil ecosystem is also an important issue (e.g. Lukow *et al.*, 2000). Although the majority of cotton cultivation in Brazil is on clay soils, a true representation of the other soil types, particularly those with low clay or high organic matter content, must also be examined.

The authors propose that the following experimental protocols are all carried out with material from the same series of incubated samples. As well as reducing the workload this will remove, or at least reduce, variations in the sample material caused by the inherent heterogeneity of the soils. Soils will be collected from cotton fields at:

1. Three sites in the Midwest region.
2. One site in the North-east region.
3. Two sites in the Meridian region.

This gives a total of six different representative soils from the three regions. Collect soil from the rooting zone or the equivalent in the field, from beneath cotton plants and unplanted areas. Take from several positions in the field, at least five points. Pass through a 4 mm sieve, bulk, homogenize and transport to the laboratories.

Amendments: Determine the amounts of residues entering the system at the sites from which the soils were taken. Then add Bt and non-Bt residues, from isolines, to the soils at the following rates:

1. 0.0 × field rate
2. 0.5 × field rate
3. 1.0 × field rate
4. 2.0 × field rate
5. 3.0 × field rate

The field rate is determined by measuring the amount of cotton crop residue in the field; it is anticipated that this will be about 8 t/ha (J.O. Siqueira, Brasília, Brazil, June 2004, unpublished data). Residues are placed on the soil surface and also mixed with the soil. Standard incubation conditions should be used for all six soils: adjust soil moisture content to 70% of soil pore space, incubate at a diurnal cycle of 25°C (12 h) and 17°C (12 h), use 5–10 kg of soil (dry weight) in a lysimeter, incubate for at least 360 days (Saxena and Stotzky, 2003). Subsamples can be taken at 0, 7, 15, 30, 60, 90, 120, 150, 180 and 360 days, depending on the requirements of specific analyses. Sample size is 5% of the total soil. At the same time, collect the leachates from the lysimeters, measure and record the volume. Incubate five replicates of each treatment.

Verification of exposure pathway: How long will Bt proteins in plant residues and Bt proteins released into the soil from plant residues and from the roots persist in the soil? Do they persist to the next cropping season?

Experiment: concentration and persistence of Bt proteins in soil, laboratory determinations

Monitor Bt protein concentrations in the incubated soils and associated leachates over time, as described above. Incubations should be continued until all the Bt proteins have apparently disappeared. Five replicates are required, using both soils that have previously cultivated either Bt cotton or the isoline and unamended soils of both types to be used as controls. Bt protein concentrations can be evaluated by several methods:

1. Bt protein presence using lateral flow quickstix. These give a qualitative or sometimes semi-quantitative result (Palm *et al.*, 1994, 1996; Head *et al.*, 2002; Herman *et al.*, 2002; Saxena and Stotzky, 2003).
2. Bt protein concentrations by ELISA (Zwahlen *et al.*, 2003a).
3. Bt protein identification by capillary electrophoresis.
4. Bt protein toxicity by bioassays; grow a target Lepidopteran larva that is highly susceptible to the Cry protein being examined, such as *M. sexta* for Cry1Ac, in the test soils at various time points during the soil incubations (e.g. Saxena *et al.*, 2004; Flores *et al.*, 2005).

Measured endpoints: Measure rates of both input and decline of the Bt proteins to estimate persistence. Determine the median lethal dose (LD_{50}) (Saxena and Stotzky, 2001a).

Testable hypothesis 1: The rate, timing or magnitude of the decomposition of Bt-cotton plant residues will be altered compared to decomposition rates of conventional cotton, due to impacts of the residues on microbial activity.

Experiment: assessment of microbial activity and decomposition rate

Microbial activity can be assessed using the same experimental procedure above and using techniques to measure: (i) substrate-induced respiration (SIR); and (ii) estimation of cellulolytic enzyme activity to indicate rates of breakdown of plant material, using techniques described in Birch *et al.* (2004) (Gilligan and Reese, 1954; Miller, 1959; Anderson and Domsch, 1978).

Measured endpoint: Rates of carbon release and cellulase activity per gram of soil in the differently treated soils. Impacts on the microbial community can also be investigated using the molecular techniques described earlier.

Testable hypothesis 2: Ammonification rates will be reduced in soils with Bt-cotton residues, with consequent effects on soil fertility and subsequent crop production.

Experiment: measurement of microbial ammonification rates

Estimate the mineralization potential by the waterlogged incubation method of Waring and Bremner (1964). Add 12.5 ml of soil to 5 g of soil in an incubation jar. Seal and incubate at 40°C for 7 days. Then add 12.5ml of 4 M KCl, shake for 1 h, filter through Whatman no. 1 paper and analyse for NH_4^+–N.

Measured endpoint: Comparative rates of NH_4^+–N formation will indicate whether the presence of the Bt proteins affect rates of organic matter breakdown and subsequent mineralization.

Testable hypothesis 3: Inorganic nitrogen-transformation rates will be reduced in soils with Bt-cotton residues, with consequent effects on soil fertility and subsequent crop production.

Experiment: measurement of nitrification rates

Potential nitrification rates can be estimated by the method of Belser and Mays (1980). Amend 25 g of each soil sample with $(NH_4)_2SO_4$ and $NaClO_3$ solutions to give a final concentration of 4 mM and 15 mM, respectively. Incubate at 20°C for 48 h.

Measured endpoint: Nitrification rates are calculated from the rate of accumulation of NO_2^-–N over time. This will show if the transgenic line has a significant effect on the rate at which nitrogen becomes available to the plant.

Testable hypothesis 4: Macrofaunal activity and interactions (e.g. earthworms) will be adversely affected with a resultant decrease in plant-residue diminution and incorporation rates.

Example experiment: impact of transgenic plant material on earthworms

Methods have been developed for the standardized testing of the impact of transgenic plants on earthworms in both field soil and laboratory incubations (Zwahlen *et al.*, 2003b). Worms are introduced into containers filled with subsamples of the various soils from the lysimeters. Mortality and individual weights are recorded every 40 days over the 360-day period.

Measured endpoint: Survival and development rates of the earthworms will indicate whether the residues of the Bt cotton had an adverse effect on the growth and development of these disseminators, which has consequences for nutrient recycling and crop growth.

Modifications of this method can be used to study the effects of Bt residues and soils on other macrofaunal components, as further information on the macrofauna found in Brazilian cotton-cultivation agroecosystems becomes available.

Testable hypothesis 5: Inputs from Bt cotton will alter the survival and efficacy of soil-borne fungal and bacterial diseases of cotton, compared to the influence of the non-transformed isoline.

Experiment: assessment of the pathogenicity of soil in which Bt cotton has been cultivated

Cultivate cotton plant seedlings from a susceptible cultivar in soils in which Bt cotton and the isoline have been grown. Transplant the young cotton seedlings into five replicate pots of each soil type and grow for 16 weeks under ideal conditions in a controlled environment chamber. Then assess the relative degree of disease development.

Measured endpoint: If there is no difference in the degree of pathogen development between plants cultivated in either soil, then the Bt proteins have had no significant effect on pathogen survival in the soil.

Discussion and Suggestions for Further Development

The cultivation of Bt cotton might have significant implications for functional dynamics in soil ecosystems. Crop production is reliant on the successful recycling of plant nutrients in the soil. Such nutrient cycling is entirely dependent on microbial and macrofaunal functional activities, which in turn obtain their energy requirements from the plants. This interdependence is partly shaped by the kind of plant residues and other inputs entering the soil. Changes in the crop type, species or cultivar, whether transgenic or not, will affect the functional dynamics of the soil ecosystem. However, those changes occurring as a consequence of the cultivation of Bt crops, which express relatively large amounts of a novel protein in their tissues, may possibly be more profound. We now understand that agricultural practices such as tillage and herbicide use, and environmental factors such as sampling date and soil type, have strong effects on microbial diversity (Fig. 10.3). Studies on changes in community structure caused by transgenic plants in temperate regions have led to the conclusion that these transient changes have no lasting impact, because they have not persisted after winter; the structural composition returns to the same state as the previous spring (Dunfield and Germida, 2004). However, the situation in a tropical climate and soil may be very different and conclusions from studies in temperate regions may have only limited relevance. A consequence of major concern was what effects the active Bt toxins may have on both the microbial and macrofaunal populations in soil ecosystems. Several hazard scenarios and hypotheses to assess the impacts of Bt cotton on soil ecosystems in Brazil were identified. The major theme in all of these was the possibility that the Bt toxins may have adverse effects on biodiversity and the functional dynamics in the soil.

Soil ecosystem populations can be described on a population constituent basis, to type, genera or species level, or on a functional basis, in which the ability to perform certain events is defined. These two population characteristics are not necessarily related; in particular, the amount of function being expressed at a particular time is most probably not related quantitatively to the numerical contribution of individuals with that ability to the total population at that time. Soil ecosystems contain tens of thousands of species, and consequently describing populations in this way is a very difficult and complex operation. An alternative approach is to define such communities by what the community and its parts do, i.e. by defining functional properties. Functions generally can be associated with many individual species, and so provide a manageable way of describing populations. Biodiversity needs to be considered here. As only a few of the many species capable of a particular function are required at any particular time to maintain that function, functional measurements alone may obscure a change in the variety of different individual species with that characteristic in the community. Although a definitive definition of the extent to which such reductions in biodiversity can be tolerated before system sustainability is compromised is difficult to determine, it is thought best practice to sustain as high a level of biodiversity in the system as possible.

We recommend that the impacts of Bt cotton on the soil ecosystem be assessed by a combination of both functional and structural assays. As key microbial and macrofaunal functional groups are responsive to plant inputs and cultural conditions that are involved in nutrient cycling, and so soil fertility, they have great potential for use in determining any impacts on soil ecosystems. The effects of transgenic plants on soil ecosystems will therefore be best studied by the application of several basic and well-proven microbial and macrofaunal activity assays. Experimental protocols were devised to assess Bt toxin persistence in soils, changes in microbial population biodiversity, and effects on plant residue decomposition and subsequent nitrogen-transformation rates. The measured endpoints from these will enable crop-management decisions that aim towards sustainable and ecologically sound Bt-cotton cultivation in Brazil.

References

Anderson, J.P.E. and Domsch, K.H. (1978) A physiological method for the quantitative measurement of microbial biomass in soils. *Soil Biology and Biochemistry* 10, 215–221.

Andow, D.A. and Hilbeck, A. (2004) Science-based risk assessment for nontarget effects of transgenic crops. *Bioscience* 54(7), 637–649.

Belser, L.W. and Mays, E.L. (1980) Specific inhibition of nitrite oxidation by chlorate and its use in assessing nitrification in soils and sediments. *Applied and Environmental Microbiology* 39(3), 505–510.

Birch, A.N.E., Wheatley, R.E., Anyango, B., Arpaia, S., Capalbo, D., Getu Degaga, E., Fontes, E., Kalama, P., Lelmen, E., Lövei, G., Melo, I.S., Muyekho, F., Ngi-Song, A., Ochiendo, D., Ogwang, J., Pitelli, R., Sétamou, M., Sithanantham, S., Smith, J., Nguyen Van Son, Songa, J., Sujii, E., Tran, Q.T., Wan, F.-H. and Hilbeck, A. (2004) Biodiversity and non-target impacts: a case study of Bt maize in Kenya. In: Hilbeck, A. and Andow, D.A. (eds) *Environmental Risk Assessment of Transgenic Organisms: A Case Study of Bt Maize in Kenya.* CAB International, Wallingford, UK.

Blackwood, C.B. and Buyer, J.S. (2004) Soil microbial communities associated with Bt and non-Bt corn in three soils. *Journal of Environmental Quality* 33, 832–836.

Blackwood, C.B., Marsh, T., Kim, S.-H. and Paul, E.A. (2003) Terminal restriction fragment length polymorphism data analysis for quantitative comparison of microbial communities. *Applied Environmental Microbiology* 69, 926–932.

Bollag, J.M. and Kurek, E.J. (1980) Nitrite and nitrous oxide accumulation during denitrification in presence of pesticide derivatives. *Applied and Environmental Microbiology* 39, 845–849.

Bruinsma, M., Veen, J.A. van and Kowalchuk, G.A. (2002) *Effects of Genetically Modified Plants on Soil Ecosystems.* Concept report for the Committee Genetic Modification (COGEM), Netherlands Institute of Ecology, Heteren, The Netherlands.

Brusetti, L., Fracia, P., Bertolini, C., Pagliuca, A.S.B. *et al.* (2004) Bacterial communities associated with the rhizosphere of transgenic Bt 176 maize (*Zea mays*) and its non-transgenic counterpart. *Plant and Soil* 266, 11–21.

Buyer, J.S., Roberts, D.P., Millner, P. and Russek-Cohen, E. (2001) Analysis of fungal communities by sole carbon utilization profiles. *Journal of Microbiology Methods* 45, 53–60.

Crecchio, C. and Stotzky, G. (1998) Insecticidal activity and biodegradation of the toxin from *Bacillus thuringiensis* subsp. *kurstaki* bound to humic acids from soil. *Soil Biology and Biochemistry* 30(4), 463–470.

Donegan, K.K., Palm, C.J., Fieland, V.J., Porteous, L.A., Ganio, L.M., Schaller, D.L., Bucao, L.Q. and Seidler, R.J. (1995) Changes in levels, species and DNA fingerprints of soil micro-organisms associated with cotton expressing the *Bacillus thuringiensis* var. *kurstaki* endotoxin. *Applied Soil Ecology* 2, 111–124.

Donegan, K.K., Seidler, R.J., Fieland, V.J., Schaller, D.L., Palm, C.J., Ganio, L.M., Cardwell, D.M. and Steinberger, Y. (1997) Decomposition of genetically engineered tobacco under field conditions: persistence of the proteinase inhibitor I product and effects on soil microbial respiration and protozoa, nematode and microarthropod populations. *Journal of Applied Ecology* 34, 767–777.

Dunfield, K.E. and Germida, J.J. (2001) Diversity of bacterial communities in the rhizosphere and root interior of field-grown genetically modified *Brassica napus*. *FEMS Microbiology Ecology* 82, 1–9.

Dunfield, K.E. and Germida, J.J. (2003) Seasonal changes in the rhizosphere microbial communities associated with field-grown genetically modified canola (*Brassica napus*). *Applied and Environmental Microbiology* 69(12), 7310–7318.

Dunfield, K.E. and Germida, J.J. (2004) Impact of genetically modified crops on soil- and plant-associated microbial communities. *Journal of Environmental Quality* 33, 806–815.

Ekschmidt, K., Wolters, V. and Weber, M. (1997) Spiders, carabids, and staphylinids: the ecological potential of predatory macroarthropods. In: Benckiser, G. (ed.) *Fauna in Soil Ecosystems. Recycling Processes, Nutrient Fluxes, and Agricultural Production.* Marcel Dekker, New York, pp. 307–362.

Escher, N., Käch, B. and Nentwig, W. (2000) Decomposition of transgenic *Bacillus thuringiensis* maize by micro-organisms and woodlice *Porcellio scaber* (Crustacea: Isopoda). *Basic and Applied Ecology* 1, 161–169.

Feng, Y., Motta, A.C., Reeves, D.W., Burmester, C.H., Santen, E. van and Osborne, J.A. (2003) Soil microbial communities under conventional-till and no-till continuous cotton systems. *Soil Biology and Biochemistry* 35, 1693–1703.

Flores, S., Saxena, D. and Stotzky, G. (2005) Transgenic Bt plants decompose less in soil than non-Bt plants. *Soil Biology and Biochemistry* 37(6), 1073–1082.

Fountain, M.T. and Hopkin, S.P. (2005) *Folsomia candida* (Collembola): a 'standard' soil arthropod. *Annual Review of Entomology* 50, 201–222.

Gilligan, W. and Reese, E.T. (1954) Evidence for multiple components in microbial cellulases. *Canadian Journal of Microbiology* 1, 90–107.

Gomes, N.C.M., Heuer, H., Schonfeld, J., Costa, R., Mendonça-Hagler, L.C. and Smalla, K. (2001) Bacterial diversity of the rhizosphere of maize (*Zea mays*) grown in tropical soil studied by temperature gradient gel electrophoresis. *Plant and Soil* 232, 167–180.

Gomes, N.C., Fagbola, O., Costa, R., Rumjanek, N.G., Buchner, A., Mendonça-Hagler, L.C. and Smalla, K. (2003) Dynamics of fungal communities in bulk and maize rhizosphere soil in the tropics. *Applied and Environmental Microbiology* 69, 3758–3766.

Grayston, S.J., Wang, S., Campbell, C.D. and Edwards, A.C. (1998) Selective influence of plant species on microbial diversity in the rhizosphere. *Soil Biology and Biochemistry* 30, 369–378.

Griffiths, B.S., Geoghegan, I.E. and Robertson, W.M. (2000) Testing genetically engineered potato, producing the lectins GNA and Con A, on non-target soil organisms and processes. *Journal of Applied Ecology* 37, 159–170.

Gupta, V.V.S.R., Roberts, G.N., Neate, S.M., McClure, S.G., Crisp, P. and Watson, S.K. (2002) Impact of Bt-cotton on biological processes in Australian soils. In: Akhurst, R.J., Beard, C.E. and Hughes, P.A. (eds) *Biotechnology of Bacillus thuringiensis and its Environmental Impact*. CSIRO Entomology, Canberra, Australia, pp. 191–194.

Head, G., Subber, J.B., Watson, J.A., Martin, J.W. and Duan, J. (2002) No detection of Cry1Ac protein in soil after multiple years of transgenic Bt cotton (Bollgard) use. *Environmental Entomology* 31, 30–36.

Herman, R.A., Scherer, P.N. and Wolt, J.D. (2002) Rapid degradation of a binary, PSI49B1, δ-endotoxin of *Bacillus thuringiensis* in soil, and a novel mathematical model for fitting curvi-linear decay. *Environmental Entomology* 31, 208–214.

Heuer, H., Kroppenstedt, R.M., Lottmann, J., Berg, G. and Smalla, K. (2002) Effects of T4 lysozyme release from transgenic potato roots on bacterial rhizosphere communities are negligible relative to natural factors. *Applied and Environmental Microbiology* 68(3), 1325–1355.

Hopkins, D.W. and Gregorich, E.G. (2003) Detection and decay of the Bt endotoxin in soil from a field trial with genetically modified maize. *European Journal of Soil Science* 54, 793–800.

Hugenholtz, P., Goebel, B.M. and Pace, N.R. (1998) Impact of culture-independent studies on the emerging phylogenetic view of bacterial diversity. *Journal of Bacteriology* 180, 4765–4774.

Koskella, J. and Stotzky, G. (1997) Microbial utilization of free and clay-bound insecticidal toxins from *Bacillus thuringiensis* and their retention of insecticidal activity after incubation with microbes. *Applied and Environmental Microbiology* 63(9), 3561–3568.

Kowalchuk, G.A. and Stephen, J.R. (2001) Ammonia-oxidising bacteria: a model for molecular microbial ecology. *Annual Review of Microbiology* 55, 485–529.

Kowalchuk, G.A., Bruinsma, M. and Veen, J.A. van (2003) Assessing responses of soil micro-organisms to GM plants. *Trends in Ecology and Evolution* 18(8), 403–410.

Lachnicht, S.L., Hendrix, P.F., Potter, R.L., Coleman, D.C. and Crossley, D.A. Jr (2004) Winter decomposition of transgenic cotton residue in conventional-till and no-till systems. *Applied Soil Ecology* 27(2), 135–142.

Lartey, R., Curl, E.A. and Peterson, C.M. (1994) Interactions of mycophagous Collembola and biological control fungi in the suppression of *Rhizoctonia solani*. *Soil Biology and Biochemistry* 26(1), 81–88.

Liu, W.T., Marsh, T.L., Cheng, H. and Forney, L.J. (1997) Characterization of microbial diversity by determining terminal restriction fragment length polymorphisms of genes encoding 16S rRNA. *Applied Environmental Microbiology* 63, 4516–4522.

Lottmann, J., Heuer, H., Smalla, K. and Berg, G. (1999) Influence of transgenic T4 lysozyme-producing potato plants on potentially beneficial plant-associated bacteria. *FEMS Microbiology Ecology* 29, 365–377.

Lottmann, J., Heuer, H., Vries, J. de, Mahn, A., During, K., Wackernagel, W., Smalla, K. and Berg, G. (2000) Establishment of introduced antagonistic bacteria in the rhizosphere of transgenic potatoes and their effect on the bacterial community. *FEMS Microbiology Ecology* 33, 41–49.

Lukow, T., Dunfield, P.F. and Liesack, W. (2000) Use of the T-RFLP technique to assess spatial and temporal changes in the bacterial community structure within an agricultural soil planted with transgenic and non-transgenic potato plants. *FEMS Microbiology Ecology* 32, 241–247.

Mendum, T.A. and Hirsch, P.R. (2002) Changes in the population structure of beta-group autotrophic ammonia oxidising bacteria in arable soils in response to agricultural practice. *Soil Biology and Biochemistry* 34, 1479–1485.

Miller, G.L. (1959) Use of dinitro salicylic acid reagent for determination of reducing sugar. *Analytical Chemistry* 31, 429.

Muyzer, G. and Smalla, K. (1998) Application of denaturing gradient gel electrophoresis (TGGE) in microbial ecology. *Antonie van Leeuwenhoek* 73, 127–141.

OGTR (2003) *Risk Assessment and Risk Management Plan Consultation Version: Commercial Release of Insecticidal (INGARD® event 531) Cotton*. Office of the Gene Technology Regulator, Woden, Australia.

Osborn, A.M., Moore, E.R.B. and Timmis, K.N. (2000) An evaluation of terminal-restriction fragment length polymorphism (T-RFLP) analysis for the study of microbial community structure and dynamics. *Environmental Microbiology* 2, 39–50.

Palm, C.J., Donegan, K.K., Harris, D. and Seidler, R.J. (1994) Quantification in soil of *Bacillus thuringiensis* var. *kurstaki* δ-endotoxin from transgenic plants. *Molecular Ecology* 3, 145–151.

Palm, C.J., Schaller, D.L., Donegan, K.K. and Seidler, R.J. (1996) Persistence in soil of transgenic plant produced *Bacillus thuringiensis* var. *kurstaki* δ-endotoxin. *Canadian Journal of Microbiology* 42, 1258–1262.

Pennanen, T., Caul, S., Daniell, T.J., Griffiths, B.S., Ritz, K. and Wheatley, R.E. (2003) Community-level responses of metabolically active soil microbes to variations in quantity and quality of substrate inputs. *Soil Biology and Biochemistry* 36(5), 841–848.

Prosser, J.I. (1989) Autotrophic nitrification in bacteria. *Advances in Microbial Physiology* 30, 125–181.

Rantalainen, M.-L., Fritze, H., Haimi, J., Kiikkilä, O., Pennanen, T. and Setälä, H. (2004) Do enchytraeid worms and habitat corridors facilitate the colonisation of habitat patches by soil microbes? *Biological Fertility of Soils* 39, 200–208.

Saxena, D. and Stotzky, G. (2000) Insecticidal toxin from *Bacillus thuringiensis* is released from roots of transgenic Bt corn in vitro and in situ. *FEMS Microbiology Ecology* 33, 35–39.

Saxena, D. and Stotzky, G. (2001a) Bt toxin uptake from soil by plants. *Nature Biotechnology* 19, 199.

Saxena, D. and Stotzky, G. (2001b). *Bacillus thuringiensis* (Bt) toxin released from root exudates and biomass of Bt corn has no apparent effect on earthworms, nematodes, protozoa, bacteria, and fungi in soil. *Soil Biology and Biochemistry* 33(9), 1225–1230.

Saxena, D. and Stotzky, G. (2001c) Bt corn has a higher lignin content than non-Bt corn. *American Journal of Botany* 88(9), 1704–1706.

Saxena, D. and Stotzky, G. (2003) Fate and effects in soil of the insecticidal toxins from *Bacillus thuringiensis* in transgenic plants. *Biosafety Reviews* 1, ICGEB, Trieste, Italy.

Saxena, D., Flores, S. and Stotzky, G. (1999) Insecticidal toxin in root exudates from Bt corn. *Nature* 402, 480.

Saxena, D., Flores, S. and Stotzky, G. (2002) Vertical movement in soil of insecticidal Cry1Ab protein from *Bacillus thuringiensis*. *Soil Biology and Biochemistry* 34, 111–120.

Saxena, D., Stewart, C.N., Altosaar, I., Shu, Q. and Stotzky, G. (2004) Larvicidal Cry proteins from *Bacillus thuringiensis* are released in root exudates of transgenic *B. thuringiensis* corn, potato, and rice but not of *B. thuringiensis* canola, cotton, and tobacco. *Plant Physiology and Biochemistry* 42, 383–387.

Scheu, S. (2001) Plants and generalist predators as links between the below-ground and above-ground system. *Basic and Applied Ecology* 2, 3–13.

Seniczak, S. and Stefaniak, O. (1981) The effect of fungal diet on the development of *Oppia nitens* (Acari, Oribatida) and on the microflora of its alimentary tract. *Pedobiologia* 21, 202–210.

Settle, W.H., Ariawan, H., Tri Asuti, E., Cahyana, W., Hakim, A.L., Hindayana, D., Sri Lestari, A. and Sartano, P. (1996) Managing tropical rice pests through conservation of generalist natural enemies and alternative prey. *Ecology* 77, 1975–1988.

Siciliano, S.D., Theoret, C.M., Freitas, J.R. de, Hucl, P.J. and Germida, J.J. (1998) Differences in the microbial communities associated with the roots of different cultivars of canola and wheat. *Canadian Journal of Microbiology* 44, 844–851.

Sims, S.R. and Ream, J.E. (1997) Soil inactivation of the *Bacillus thuringiensis* subsp. *kurstaki* CryIIA insecticidal protein within transgenic cotton tissue: laboratory microcosm and field studies. *Journal of Agricultural and Food Chemistry* 45, 1502–1505.

Smalla, K., Wieland, G., Buchner, A., Zock, A., Parzy, J., Kaiser, S., Roskot, N., Heuer, H. and Berg, G. (2001) Bulk and rhizosphere soil bacterial communities studied by denaturing gradient gel electrophoresis: plant-dependent enrichment and seasonal shifts revealed. *Applied and Environmental Microbiology* 67, 4742–4751.

Stotzky, G. (2000) Persistence and biological activity in soil of insecticidal proteins from *Bacillus thuringiensis* and of bacterial DNA bound on clays and humic acids. *Journal of Environmental Quality* 29, 691–705.

Tapp, H. and Stotzky, G. (1995) Dot blot enzyme-linked immunosorbent assay for monitoring the fate of insecticidal toxins from *Bacillus thuringiensis* in soil. *Applied and Environmental Microbiology* 61, 602–609.

Tapp, H. and Stotzky, G. (1998) Persistence of the insecticidal toxin from *Bacillus thuringiensis* subsp. *kurstaki* in soil. *Soil Biology and Biochemistry* 30, 471–476.

Tapp, H., Calamai, L. and Stotzky, G. (1994) Adsorption and binding of the insecticidal proteins from *Bacillus thuringiensis* subsp. *kurstaki* and subsp. *tenebrionis* on clay minerals. *Soil Biology and Biochemistry* 26(6), 663–679.

van Elsas, J.D., Trevors, J.T. and Wellington, E.M.H. (1997) *Modern Soil Microbiology*. Marcel Dekker, New York.

Venkateswerlu, G. and Stotzky, G. (1992) Binding of the protoxin and toxin proteins of *Bacillus thuringiensis* subsp. *kurstaki* on clay minerals. *Current Microbiology* 25, 225–233.

Waring, S.A. and Bremner, J.M. (1964) Ammonium production in soil under waterlogged conditions as an index of nitrogen availability. *Nature* 201, 951–952.

Wheatley, R.E., Hackett, C.A., Bruce, A. and Kundzewicz, A. (1997) Effect of substrate composition on production and inhibitory activity against wood decay fungi of volatile organic compounds from *Trichoderma* spp. *International Biodeterioration and Degradation* 39, 199–205.

Wheatley, R.E., Ritz, K., Crabb, D. and Caul, S. (2001) Temporal variations in potential nitrification dynamics related to differences in rates and types of carbon inputs. *Soil Biology and Biochemistry* 33, 2135–2144.

Wheatley, R.E., Caul, S., Ritz, K., Daniell, T.J., Crabb, D. and Griffiths, B.S. (2003) Microbial population dynamics related to temporal variations in nitrification function in a field soil. *European Journal of Soil Science* 54, 707–714.

Yu, L., Berry, R.E. and Croft, B.A. (1997) Effects of *Bacillus thuringiensis* toxins in transgenic cotton and potato on *Folsomia candida* (Collembola: Isotomidae) and *Oppia nitens* (Acari: Oribatida). *Journal of Economic Entomology* 90(1), 113–118.

Zwahlen, C., Hilbeck, A., Gugerli, P. and Nentwig, W. (2003a) Degradation of the Cry1Ab protein within transgenic *Bacillus thuringiensis* corn tissue in the field. *Molecular Biology* 12, 765–775.

Zwahlen, C., Hilbeck, A., Howald, R. and Nentwig, W. (2003b) Effects of transgenic Bt-corn litter on the earthworm *Lumbricus terrestris*. *Molecular Ecology* 12, 1077–1086.

11 Assessing Gene Flow from Bt Cotton in Brazil and its Possible Consequences

J.A. Johnston, C. Mallory-Smith, C.L. Brubaker, F. Gandara, F.J.L. Aragão, P.A.V. Barroso, Vu Duc Quang, L.P. de Carvalho, P. Kageyama, A.Y. Ciampi, M. Fuzatto, V. Cirino and E. Freire

Corresponding author: Dr Jill Johnston, Plant Biology, University of Minnesota, 250 Biological Sciences Centre, 1445 Gortner Avenue, St Paul, MN 55108, USA. e-mail: jillwest@utah.gov

Introduction

The potential environmental risks of *Bacillus thuringiensis* (Bt) cottons arise from the Bt gene that has been genetically engineered into these cottons and is absent from the cotton cultivars currently grown in Brazil. This chapter focuses on whether environmental harm could result if the Bt gene was unintentionally transferred to wild *Gossypium* species, cotton landraces, dooryard cottons, feral cottons or crop volunteers.

In the first instance, introgression of the Bt gene into non-transgenic cottons could possibly exacerbate the threat to the genetic integrity of Brazil's rich diversity of cotton landraces and dooryard cottons as well as the endemic wild species, *Gossypium mustelinum*. The indigenous landraces and dooryard cottons are critical components of low-input cotton cultivation in some regions of Brazil, and are invaluable resources for Brazilian cotton breeders who seek to develop locally adapted cultivars. Their genetic integrity is already at risk due to gene flow from elite cotton cultivars, and it is critical to ascertain whether Bt cottons will increase this risk. *G. mustelinum* is a wild cotton relative endemic to Brazil, and is already at risk of extinction. Secondly, introgression of the Bt gene to organic and other non-transgenic cotton cultivars could have significant socio-economic consequences. While the lint of the initial pollen recipients is arguably non-transgenic (being a maternal tissue), market perceptions will probably be insensitive to this scientific nuance. More importantly, the seed planted the following year will carry the Bt gene. There is evidence that there is a low level of transgene contamination already present in cottonseed that is currently being planted in Brazil (see Andow *et al.*, Chapter 1, this volume) and the concern would increase if Bt cotton were planted on a wide scale in the country. Seed contamination would be particularly damaging to the organic cotton industry, where planting seed is obtained from the previous year's crop. Finally, there is the real concern that the Bt gene would impart a selective advantage to introgressed populations or Bt crop volunteers and increase their colonizing ability. This, in turn, may allow these introgressed populations or crop volunteers to replace other feral, landrace or dooryard cottons and most critically, *G. mustelinum*.

We applied a general scientific risk-assessment model to the question of gene-flow risks in the event that Bt cotton were introduced (Box 11.1). The framework has three parts that examine: (i) factors that affect the likelihood of gene flow; (ii) factors that affect the likelihood of transgene establishment and spread; and (iii) the potential adverse consequences of gene flow.

Box 11.1. The general form of the GMO Guidelines gene flow section.

1. *What factors affect the likelihood of intra- and interspecific gene flow?*

    1.1. Are related taxa present in the region?
        1.1.1. In what regions is the crop likely to be cultivated?
        1.1.2. What is the frequency and distribution of closely related species within this area?

    1.2. To what extent does the crop species cross-pollinate with other plantings of the same crop?
        1.2.1. To what extent does the crop outcross?
            1.2.1.1. Are male-sterile lines used to produce hybrid seed? (Male-sterile plants will always cross-pollinate.)
        1.2.2. Over what distances does cross-pollination occur?

    1.3. To what extent do fertile hybrids occur between the crop and nearby relatives?
        1.3.1. Are hand-crosses successful?
            1.3.1.1. Does the direction of pollination affect the success of cross-pollination (e.g. crop-to-wild vs. wild-to-crop)?
        1.3.2. Does hybridization occur spontaneously under natural or experimental conditions?
            1.3.2.1. Have putative hybrid or backcrossed plants been observed?
            1.3.2.2. Have hybrid or backcrossed plants been confirmed using genetic markers?
        1.3.3. Over what distances can cross-pollination occur?

Box 11.1. The general form of the GMO Guidelines gene flow section – *cont'd.*

1.4. How fit are crop–weed hybrids?

(The fitness of hybrids could be lower, similar or higher; here, the term hybrid includes crosses between cultivated and wild forms of the same species, as well as interspecific crosses.)

1.4.1. Are F_1 hybrids fit enough to survive and reproduce while growing among wild relatives or crop plants?

1.4.2. Are F_2 or BC_1 hybrids fit enough to survive and reproduce while growing among wild relatives or crop plants?

1.4.3. What proportion of F_2 or BC_1 plants exhibit lifetime fecundity that is equal to or greater than the average fecundity of wild plants grown in the same experimental conditions?

1.4.4. How does hybrid fitness compare among different field locations, growing conditions and years?

1.4.5. Does hybrid fitness differ depending on whether the hybrid seed formed on a crop or wild plant?

1.4.6. Can interspecific hybrids persist as perennial plants without sexually reproducing?

    1.4.6.1. Could these perennial plants spread clonally?

1.5. Does the crop produce volunteers or establish feral populations?

1.5.1. Can feral populations persist for more than one or two growing seasons?

1.5.2. Do feral populations establish a seed bank that could potentially be a source of weedy plants for many years?

1.5.3. Are the plants perennial and, if so, are they clonal (e.g. grasses)?

1.5.4. Are volunteer or feral populations considered to be weedy in agricultural or unmanaged areas?

1.5.5. Do feral populations act as a 'genetic bridge', increasing the chances that crop genes will escape into populations of wild relatives?

2. *What is the likelihood that a specific transgene from the crop will increase in frequency following gene flow or establishment of feral populations?*

2.1. If the transgene spreads to non-transgenic crops, is it likely to persist in an agricultural setting?

2.1.1. Do crops with introgressed transgenes have higher yields or better quality than non-transgenic crops (e.g. due to resistance traits) that make them more likely to be selected and propagated by farmers?

2.1.2. Do transgenic crops produce more pollen than non-transgenic versions of the same crop?

2.1.3. Is the largest proportion of a cultivated area planted to transgenic crops, such that transgenic pollen will be more abundant and sire more seeds than non-transgenic plants growing nearby?

2.1.4. Is the transgene stably inherited over many generations and many crop backgrounds?

2.1.5. Are seeds or vegetative propagules dispersed from the crop, either naturally or by people?

    2.1.5.1. To what extent will transgenic seeds or vegetative propagules be transported among regions?

Continued

Box 11.1. The general form of the GMO Guidelines gene flow section – *cont'd.*

2.2. Is the transgene likely to spread and persist in free-living intra- or interspecific hybrid populations?

(Compare populations in farmers' fields, field margins and unmanaged areas, including habitats of any rare species that hybridize with the crop.)

2.2.1. If the transgene construct under consideration is available for research purposes, what are the fitness effects of the transgene when tested empirically in hybrid or backcrossed plants in comparison to appropriate control plants?

2.2.1.1. Is a fitness benefit associated with the transgene under natural or artificial selective pressures (e.g. natural levels of disease, herbivory, or herbicide use)?

2.2.1.1.1. What ecological factors have the greatest effect on components of fitness in wild populations, and how is the transgene likely to alter these effects?

2.2.1.1.2. Will the factors limiting the population size of wild relatives (herbivores, pathogens, etc.) limit populations that have introgressed the transgene?

2.2.1.2. Is a fitness cost associated with the transgene in the absence of selective pressures?

2.2.1.3. How do fitness comparisons between transgenic and non-transgenic wild relatives vary among different field locations, growing conditions and years?

2.2.1.4. Is the transgene stably inherited over many generations?

2.2.1.4.1. Is transgene expression consistent among different genetic backgrounds, generations and environmental conditions?

2.2.2. If transgenic crop plants are not available for experimental studies, can the effect of the transgene be mimicked (e.g. using insecticidal sprays)?

2.3. Is the transgene likely to spread and persist in feral crop populations?

2.3.1. Can feral populations redevelop seed dormancy that is often bred out of cultivated varieties?

2.3.1.1. Will seed banks establish, creating a source of feral transgenic plants for many years to come?

2.3.2. Do feral populations possess mechanisms for dispersing seed further than their relatives in cultivation?

2.3.2.1. If seeds have enhanced dispersal, are feral populations likely to invade new habitat?

2.3.2.2. Could spread of feral populations allow transgenes to come into contact with wild relatives that don't occur immediately in agricultural areas?

2.3.3. Could feral populations create a 'genetic bridge' through which crop genes are more likely to flow into wild populations than from direct crop–weed pollen transfer?

2.4. What possible variations on the direct, expected consequences of gene flow could affect the persistence and spread of transgenes in the environment?

2.4.1. Will the introduction of the transgenic promoter alone alter the phenotype of introgressed crops or weeds?

2.4.2. How will genetic background affect the likelihood of transgene silencing following introgression?

2.4.3. Could the transgene cause an increase in one fitness component, but overall reduction in fitness (e.g. the case with salmon growth hormone), creating extinction concerns for related taxa?

2.4.4. If multiple transgenes are 'stacked', can you assume that fitness effects will be additive?

3. *What potential ecological and agronomic effects could result from the spread and persistence of transgenes?*

3.1. How could the crop-to-crop spread of transgenes affect local agricultural production?

3.1.1. Are there possible effects of transgenic insect resistance on resistance management, integrated pest management or beneficial non-target species?

3.1.2. Are there possible effects of transgenic herbicide resistance on the efficacy of locally used herbicides (e.g. glyphosate)?

3.1.2.1. Could these types of effect reduce or enhance options for crop rotation or rotations of herbicides with different modes of action?

3.1.3. Are there other ways in which transgenic crops will alter local agricultural practices?

3.1.3.1. For example, will more growers use 'no-till' methods in conjunction with herbicide-resistant crops?

3.1.4. Will use of transgenic crops reduce negative environmental impacts of traditional agricultural practices using non-transgenic crops?

3.1.5. Could the spread of any type of transgene compromise the market value of non-transgenic crops?

3.1.6. Could transgenes that are not approved for human consumption spread to other plantings of the same crop?

3.2. How could the crop-to-crop spread of transgenes affect the genetic diversity of local landraces?

3.2.1. Is the area known to be a centre of origin for the crop?

3.2.2. Is there already gene flow from modern cultivars into non-transgenic landraces?

3.2.3. Could a strongly selected transgene or a highly popular transgenic crop lead to more rapid displacement of landrace gene pools in comparison to displacement that may already occur?

3.3. How could the crop-to-wild spread of transgenes affect the biology of wild relatives of the crop?

3.3.1. Could a strongly selected transgene reduce the genetic diversity of rare and non-weedy species, beyond effects that may already occur due to gene flow from non-transgenic crops?

Continued

Box 11.1. The general form of the GMO Guidelines gene flow section – *cont'd.*

    3.3.2. Could the transgenic trait allow free-living relatives to become more abundant within their typical habitats?
    3.3.3. Could the transgenic trait allow free-living relatives to occupy new ecological niches (e.g. due to cold or drought tolerance)?

3.4. How could the crop-to-wild spread of transgenes affect non-target species and biodiversity, either directly or indirectly?
    3.4.1. What is the transgene product, and what are the likely effects of this trait on other plants, herbivores and beneficial organisms in the region?
        3.4.1.1. Is the transgene product likely to affect multiple trophic levels?
        3.4.1.2. Will below-ground food webs be affected?
            3.4.1.2.1. Is there enough baseline data to make this determination?
    3.4.2. Does the transgene confer any unintentional but ecologically significant changes in the chemical composition of hybrid and backcrossed progeny?
        3.4.2.1. If so, are these effects likely to affect plant competition, plant–insect or plant–soil interactions?
    3.4.3. If free-living relatives of the crop become more widespread due to the transgene, are they likely to competitively displace any native plant species in their typical habitat?
        3.4.3.1. Could the increase in dominance by the transgenic weed destabilize the whole food web?
    3.4.4. If free-living relatives of the crop become more widespread due to the transgene, are they likely to competitively displace any native plant species by extending their native range?
        3.4.4.1. Will the food web or ecosystem processes in the new habitat be severely altered?

Factors that Affect the Likelihood of Gene Flow

Evolutionary and taxonomic background

The cotton genus, *Gossypium*, comprises approximately 44 diploid (2N = 26) and 5 allotetraploid (2N = 52) species (Fryxell, 1992), whose aggregate geographical range spans most tropical and subtropical regions of the world. There are three primary centres of diversity for the diploid *Gossypium* species in Australia, north-eastern Africa with Arabia and west-central and southern Mexico. This global radiation was accompanied by diversification in morphology, ecology and chromosomal homology. As a result, the diploid *Gossypium* species can be grouped into eight morphologically and cytogenetically distinct groups, called genomes, based on similarities in chromosome size and structure designated A through G, and also K (Beasley, 1940, 1942; Phillips and Strickland, 1966; Edwards and Mirza, 1979; Endrizzi *et al.*, 1985; Stewart, 1995). Chromosome homology within genomes is reasonably high and this is reflected in the normal meiotic pairing and high F_1 fertility observed in intragenomic hybrids. In contrast, intergenomic crosses are difficult or impossible to effect, and the rare F_1 plants are characterized by meiotic abnormalities and sterility.

Modern, improved varieties of cotton are all members of the tetraploid species *Gossypium hirsutum*. Several decades of study and accumulated evidence established that the tetraploid *Gossypium* species are allopolyploids that have recombined the Old World A genome and the New World D genome (reviewed in Endrizzi *et al.*, 1985; see also Wendel, 1989). Molecular data indicate that all allopolyploids in *Gossypium* share a common ancestry, i.e. that polyploid formation occurred only once (Brubaker *et al.*, 1999). As a result, crosses among tetraploid *Gossypium* species can be successful, while crosses between tetraploid and diploid *Gossypium* species are essentially unknown without human intervention.

The allotetraploid *Gossypium* species in Brazil, *Gossypium barbadense*, *G. hirsutum* and *G. mustelinum*, are robust shrubs with arborescent tendencies (Fryxell, 1979, 1992). *G. barbadense* is indigenous to western South America with range extensions into Mesoamerica, the Caribbean and eastern South America. The centre of morphological diversity for *G. hirsutum* is Mesoamerica, but its indigenous range includes the Caribbean, northern South America and some Pacific islands. *G. mustelinum*, the sole descendant of one branch of this radiation, is restricted to a small region of north-eastern Brazil (Wendel *et al.*, 1994).

Taxonomy of the *Gossypium* species indigenous to Brazil

Gossypium mustelinum

Originally described by Watt (1907), *G. mustelinum* has been treated at various ranks within *G. hirsutum* based on the superficial similarity of *G. mustelinum* to *G. hirsutum* var. *marie-galante*. However, the genetic evaluation clearly indicates that *G. mustelinum* represents a unique gene pool among the tetraploid *Gossypium* species and thus warrants specific status (Wendel *et al.*, 1994).

Gossypium barbadense

G. barbadense occurs as two distinct entities: var. *barbadense* and var. *braziliensis*. Variety *braziliensis* comprises the 'kidney' cottons of the Amazon basin. The definitive feature of the 'kidney' cottons is that the seeds in each locule of the capsule are fused into a solid kidney-shaped mass. This unique feature, in conjunction with a degree of morphological differentiation from the rest of *G. barbadense*, has been used to justify taxonomic recognition of these unique cottons. They are not, however, genetically distinct from *G. barbadense* var. *barbadense* (Percy and Wendel, 1990). They have no unique alleles and no unique multi-locus genotypes. The 'kidney' seed trait itself is monogenic, being encoded by a recessive allele (Turcotte and Percy, 1990).

Under low-input agriculture, the 'kidney' seed trait has two advantages. It facilitates hand-ginning by allowing all the seeds of a locule to be ginned simultaneously rather than individually, and when entire 'locks' of fused seeds are planted, the probability that all the seeds are properly oriented and placed in the soil at the proper depth is greater (Fryxell, 1979). The 'kidney' seed cottons show evidence of considerable agronomic improvement and it is likely that the 'kidney' seed trait and its complex of associated characteristics arose under domestication (Fryxell, 1979). Therefore, *G. barbadense* var. *braziliensis* is more properly considered a geographically restricted domesticated form of *G. barbadense* than a natural taxonomic entity.

Gossypium hirsutum

In their first treatment of *G. hirsutum*, Hutchinson *et al.* (1947) recognized three varieties. Variety *hirsutum* that includes Upland cotton and other early-cropping or annualized forms, has a centre of diversity in Guatemala but extends throughout Mesoamerica. The other two, mostly perennial forms, are the primarily Mesoamerican var. *punctatum* and var. *marie-galante* from northern and north-eastern South America and the Caribbean. Later, Hutchinson (1951) replaced this formal circumscription with an informal system in which he recognized seven geographical races. The Hutchinson *et al.* (1947) circumscription will be used here to conform to Brazilian practice, with the note that *G. hirsutum* var. *hirsutum* is typically designated *G. hirsutum* var. *latifolium*.

G. hirsutum in Brazil can be attributed to either var. *latifolium* or var. *marie-galante*. *Gossypium* var. *latifolium* was introduced to Brazil as a cultigen and does not exist as a truly wild plant in Brazil. The *latifolium* cottons are sub-shrubs with medium large to very large bolls and exhibit a range of day-length sensitivities. The *marie-galante* cottons possess a dominant central stem that often approaches 15 cm in diameter. Although *marie-galante* cottons do not occur as truly wild plants in Brazil, they almost certainly arrived through intentional germplasm dispersal in pre-Columbian times.

Non-*Gossypium* relatives of cotton and their potential role in Bt gene escape

There are numerous non-*Gossypium* Malvaceae species that occur in Brazil. Of these, the most relevant to assessing the risk of Bt gene escape are the closest non-*Gossypium* species found in Brazil (*Thespesia* spp. and *Cienfugosia* spp.) and the Malvaceous species that occur as weeds in cotton fields, e.g. *Sida* spp. and, to a lesser extent, *Malva* spp.

The closest relatives within the botanical family are *Cienfugosia* (2n = 20, 22) and *Thespesia* (2n = 26). While there are no known studies assessing the possibility that *Gossypium* species can hybridize with any *Cienfugosia* or *Thespesia* species, the tremendous evolutionary distance between *Cienfugosia* or *Thespesia* and *Gossypium* (10–15 million years before present; Seelanan *et al.*, 1997), the difference in chromosome numbers and the profound differences in reproductive architecture undermine any expectation that *Cienfugosia* × *Gossypium* or *Thespesia* × *Gossypium* hybrids are possible.

If there was any possibility of fertile hybrids arising between Bt cottons and weedy Malvaceous species (e.g. *Sida* spp. and *Malva* spp.), then the environmental consequences would be much greater than for Bt gene flow among *Gossypium* species, none of which can be characterized as weedy. However, the evolutionary distances between *Sida* or *Malva* and *Gossypium* are much greater than for the *Cienfugosia* or *Thespesia*, coupled with similar differences in chromosome numbers. Even in the absence of experimental data, there is no reasonable expectation that *Sida* × *Gossypium* or *Malva* × *Gossypium* hybrids are possible.

Gossypium populations in Brazil

The three *Gossypium* species that occur in Brazil are found as a variety of forms that span the wild to domesticated continuum, including the endemic wild species (*G. mustelinum*), *G. barbadense* and *G. hirsutum* landraces, dooryard cottons, and feral cottons and modern elite cultivars of *G. hirsutum*. In this chapter, the authors present our consensus regarding the types of populations in which each species can be found in Brazil (Table 11.1).

G. hirsutum var. *latifolium*, or upland cotton, occurs within Brazil exclusively as an elite cultigen with its subsequent field and roadside volunteers. *G. hirsutum* var. *latifolium* cultivars are grown in all four cotton areas, and

Table 11.1. The types of populations that *Gossypium* species and varieties can form in Brazil. Each recipient population type is a potential avenue of Bt gene escape from cotton in Brazil.

	Wild	Feral	Volunteer	Dooryard	Landrace	Cultivated
G. hirsutum var. *latifolium*			×	×		×
G. hirsutum var. *marie-galante*		×	×	×	×	×
G. barbadense var. *barbadense*			×	×	×	
G. barbadense var. *braziliensis*			×	×	×	
G. mustelinum	×					

are the dominant forms found in the Midwest, Meridian and North-east regions (see Fig. 2.2 in Fontes *et al.*, Chapter 2, this volume). The *G. hirsutum* var. *latifolium* cultivars grown in Brazil are mostly locally adapted cultivars but overseas germplasm is commercially available.

G. hirsutum var. *marie-galante*, also known as mocó cotton, occurs as dooryard cotton in the Midwest and the Amazon Basin. In the North-east, *G. hirsutum* var. *marie-galante* occurs as dooryard cotton, in feral populations, as a landrace, and as improved cultivars (see Fig. 2.2 in Fontes *et al.*, Chapter 2, this volume). Volunteers are associated with cultivated fields and on roadsides (between farms and gin). Cottons ascribed to *G. hirsutum* var. *marie-galante* are genetically diverse (Moreira *et al.*, 1995). The mocó cottons represent a distinct subset of the *G. hirsutum* var. *marie-galante* gene pool, and in the North-east, two other distinct forms occur. Verdões or 'green types' are thought to be a hybrid between *G. hirsutum* var. *marie-galante* and *G. hirsutum* var. *latifolium*. Brown-linted forms are the second type. Some brown-linted feral populations present in the North-east are thought by some to reflect hybridization between *G. hirsutum* var. *marie-galante* and *G. mustelinum*. However, brown-linted forms of mocó cotton are common, and the geographical distance between *G. mustelinum* and most mocó cottons is so large that the probability that these brown-linted forms arose from introgression is low.

G. barbadense var. *barbadense*, also known as 'quebradinho' cotton, is assumed to have been moved into Brazil as a landrace via human-mediated germplasm diffusion either via the Caribbean or across the Andes (Percy and Wendel, 1990). Although landraces still persist, in some indigenous reserves, *G. barbadense* var. *barbadense* primarily occurs as dooryard cotton. Dooryard forms occur in all four cotton districts, although in the Meridian district they are not found in the primary cotton-cultivation areas. Elite *G. barbadense* cultivars are grown in Brazil on a very small scale.

G. barbadense var. *braziliensis*, also known as 'kidney' cotton ('Rim de boi'), occurs mainly as dooryard cottons and is present throughout the Midwest and Amazon Basin, including in the cotton districts. As for *G. barbadense* var. *barbadense*, some Brazilian Indian ethnic groups maintain landraces. Some dooryard forms are found in the Meridian district but not in the primary areas of cotton cultivation. *G. barbadense* var. *braziliensis* is presumed to have arisen within Brazil as a local cultigen developed from introduced *G. barbadense* var. *barbadense* landraces.

The data available regarding *G. barbadense* and *G. hirsutum* var. *marie-galante* are inadequate to reliably describe the geographical distribution and abundance of these two species. Scientists at Embrapa have identified this gap in knowledge as a priority that requires intense sampling for a robust understanding of where these species occur in Brazil.

G. mustelinum is known only as a wild species, and is thought to be restricted to discrete areas in the Brazilian states of Rio Grande do Norte and Bahia in the North-east cotton-growing area (Plate 2 – see colour frontispiece) (Freire, 2000). *G. mustelinum* has never been known as a cultigen, and, in fact, the populations are considered under threat of extinction.

Occurrence of *Gossypium* in the three cotton-growing regions of Brazil

There are three main geographical regions in which cotton is grown in Brazil: Midwest, Meridian and North-east. They differ significantly in the level of agricultural inputs, insect and disease complexes and in ecology. Historically, the North-east and Meridian were the primary cotton-growing area, but as production declined with the advent of the boll weevil, the Midwest region has become the primary region for large-scale cotton production. The density and forms of cotton vary from district to district, and in each area different forms of cotton would co-occur with Bt cottons (Table 11.2).

In the Midwest, *G. hirsutum* var. *latifolium* cultivars are grown on large-scale farms. Farm sizes range from 5 ha to 10,000 ha, with an average size of 500+ ha. In the state of Mato Grosso, where more than 50% of the cotton fibre is produced, most farms are larger than 1000 ha. Dooryard forms of *G. hirsutum* var. *marie-galante*, *G. barbadense* var. *barbadense* and *G. barbadense* var. *braziliensis* also occur. Both varieties of *G. barbadense* occur as landraces, restricted to some indigenous reserves. The Midwest region will be the primary region in which Bt cottons will be grown.

The North-east region comprises the semiarid cotton-production areas in north-eastern Brazil. The area is characterized by small-scale, low-input farming on family holdings that average 3 ha in size. Most of the area is planted with *G. hirsutum* var. *latifolium*, but around 9000 ha in the Seridó regions of Paraiba and Rio Grande do Norte are planted with landrace and elite mocó cottons (*G. hirsutum* var. *marie-galante*) (see Fig. 2.2 in Fontes *et al.*, Chapter 2, this volume). Feral and dooryard forms of mocó cottons are also found. *G. barbadense* var. *braziliensis* and var. *barbadense* can occur as landraces and dooryard cottons. *G. mustelinum* occurs in patches that are generally away from cultivated cottons and can be fairly remote. Due to low productivity, it may be less likely that Bt cottons will be grown commercially in the North-east. However, the possibility that seed will be imported into the area cannot be precluded.

The Meridian cotton-production region includes the south-eastern states of São Paulo and Paraná, where 90,000 ha of cotton are planted. The level of agricultural input is intermediate between the North-east and the Midwest. As with the other districts, the primary cultivated cottons are elite *G. hirsutum* var. *latifolium* cultivars; however, some dooryard forms of *G. barbadense* var. *braziliensis* and var. *barbadense* do occur.

Cotton production in the Amazon Basin occurs on a very limited basis, but the full extent of cultivation is unknown (see Fontes *et al.*, Chapter 2, this volume). It is believed that *G. hirsutum* var. *latifolium* is the primary cultivated variety and that *G. barbadense* occurs as landraces and dooryard cottons. *G. hirsutum* var. *marie-galante* was believed to have been introduced from the North-east as dooryard cotton. Efforts are currently underway to increase and actualize the knowledge about *Gossypium* genera in Brazil. The results should be available sometime in 2005. In a full risk assessment of gene flow in Brazil, if cotton production increases in the Amazon, the potential impacts of gene flow in this region should be re-evaluated.

Table 11.2. Regional differences in occurrence of cotton populations in cotton-production regions of Brazil and in the Amazon region. Each type of recipient population is a potential avenue for gene flow from Bt cotton. White boxes indicate the presence of a population; shaded boxes indicate the absence of a population.

	G. hirsutum var. *latifolium*						*G. hirsutum* var. *marie-galante*[1]						*G. barbadense* var. *barbadense*						*G. barbadense* var. *braziliensis*						*G. mustelinum*					
	W	F	V	D	L	E	W	F	V	D	L	E	W	F	V	D	L	E	W	F	V	D	L	E	W	F	V	D	L	E
MW					a												a						a							
NE									BL		v	a					b						a				a			
MR																	b						a,b							

aRegion where the *Gossypium* variety is dominant.
bNo overlap with other forms of cotton (populations occur at > 300 m distances).
Key: MW, Midwest region; NE, North-east region; MR, Meridian region; W, Wild; F, Feral; V, Volunteer; D, Dooryard; L, Landrace; E, Elite.
[1]v: Verdao: Hybrid landrace (*G. hirsutum* var. *marie-galante* × *G. hirsutum* var. *latifolium*); BL: brown-lint cotton; possible *G. hirsutum* var. *marie-galante* × *G. mustelinum*); boxes without symbols are 'pure' *G. hirsutum* var. *marie-galante*.

Reproductive isolation among Brazilian *Gossypium* species

Cytological assessments demonstrate that the tetraploid *Gossypium* species are not separated by any large-scale cytological rearrangements (Gerstel and Sarvella, 1956), and can be hybridized to produce fertile F_1 progeny (Stephens, 1967). There is some 'incipient chromosome structural differentiation' evident between *G. mustelinum* and *G. barbadense* and *G. hirsutum* (Hasenkampf and Menzel, 1980), although this does not disrupt meiosis sufficiently to render hybrids infertile. However, unfit genotypes and transgressive segregation are evident in the F_2 and succeeding generations (Stephens, 1950).

From the perspective of cotton breeding and risk assessment, the *Gossypium* species can be subdivided into three germplasm pools, based on two criteria (Stewart, 1995): (i) the ability to form fertile hybrids with Bt cottons; and (ii) the frequency of genetic recombination between the Bt-cotton chromosomes and the chromosomes in other Brazilian cottons. The primary gene pool contains the species that should hybridize readily with the Bt cottons to produce hybrids whose donor and recipient chromosomes readily exchange genes. The data would suggest that all the indigenous Brazilian cottons belong to the primary germplasm pool for Bt cotton. Thus, conservatively one must assume that there are no significant genetic or cytogenetic barriers to gene flow among the three tetraploid *Gossypium* species found in Brazil.

The secondary germplasm pool includes species that cannot form fertile hybrids with the cultivated cottons without human intervention, but whose chromosomes are still sufficiently similar to those of the cultivated species to permit gene exchange once these fertile hybrids have been obtained. This germplasm pool includes the diploid species of the A, B, D and F genomes. None of the secondary germplasm pool species occurs in Brazil. The tertiary germplasm pool contains those species that can be crossed with the cultivated cottons to form fertile hybrids only with extensive human intervention. Even when fertile hybrids are obtained, the chromosome structures are so dissimilar that gene exchange occurs only at very low frequencies. This germplasm pool contains the African–Arabian E genome species and the Australian C, G and K genome species. Thus, among the wild *Gossypium* species, the Australian species (along with the African–Arabian E genome species) are those least likely to provide an avenue of escape for the Bt gene. None of the tertiary germplasm species occurs in Brazil. Species outside the *Gossypium* genus fall outside the tertiary germplasm pool, and therefore present a near-zero likelihood of crossing with cultivated cotton.

A limited species-isolation mechanism does exist. Hybrids among *G. barbadense*, *G. hirsutum* and *G. mustelinum* can give rise to 'corky' plants (Stephens, 1946, 1950; Stephens and Phillips, 1972). The 'corky' phenotype is characterized by mosaic yellowing of the leaf mesophyll and excessive cork production on stems, leaf petioles and midribs. Because this cork production begins very early, growth is stunted. 'Corky' plants will flower and the pollen is viable, but female fertility is severely reduced. Although the corky gene is not a complete barrier to gene flow between the species, this locus has probably contributed to maintaining the distinctions among the three species where they occur together. Other relevant variables contributing to maintaining genetic distinctions among sympatric allotetraploid *Gossypium* species include differences in the timing of pollen shedding (Stephens and Phillips, 1972), pollen competition (Kearney and Harrison, 1932) and differences in ecological preferences (Mauer, 1930).

Based on the considerations listed above, all forms of cotton found in Brazil (Table 11.2) must be considered members of the primary gene pool relative to Bt cottons. Consequently, there is no *a priori* expectation that there are any genetic or cytological barriers among the species sufficient to halt gene flow from Bt cottons to any of the various forms of *G. barbadense* and *G. hirsutum* and the wild species *G. mustelinum*. Genetic studies have clearly established that where *G. barbadense* and *G. hirsutum* occur sympatrically, introgression has occurred (Brubaker *et al.*, 1993; see also Brubaker *et al.*, 1999). In fact, gene flow from current commercial cotton cultivars into indigenous dooryard and landrace cottons is already considered an ongoing issue of concern (P. Barroso, Brasília, 2004, unpublished data). While there is little direct evidence for introgression between *G. mustelinum* and either *G. barbadense* or *G. hirsutum* (Wendel *et al.*, 1994), this most likely reflects the geographic isolation of *G. mustelinum* populations rather than any effective species-isolation mechanism (Wendel *et al.*, 1994). It must be assumed that when any of the current forms of cotton in Brazil occur within pollen vector range of Bt cotton, there is a sufficient probability of gene flow that this may require active management.

We evaluated several additional factors that could affect gene flow from Bt cotton to recipient populations in Brazil such as geographic proximity, pollination biology and pollen dispersal distance. A summary of our full evaluation is shown in Table 11.3, including information on likely pollination distances, the success of natural versus hand crosses between crop and recipient populations, and the likelihood that populations will overlap. These evaluations support our conclusion that there are no barriers to halt gene flow from Bt cottons to any of the other tetraploid *Gossypium* species in Brazil.

What would it take to isolate crops and prevent gene flow?

It will be impossible to eliminate the possibility of gene flow from Bt cotton in Brazil, because there are no genetic barriers capable of isolating cotton cultivars from recipient populations with overlapping distributions. It may be possible to reduce the chances of gene flow by restricting the area over which Bt cotton is planted, excluding potential gene-flow 'hotspots'. A century ago, in the USA, it was suggested that 5–10 miles would be necessary to prevent all natural crossing (Weber, 1902, as cited in Green and Jones, 1953). More recent data from the USA show that 7–10 rows of non-transgenic cotton can trap transgenic pollen, allowing less than 1% to escape beyond the rows (Umbeck *et al.*, 1991; Berkey *et al.*, 2002). However, this was done with North American pollinators in an intensive agricultural setting. Pollination

Table 11.3. Likelihood that gene flow will occur from Bt cotton to recipient populations in Brazil (Part 1 of general scientific risk-assessment model in Box 11.1). White boxes indicate a case where the question was relevant to the population in question. Shaded boxes indicate that the question did not apply to that population.

	G. mustelinum		*G. barbadense*				*G. hirsutum*				
	Wild	Feral	Volunteer	Dooryard	Landrace	Feral	Volunteer	Dooryard	Landrace	Elite (non-Bt)	Bt cotton crop
Will transgenic crop overlap (< 300 m between populations) with recipient populations?	Not likely	Yes	Yes	Not likely	Yes	Yes	Yes	Yes	Yes	Yes	
To what extent does taxon outcross?	Very little	More pollinators than crop field but fewer plants	Very little	Very little	More pollinators than crop field but fewer plants	More pollinators than crop field but fewer plants	Very little	Very little	Very little	10–30%?	10–30%?
Over what distance can cross-pollination occur?	Probably less than 1% after 30 m, but rare long-distance events can occur.										
Are volunteer crops or wild plants considered weeds?	No	No	No	No	No	No	No	No	No. Cattle graze volunteers after harvest		No. Efforts are made to destroy volunteers
Do hand crosses with Bt cotton result in fertile hybrids?	Yes	Yes	Yes	Yes	Yes	Yes	Yes	Yes	Yes	Yes	

Continued

Table 11.3. Likelihood that gene flow will occur from Bt cotton to recipient populations in Brazil (Part 1 of general scientific risk-assessment model in Box 11.1). White boxes indicate a case where the question was relevant to the population in question. Shaded boxes indicate that the question did not apply to that population – *cont'd.*

	G. mustelinum		G. barbadense				G. hirsutum				
	Wild	Feral	Volunteer	Dooryard	Landrace	Feral	Volunteer	Dooryard	Landrace	Elite (non-Bt)	Bt cotton crop
Does spontaneous hybridization occur in the greenhouse?	No, requires insects	No, requires insects	No, requires insects	No, requires insects	No, requires insects	No, requires insects	No, requires insects	No, requires insects	No, requires insects	No, requires insects	
Does spontaneous hybridization occur in nature?	It can	It can	It can	It can	It can	It can	It can	It can	It can	It can	
Have hybrids with crop been confirmed with molecular markers?	Yes	Yes	Yes	Yes	Yes	Yes	Yes	Yes	Yes	Yes	
Does direction of cross-pollination affect the success of crossing?	No data, probably not	No data, probably not	No data, probably not	No data, probably not	No data, probably not	No data, probably not	No data, probably not	No data, probably not	No data, probably not	No data, probably not	
Are F_1 hybrids viable and vigorous?	Yes	Yes	Yes	Yes	Yes	Yes	Yes	Yes	Yes	Yes	
Do F_1 hybrids reproduce successfully?	Yes	Yes	Yes	Yes	Yes	Yes	Yes	Yes	Yes	Yes	
Are later generation hybrids vigorous and reproductively fit?	Some, F_2 break-down e.g. Corky locus	Some, F_2 break-down e.g. Corky locus	Some, F_2 break-down e.g. Corky locus	Some, F_2 break-down e.g. Corky locus	Some, F_2 break-down e.g. Corky locus	Yes	Yes	Yes	Yes	Yes	

Question	1	2	3	4	5	6	7	8	9	10	11
How is fitness of hybrids tied to field locations and growing conditions?	No data. See 'regional factors' and 'hybrid fitness' sections for our predictions.										
Can hybrids have seed dormancy or form seed banks?	Possibly, short dormancy (1–3 years)	Possibly, short dormancy (1–3 years)	Possibly, short dormancy (1–3 years)	Possibly, short dormancy (1–3 years)	Possibly, short dormancy (1–3 years)	No	No	No	No	No	No
Do crop plants volunteer in years following cultivation?					Yes	Yes	Yes	Yes	Yes	Yes	Yes
Can crop volunteers establish feral populations?					Not observed in 100 years. Theoretically possible	Not observed in 100 years. Theoretically possible	Not observed in 100 years. Theoretically possible	Not observed in 100 years. Theoretically possible	Not observed in 100 years. Theoretically possible	Not observed in 100 years. Theoretically possible	Theoretically possible
Could transgene persist as a perennial for more than one year in volunteer or hybrid populations?	Yes	Yes	Yes	Yes	Yes	Yes	Yes	Yes	Yes	Yes	Yes
Could transgene spread clonally?	No	No	No	No	No	No	No	No	No	No	No
Are populations of feral or wild plants considered weeds?	No	No	No	No	No	No	No	No	No	No	Yes

distance may be different at field edges and near dooryard or low-density feral cottons. Also, Brazilian pollinators may behave differently or occur in different densities than pollinators in the USA, making it difficult to estimate relevant pollination distances from existing data. To the best of our knowledge, there have been no studies of cotton pollination distance completed in Brazil.

Potential pollination distance for Bt cotton in Brazil is the first major knowledge gap we identified during the workshop. Pollination distance should be tested in Brazilian cotton varieties in Brazilian cotton fields. Neutral genetic markers such as fibre colour (Arriola and Ellstrand, 1996) or neutral transgenic markers such as green fluorescent protein (GFP) or kanamycin resistance (Harper *et al.*, 1999) may be helpful for such a study. The procedure in Umbeck *et al.* (1991) would be adequate for assessing pollen movement among adjacent cotton fields or within the agricultural setting. The movement of pollen outside the agricultural setting is much more difficult to predict. Both Umbeck *et al.* (1991) and Berkey *et al.* (2002) point out that high-density boundary crops are necessary for effective pollen trapping. Thus, it may be necessary to perform additional tests to predict the pollination distances possible at field edges, and in small or widely spaced populations of *Gossypium* sp. in Brazil.

Other considerations about the likelihood of gene flow

Clearly, there are many opportunities for gene flow from Bt cotton to recipient populations in Brazil. However, viable hybrids have been shown to form spontaneously and be reproductively successful. Volunteers of non-transgenic cotton have been heavily managed around agricultural fields, but are likely to escape occasionally. No *Gossypium* species are currently considered to be weeds, so gene flow from Bt cotton will not exacerbate an existing weed problem. However, some recipient populations of *Gossypium* are perennials, creating an avenue of escape for the transgene that might allow it to persist in the environment for several years. Volunteer crops and perennial hybrids that escape the agricultural environment to form feral populations may act as genetic bridges to further spread the transgene to dooryard or wild cottons.

Likelihood of Transgene Establishment and Spread

There are two ways that a transgene in cotton might escape its cultivated state: pollen- or seed-mediated gene flow. In the case of pollen-mediated gene flow, pollen from the crop fertilizes individuals in a recipient population, resulting in a hybrid organism that is a mixture of the two. Alternatively, gene flow could occur via mature transgenic crop seeds that are dispersed to germinate the following year. When cottonseed and fibres are transported for ginning and processing, inadvertent loss of seeds and spread of volunteer plants along transport routes are possible.

For transgene flow to occur via pollen, an intra- or interspecific hybridization must occur. There are three main factors that will determine the effect of the transgene on hybrid fitness: genetic isolation mechanisms, transgene expression in a new genetic background and the transgene effect on fitness in the recipient population. If transgene flow occurs via seed, volunteer or feral plants could arise from crop plants, establishing transgenic populations outside the agricultural environment. If hybrids formed successfully, the primary factor that will determine the extent of transgene spread is its fitness effect.

The second major knowledge gap identified during the workshop is the lack of data on the Bt gene fitness effect on recipient *Gossypium* populations. For Bt cotton grown in Brazil, genetic barriers between crops and recipient populations are weak, so hybrid fitness is predicted to be high. It is unknown how Bt gene expression in the genetic background of a recipient population will compare to expression in the Bt cotton crop. The genetic background presented by *G. mustelinum* or mocó cotton may interfere with or silence expression of the Bt gene. Additionally, it is not known whether recipient populations would experience a benefit from resistance to target pests. To fill this critical knowledge gap, we propose two studies, one to investigate transgene expression in a variety of genetic backgrounds and another to evaluate insect pressure on *Gossypium* species outside the agricultural environment.

There are currently few examples of studies that examine the fitness of a transgene following a wide outcross. One experimental model for testing the Bt fitness effect in Brazilian cotton can be found in Snow *et al.* (2003). Also, a previously suggested protocol for investigating fitness effect of a Bt gene following outcrossing between elite and landrace maize varieties might serve as a useful example (Box 6.2; Johnston *et al.*, 2004).

It is widely predicted that the fitness effect of an escaped transgene will be a primary determinant of its rate of spread (Snow and Palma, 1997; Ellstrand *et al.*, 1999; Burke *et al.*, 2002). The model in Box 11.2 illustrates the changes in gene frequency and genetic diversity when a population receives a novel allele that has a beneficial, neutral or detrimental effect on fitness. If the Bt gene is capable of reducing the amount of herbivore damage suffered by *Gossypium* sp., it is reasonable to expect that Bt will most likely have a positive effect on fitness of recipient populations. If that is the case, selection will cause an increase in Bt gene frequency in the population. If the Bt gene is neutral and has no effect on the fitness of hybrids, the transgene is still likely to spread into recipient populations, but not as quickly. In the neutral case, the number of introductions of Bt into a population will have a large effect on the eventual frequency of the gene in a population.

If the Bt gene has a negative effect on the fitness of recipient individuals, the gene is unlikely to spread into the population, unless gene flow is recurrent and high. If all resulting hybrids are at a selective disadvantage, the gene will remain at low frequencies or be removed from the population completely (Ellstrand *et al.*, 1999; Bergelson and Purrington, 2002). The recurrent introduction of transgenes with negative fitness effects may generate problems in recipient populations that are small or endangered. If the recipient populations are producing a small number of gametes and a large proportion of them are 'wasted' by making hybrid seeds with low fitness, the population may be put at an increased risk of extinction.

Box 11.2. Potential population genetic consequences of unilateral gene flow of a transgene following single versus recurrent gene-flow events.

Assumptions:
Gene flow unilateral, system isolated from other populations.
Single dominant or additive novel allele fixed in source population.
Novel allele initially absent in recipient population.
Recipient population large.
Mutation rate « hybridization rate « 100%

MODEL 1. A single to a few gene-flow events.

Relative fitness of novel immigrant allele	Initial change in recipient population genetic diversity	Long-term change in genetic diversity (~100 generations)	Persistence of novel allele (~100 generations)
Beneficial	Increase	None or decrease	Fixation
Neutral	Increase	Increase or none	Coexistence
Detrimental	Increase	None	No

MODEL 2. Recurrent gene-flow events.

Relative fitness of novel immigrant allele	Initial change in recipient population genetic diversity	Long-term change in genetic diversity (~100 generations)	Persistence of novel allele (~100 generations)
Beneficial	Increase	None or decrease	Fixation
Neutral	Increase	None or decrease	Fixation or coexistence
Detrimental	Increase	Increase, none or decrease	Coexistence

Regional differences in insect pressure will change fitness effect of Bt gene

Since the Bt gene is intended to offer some protection against insect herbivores, it is likely that following gene flow the effects of herbivore pressure on recipient populations would be altered as well. Within the agricultural setting, we know there are large regional differences in cotton herbivore and disease pressure (Table 11.4). However, we currently do not know if or how much the target insects *Alabama argillacea*, *Pectinophora gossypiella* and *Heliothis virescens* are feeding on feral, dooryard and wild cotton populations. It is possible that herbivory is inconsequential to the fitness and ecology of these populations. The most robust studies of the Bt gene effect on fitness of cotton in Brazil would ultimately need to be carried out in several genetic backgrounds and under realistic herbivory levels that reflect regional variation.

Table 11.4. Insect and disease pressure in each region of cotton cultivation in Brazil. Within each region, there will be a considerable amount of variation for specific areas.

Agronomic environment	Insect impact			Disease		
	High	Medium	Low	High	Medium	Low
Midwest region Natural enemies present at low levels, high levels of input and technology available in these areas, many farms are large and mechanized	Pink bollworm,[a] boll weevil, tobacco budworm,[a] aphids (as vector), cotton leafworm, fall armyworm	Thrips, cotton stem-borer, cotton stainer, mites	Costalimaita ferruginea, grasshopper, leaf miner	Fusarium wilt (in some areas, but spreading with nematodes), nematodes, blue disease,[a] bacterial blight, witches broom, areolate mildew	Leaf spot (Alternaria, Cercospora, Stemphylium)	Anthocyanosis, verticillium wilt
North-east region Natural enemies more abundant. Upland and móco cotton are both grown in this region, inputs are lower than in the Midwest and many farms are small	Cotton leafworm,[a] Pink bollworm,[a] aphids Boll weevil is the key pest in some areas, controlled by IPM in others	Cotton stemborer	Mites, Costalimaita ferruginea, leaf miner	(None)	Witches broom, areolate mildew	Fusarium wilt, nematodes, blue disease,[a] bacterial blight, Anthocyanosis, verticillium wilt, leaf spot (Alternaria, Cercospora, Stemphylium)
Meridian region Intermediate level of mechanization. Farms range in size from 10 ha to 500 ha. The most useful varieties are adapted to local diseases and conditions	Tobacco budworm,[a] pink bollworm,[a] cotton leafworm, boll weevil	Pink bollworm,[a] cotton stem-borer, cotton stainer, mites (some spp.)	Mites (some spp.), Costalimaita ferruginea, leaf miner	Blue disease,[a] witches broom, fusarium wilt, nematodes	Areolate mildew, bacterial blight, verticillium wilt, leaf spot (Alternaria, Cercospora, Stemphylium)	Anthocyanosis

[a]Blue disease is caused by a luteovirus.

Other factors that affect establishment and spread of transgene

There are several factors that might affect the rate and extent of Bt gene establishment and spread in Brazilian *Gossypium* populations. However, the fitness effect of the transgene will be the driving factor and we are unable to evaluate fitness effect without more information on transgene expression in a new genetic background, the strength of target pest herbivore pressure outside cultivation and the ecological factors that constrain the fitness of free-living populations of *Gossypium* species. We will, however, raise a few other, probably less important factors without discussing them in detail. A summary of our evaluation of transgene establishment and spread is shown in Table 11.5.

If the Bt gene were to become incorporated into cotton landraces, farmers who save seeds and plant landraces of cotton may select for the Bt gene unintentionally. Plants that get the Bt gene via pollen-mediated gene flow may be more vigorous or have higher productivity, and be inadvertently selected for planting the following year. Similarly, contaminated seed lots may produce a few vigorous individuals that contain the Bt construct without the farmer's knowledge. If these individuals are more vigorous, their seeds may be saved, increasing the frequency of the transgene in subsequent years. While there may be no inherent harm in spreading the Bt gene to small-scale farmers and landraces, there may eventually be intellectual property or organic standard issues caused by the spread. Additionally, movement of the Bt gene into landraces might have negative effects on resistance-management efforts or create genetic bridges that could further spread the gene into wild populations.

Seed-mediated gene flow may be difficult to predict or follow, once the gene has escaped the agricultural environment. The mechanisms of seed dispersal (birds, mammals, water, humans) are not likely to be different outside an agricultural field, but the opportunity for seed dispersal may increase in disturbed or natural areas. For example, birds and mammals are likely to be more numerous in areas with tree cover than in an open field of cotton. Therefore, current estimates of seed gene flow may not adequately account for dispersal outside the cotton-production fields.

While the wild cotton *G. mustelinum* is the least likely type of recipient population to directly overlap with Bt-cotton cultivars, there are several ways that genetic bridges between Bt cotton crops and wild cotton populations could develop. The distribution of Bt cotton would likely overlap with plantings of *G. hirsutum* var. *marie-galante* landraces in the North-east. The *marie-galante* cotton is also grown in areas that overlap with *G. mustelinum* (P. Barroso, Brasília, 2004, unpublished data). If the Bt gene became integrated into the landrace, unknowing farmers might share seeds, putting the wild cotton at risk. Feral populations, if formed, would be unmonitored and could act as a genetic bridge between Bt crops and dooryard or landrace cottons. Transgenic hybrids might have a perennial growth form and could serve as an undetected source of transgenic pollen for several years if established. Some *Gossypium* varieties have seed dormancy as well, so it is possible that transgenic hybrid seed could persist in the environment for more than a year, germinating perhaps after removal of volunteers had ceased.

Table 11.5. Likelihood of transgene establishment and spread from Bt cotton to recipient populations in Brazil (Part 2 of general scientific risk-assessment model in Box 11.1). White boxes indicate a case where the question was relevant to the population in question. Shaded boxes indicate that the question did not apply to that population.

	Hybrids with *G. mustelinum*	Hybrids with *G. barbadense*				Hybrids with *G. hirsutum*				
	Wild	Feral	Volunteer	Dooryard	Landrace	Feral	Volunteer	Dooryard	Landrace	Cultivated (non-Bt)
If the transgene spreads to non-transgenic crops is it likely to persist in an agricultural setting?	Probability near 0 because populations are very isolated (for now)	Yes	Yes	Yes	Yes	Yes	Yes	Yes	Yes	Yes
Will introgressed transgene increase yield, making farmers more likely to select for it?		Probably not			Yes				Yes	
Is transgene stably inherited over several generations?	Expression in F_2 hybrids	Expression in F_2 hybrids in controlled conditions	Expression in F_2 hybrids in controlled conditions	Expression in F_2 hybrids in controlled conditions	Expression in F_2 hybrids in controlled conditions	No data (probably yes)	No data (probably yes)	No data (probably yes)	No data (probably yes)	No data (probably yes)
Can the transgene be dispersed by vegetative propagules?	Probability near 0	Probability near 0	Probability near 0	Probability near 0	Probability near 0	No	No	No	No	No

Continued

Table 11.5. Likelihood of transgene establishment and spread from Bt cotton to recipient populations in Brazil (Part 2 of general scientific risk-assessment model in Box 11.1). White boxes indicate a case where the question was relevant to the population in question. Shaded boxes indicate that the question did not apply to that population – *cont'd.*

	Hybrids with G. mustelinum		Hybrids with G. barbadense				Hybrids with G. hirsutum			
	Wild	Feral	Volunteer	Dooryard	Landrace	Feral	Volunteer	Dooryard	Landrace	Cultivated (non-Bt)
Are seeds dispersed from the crop naturally or moved by humans?	Birds, water, and mammals, humans unlikely	Birds, water and mammals, humans	Birds, water and mammals, humans	Birds, water and mammals, humans	Birds, water and mammals, humans	Birds, water and mammals, humans	Birds, water and mammals, humans	Birds, water and mammals, humans	Birds, water and mammals, humans	Birds, water and mammals, humans
Is it possible for the transgene to spread and persist in 'natural' areas beyond the agricultural setting?	Yes	Yes	Yes	Yes		Yes	Yes	Yes		
What are the fitness effects of the transgene in hybrids? Can they be tested empirically?	No data; can be tested	No data; can be tested	No data; can be tested	No data; can be tested	No data; can be tested	No data; can be tested	No data; can be tested	No data; can be tested	No data; can be tested	No data; can be tested

Continued

Question												
Is the fitness effect associated with specific artificial or natural selective pressures?	Natural selection by target insect herbivory	Natural selection by target insect herbivory	Natural selection by target insect herbivory	Natural selection by target insect herbivory	Natural selection by target insect herbivory	Natural selection by target insect herbivory	Natural selection by target insect herbivory	Natural selection by target insect herbivory	Natural selection by target insect herbivory	Natural selection by target insect herbivory	Natural selection by target insect herbivory	Natural selection by target insect herbivory
What ecological factors have the greatest effect on fitness components of the wild population?	No data	No data	No data	No data	No data	No data	No data	No data	No data	No data	No data	No data
Is there a fitness cost of Bt if there is no insect pressure?	No data	No data	No data	No data	No data	No data	No data	No data	No data	No data	No data	No data
How does the fitness effect of the Bt transgene components are across field locations, conditions and years?	No data	No data	No data	No data	No data	No data	No data	No data	No data	No data	No data	No data
Could hybrid seed have dormancy or establish seed banks?	Possibly short dormancy (1–2 years)	Small possibility of short dormancy (1–2 years)	Small possibility of short dormancy (1–2 years)	Small possibility of short dormancy (1–2 years)	Small possibility of short dormancy (1–2 years)	Small possibility od short dormancy (1–2 years)	Not likely, but no data	Not likely, but no data	Not likely, but no data	Not likely, but no data	Not likely, but no data	Not likely, but no data

Table 11.5. Likelihood of transgene establishment and spread from Bt cotton to recipient populations in Brazil (Part 2 of general scientific risk-assessment model in Box 11.1). White boxes indicate a case where the question was relevant to the population in question. Shaded boxes indicate that the question did not apply to that population – *cont'd.*

	Hybrids with *G. mustelinum*		Hybrids with *G. barbadense*			Hybrids with *G. hirsutum*				
	Wild	Feral	Volunteer	Dooryard	Landrace	Feral	Volunteer	Dooryard	Landrace	Cultivated (non-Bt)
Do hybrid and feral populations have mechanisms for dispersing seeds further than cultivated varieties?	No special mechanisms, but plants in wild areas may have greater opportunity for dispersal	No special mechanisms, but plants in wild areas may have greater opportunity for dispersal	No special mechanisms, but plants in wild areas may have greater opportunity for dispersal	No special mechanisms, but plants in wild areas may have greater opportunity for dispersal		No special mechanisms, but plants in wild areas may have greater opportunity for dispersal	No special mechanisms, but plants in wild areas may have greater opportunity for dispersal	No special mechanisms, but plants in wild areas may have greater opportunity for dispersal		
Could populations invade new habitat after transgene introgression?	Not likely, other ecological limitations will restrict populations	Not likely	Not likely	Not likely	Not likely	Not likely	Not likely	Not likely	Not likely	Not likely

Consequences of Gene Flow

As we conducted our analysis of the factors that could affect gene flow, we considered several of the potential ecological consequences of gene flow. Foremost was the concern that existing geographic and reproductive barriers would not prevent gene flow to the endangered *G. mustelinum*, a species at risk of extinction or genetic swamping. The incorporation of the Bt gene into *G. mustelinum* populations could potentially change the ecological character of the species, with unpredictable consequences. As a worst-case scenario, repeated gene flow from Bt cotton (or other elite cotton varieties) could replace the unique genetic character of *G. mustelinum* with crop genes, essentially hybridizing the species out of existence. The relative geographic isolation of *G. mustelinum* may slow or even prevent gene flow from Bt cotton; however, we cannot now categorize the risk of transgene flow to the wild species as negligible.

Wide outcrosses between Bt cotton and *G. mustelinum* or *G. barbadense* may have low hybrid fitness. If there is a significant loss of fitness, small *G. mustelinum* populations may have low rates of seed production as a result. For *G. mustelinum* populations, ovules 'wasted' by fertilization by crop pollen may result in population decline and even extinction. For *G. barbadense*, if reduced fitness leads to the loss of landraces, valuable germplasm might be lost as well.

Our second concern was for the purity of the germplasm contained in *G. mustelinum*, *G. barbadense* and *G. hirsutum* var. *marie-galante*. Breeders use cotton landraces and dooryard cottons as sources of genetic diversity in the production of new locally adapted varieties. Any loss of genetic diversity could result in loss of germplasm for production of new cotton varieties. This is of special concern because northern Brazil is part of the Caribbean centre of diversity for *Gossypium* species. Much of the genetic variation contained within the landrace, dooryard, feral and wild populations is potentially truly unique to that region of the world.

Much less likely, but still a potential concern is that increased fitness of feral and dooryard populations could lead to increased weediness of non-cultivated *Gossypium* species. Although feral cotton is not considered a weed at this time, the introduction of the Bt gene might provide an advantage that could increase invasiveness and colonization ability. Feral populations that possess the Bt gene could conceivably have decreased seed predation, and, therefore, increased seed production and/or increased biomass production due to release from herbivore pressure. These populations would have a competitive advantage that might allow them to increase in number and expand the range in which they occur. Feral populations may act as a genetic bridge by facilitating movement of the Bt gene from Bt cotton into other feral, dooryard or landrace cottons.

In addition, we raise two issues related to the loss of crop purity in non-transgenic cotton. If gene flow occurs into organic cotton fields, then the cotton cannot be sold as organic and growers will lose the opportunity to participate in a potentially profitable niche market. Producers of non-transgenic cotton also would be subject to possible loss of credibility and market share if they attempt to guarantee delivery of a non-transgenic product and cross-pollination occurs. If small-scale farmers save seeds from one year's crop to plant the following year, they will be at higher risk of accidentally incorporating transgenes into their landraces or cultivars than farmers who purchase seeds each year. Zones of exclusion or buffering from Bt cotton would be particularly important in regions where seed cotton is commercially produced. Reports that seed sources are already contaminated with transgenes (see Andow *et al.*, Chapter 1, this volume) are especially troubling; our analyses would suggest that wide-scale release of transgenic cotton is likely to exacerbate the problem.

The ecological consequences and environmental risks of Bt-cotton introductions in Brazil are difficult to predict. It is critical that monitoring systems be put in place before the widespread introduction of transgenic cotton. Baseline information must be collected on the distributions of all types of potential gene flow-recipient populations so that changes can be quantified. Monitoring is needed for all *Gossypium* spp. to determine if changes have occurred. Special consideration should be made for the endangered status of the wild species *G. mustelinum* before the deployment of this crop.

If monitoring reveals that transgenes are spreading from the crop to recipient populations, a more thorough analysis of gene-flow consequences would need to be completed. Some of the consequences of gene flow that were not given much consideration at the workshop due to time constraints were ways in which gene flow could worsen some of the adverse effects discussed by other workshop groups. For instance, if the Bt gene becomes incorporated into wild or feral populations, a broader range of non-target species would likely be exposed to the Bt toxin than is found in the agricultural setting alone. Furthermore, estimates of the risk of resistance evolution would probably be increased by the spread of the Bt gene throughout the landscape. Most of what was discussed by all subgroups during the course of the workshop was the risks of Bt cotton within the cotton agroecosystem. Considerable movement of the Bt gene via gene flow might change many of these evaluations.

At the time of publication, Embrapa scientists in Brazil are in the process of collecting baseline data on population distribution and pollination biology of potential recipient populations of cotton relatives (A.Y. Ciampi, Brasília, 2003, unpublished data). It is possible that this information will be available and published sometime in 2005. The results will be used to design or adapt more efficient strategies to prevent undesirable gene flow.

Staging the Risk-Assessment Process

Our testing and monitoring recommendations are organized here into a time-ordered sequence that reflects the stages of developing a genetically engineered organism (see Andow *et al.*, Chapter 1, this volume). The information that we suggest gathering begins with general crop-specific background information and culminates with transformation-specific information gathered in an ecologically relevant context. Transgene exposure to the environment increases with each testing stage, culminating with release of the crop for commercial use. The authors agree that all relevant data for evaluating the likelihood of gene flow must be gathered by the end of the small-scale field-testing stage. Once large-scale field testing has begun, the environment will experience significantly higher exposure and potential risks may be difficult to avoid; only monitoring and mitigation of gene flow should be left to be done.

Pre-transformation: consideration of crop biology and distribution of recipient populations

Prior to design of a transgenic crop, much can be learned about potential gene-flow rates by evaluating the biology of the crop and the distribution of its wild relatives. The information to be gathered in this early stage will often be available in existing databases or scientific literature. Fine-scale data on population presence in the specific region under consideration may not be available. To evaluate the overlap in crop and recipient population distributions, existing surveys should be consulted, or performed if unavailable. It may be important to consider the density of sample points, how dynamic populations are and whether volunteer or feral plants were surveyed at all. To evaluate genetic diversity of potential recipient populations, consult germplasm characterization and seed-bank efforts if available. If not, this is a time consuming, relatively expensive step to take. However, if genetic resources found in landrace and wild populations are not catalogued, they could be lost. When evaluating phenological overlap, it is important to look at variation among years, regions and crop varieties. Are there long- and short-season varieties of the crop that will likely be transformed? Do altitudinal or regional climate differences affect the timing of flowering in the crop or recipient populations? Do plants perform differently in very wet or dry years?

At this stage, it is important to identify potential recipient populations that are of special concern, due to their rarity, their invasiveness or their particular economic or cultural importance. The value placed on the recipient population will guide the evaluation of adverse consequences from gene flow.

In the event that the answers to questions in any of these areas are not known, data can be collected without working with transgenic crops. When using data collected for another purpose or from a third-party source, it is important to consider where data were collected and how intensely, to evaluate how broadly the results can be generalized.

Posttransformation, prerelease: gene flow and introgression for crop and local recipient populations

Once a crop has been transformed, more detailed information about its local biology will be useful in predicting gene flow and introgression. It is important to have robust estimates of pollen-dispersal distances under local conditions with local pollinators and climate conditions. Studies of pollen movement could be evaluated in the field with non-transgenic varieties of the crop variety. The protocol used in Umbeck *et al.* (1991) is useful for evaluating cotton gene flow within a high-density crop environment. The procedure used in Arriola and Ellstrand (1996) might be more helpful for evaluating long-distance dispersal events or working with wind-pollinated species. Regardless, pollination estimates should be done in both high- and low-density stands to better simulate populations that might occur inside or outside of an agricultural environment.

Stable inheritance of transgene and gene expression in different genetic backgrounds should be evaluated as a matter of course in creating genetically engineered crops (see Grossi *et al.*, Chapter 4, this volume). However, the value of this information as an indicator of potential gene flow may be overlooked. In the case of Bt crops, if the gene has little or no expression in the genetic background of a wild relative, the likelihood of the gene spreading via gene flow is different from the case of gene expression similar to the crop. Presumably, the fitness benefit of the Bt gene can only be realized if the gene is expressed, the protective protein is made and insect herbivory decreases as a result. Ideally, each step of this interaction needs to be evaluated for each genetic background. At a minimum, the interaction of gene expression and insect herbivory should be evaluated for the most sensitive recipient populations, such as rare or invasive relatives, under a natural level of insect herbivory.

In order to evaluate the likelihood of successful outcrossing of the crop, a small field test could be performed to evaluate the fitness of hybrids between non-transgenic crops and recipient populations. By evaluating fitness without the transgene, the risks associated with transgene escape are zero. In some cases, it might be found that genetic or phenotypic barriers are sufficient to prevent successful outcrossing and therefore gene flow. Estimates of hybrid fitness would ideally involve as many crop-recipient population crosses in as many environmental conditions as possible. Fitness comparisons should include the original transgenic cultivar(s) and the unaltered members of the original recipient population(s). A robust fitness evaluation would include vigour as well as reproductive output.

At this stage, if transgenic crops were available, it would be useful to evaluate pleiotropic and unintended phenotypic effects of the transgene. Unintended effects on non-yield related traits might go unnoticed by breeders, but have significant ecological effects. We suggest that these tests be done in a controlled setting, where plant phenotypes can be observed closely with as little environmental variation as possible. If an unintended effect is observed at this stage, its potential environmental impact should be evaluated prior to field trials.

Small-scale field trials

Once a crop has been successfully transformed and gene-flow rates and probabilities have been characterized in a controlled setting, small-scale field trials can begin. During small field trials of transgenic crops, there is some risk associated with transgene escape. A single exposure of the environment to a transgene carries a much smaller probability of gene flow occurring successfully than repeated exposure (Table 11.6) that will be created when crops are commercially released and grown year after year (Crow and Kimura, 1970; Nagylaki, 1977; Haygood *et al.*, 2003). These small-scale field trials should provide sufficient data to model and assess the risks associated with recurrent release.

Before field testing can begin, intensive surveys should be performed, in order to rule out the chance that there are any recipient populations in the area. For Bt cotton, the most conservative estimate given of a necessary isolation distance is 10 miles (Green and Jones, 1953).

The fitness effect of the transgene should be evaluated in as many natural conditions as possible. As herbivore identity and pressure change across regions, so will the fitness effect of the Bt gene. It is likely that there will be some conditions in which there is a large positive fitness effect of the transgene, and others where there is little or no effect (Burke and Rieseberg, 2003). Ideally, small-scale field tests should be performed in multiple years, several sites and with as many relevant hybrid crosses as possible. Variation in environmental conditions (temperature, humidity, precipitation, wind speed and direction) will produce variation in phenology of crop and wild species (Snow *et al.*, 2003). Pollinators and herbivores are likely to vary regionally and annually as well, affecting both pollination distances and fitness effects of the transgene.

Large-scale testing and release: monitoring and mitigating gene flow

Once the transgenic crop is released on a larger scale, the general and situation-specific questions regarding gene flow should be answered. As the scale of release increases, broader surveys would help to minimize the chances of gene flow occurring. Once the crop has been released commercially, zones of exclusion or zones with active removal of volunteers and other recipient populations might be helpful in slowing the escape of the transgene. However, in the case of Bt cotton in Brazil, it is essentially a foregone conclusion that as there are no substantial genetic or physical barriers to gene flow, it will happen eventually. The real questions then revolve around the impact of that gene flow.

Overall Conclusions

If Bt cotton is introduced in Brazil, there will be opportunity for gene flow to occur. Natural reproductive barriers will not stop gene flow from Bt cotton to other *G. hirsutum* varieties, including the local cultivar *marie-galante*. The other two *Gossypium* species present in Brazil, *G. barbadense* and *G. mustelinum*, can also successfully cross with Bt cotton. If there is a fitness advantage of the Bt gene to other *Gossypium* species and varieties, the

Table 11.6. Proposed testing stages for assessing possible risks associated with gene flow from transgenic crops to their wild or cultivated relatives.

Stage	Questions to be answered	Setting for test

1. Consideration of the biology of crop to be transformed and site where it will be released	• What crop will be transformed?
	 • Over what area will the crop be released?
	 • Are there crop relatives with overlap in distribution?
	   • Is there phenological overlap between crop and relative populations?
	   • How much genetic diversity is found in populations of relatives?
	 • Are any of the potential recipient populations endangered or rare?
	 • How far can pollen move from the crop?
	 • Over what distance can outcrossing occur?
	 • How fit are hybrids between crop and recipient populations?
		Existing knowledge from literature or experts. If information is not available, perform experiments. If not in literature, test in greenhouse/ field using non-transgenic crop.

2. Post-transformation, prerelease: gene flow and introgression for crop and local recipient populations	• Will transgene be expressed similarly in different genetic backgrounds? • Will inheritance of the transgene be stable over multiple generations? • What is the maximal fitness effect of transgene in optimal environments? • What pleiotropic effects are observed when transgene is present?	*In greenhouse using transgenic crop.*
3. Small-scale field testing	*Assessing and managing possible risks associated with small-scale test* • What related populations occur in area where test will be done? *Data for assessing the risk of commercial release*	*In areas where crop is being tested in both agricultural and natural environments.*

4. Large-scale field testing	• How does fitness effect of the transgene and phenology of crop and relatives vary in different years, sites and hybrid crosses? • How does fitness of transgenic hybrids compare to crop and weed populations in a realistic pest environment? Continue experiments in test area. Monitor for volunteers, test for gene flow in surrounding fields or populations.	*In area where crop is being tested in both agricultural and natural environments.*
5. Commercial use	Monitor for volunteers, intensify efforts to remove escape plants in areas of special concern. Continue to test for gene flow and quantify the rate of gene movement.	*In areas where crop is released in both agricultural and natural environments.*

transgene will establish and will likely spread. Once the transgene is incorporated into a recipient population, it will be subject to natural selection pressure
and the outcome is difficult to predict. Where Bt cotton is introduced and how
it is managed will influence the extent and type of consequences.

Due to the knowledge gaps in the information needed to assess the potential for gene flow, the identification of potential adverse consequences could
not be thoroughly addressed in this chapter. As more information becomes
available on the biology and ecology of the species involved, a more in-depth
evaluation of the ecological and agronomic risks should be conducted.

References

Arriola, P.E. and Ellstrand, N.C. (1996) Crop-to-weed gene flow in the genus
    Sorghum (Poaceae): spontaneous interspecific hybridization between johnson
    grass, *Sorghum halepense*, and crop sorghum, *S. bicolor. American Journal of
    Botany* 83, 1153–1160.
Beasley, J.O. (1940) The origin of American tetraploid *Gossypium* species. *American
    Naturalist* 74, 285–286.
Beasley, J.O. (1942) Meiotic chromosome behaviour in species, species hybrids, hap-
    loids and induced polyploids of *Gossypium. Genetics* 27, 25–54.
Bergelson, J. and Purrington, C.B. (2002). Factors affecting the spread of resistant
    Arabidopsis thaliana populations. In: Letourneau, D.K. and Burrows, B.E. (eds)
    *Genetically Engineered Organisms: Assessing Environmental and Human
    Health Effects.* CRC Press, Boca Raton, Florida, pp. 33.
Berkey, D.A., Savoy, B.R., Miller, S.R. and Johnson, P.G. (2002) Pollen dissemination
    from adjacent fields of genetically enhanced cotton in the Mississippi delta.
    Proceedings of the 2002 Beltwide Cotton Conferences. National Cotton
    Council, Memphis, Tennessee.
Brubaker, C.L., Koontz, J.A. and Wendel, J.F. (1993) Bidirectional cytoplasmic and
    nuclear introgression in the New World cottons, *Gossypium barbadense* and
    G. hirsutum. American Journal of Botany 80, 1203–1208.
Brubaker, C.L., Bourland, F. and Wendel, J.F. (1999) The origin and domestication of
    cotton. In: Smith, C. and Cothren, J. (eds) *Cotton: Origin, History, Technology,
    and Production.* John Wiley & Sons, New York, pp. 3–31.
Burke, J.M. and Rieseberg, L.H. (2003) Fitness effects of transgenic disease resistance
    in sunflowers. *Science* 300, 1250–1250.
Burke, J.M., Gardner, K.A. and Rieseberg, L.H. (2002) The potential for gene flow
    between cultivated and wild sunflower (*Helianthus annuus*) in the United States.
    American Journal of Botany 89, 1550–1552.
Crow, J.F. and Kimura, M. (1970) *An Introduction to Population Genetics Theory.*
    Harper & Row, New York.
Edwards, G.A. and Mirza, M.A. (1979) Genomes of the Australian wild species of cot-
    ton. II. The designation of a new G genome for *Gossypium bickii. Canadian
    Journal of Genetics and Cytology* 21, 367–372.
Ellstrand, N.C., Prentice, H.C. and Hancock, J.F. (1999) Gene flow and introgression
    from domesticated plants into their wild relatives. *Annual Review of Ecology and
    Systematics* 30, 539–563.
Endrizzi, J.E., Turcotte, E.L. and Kohel, R.J. (1985) Genetics, cytology, and the evo-
    lution of *Gossypium. Advanced Genetics* 23, 271–375.

Freire, E.C. (2000) *Distribuicao, coleta, uso e preservacao das especies silvestres de algodao no Brasil*. Embrapa, Centro Nacional de Pesquisa de Algodao, Campina Grande, PB, Brazil.

Fryxell, P.A. (1979) *The Natural History of the Cotton Tribe*. Texas A&M University Press, College Station, Texas.

Fryxell, P.A. (1992) A revised taxonomic interpretation of *Gossypium* L. (Malvaceae). *Rheedea* 2, 108–165.

Gerstel, D.U. and Sarvella, P.A. (1956) Additional observations on chromosomal translocations in cotton hybrids. *Evolution* 10, 408–414.

Green, J.M. and Jones, M.D. (1953) Isolation of cotton for seed increase. *Agronomy Journal* 45, 366–368.

Harper, B.K., Mabon, S.A., Leffel, S.M., Halfhill, M.D., Richards, H.A., Moyer, K.A. and Stewart, C.N. (1999) Green fluorescent protein as a marker for expression of a second gene in transgenic plants. *Nature Biotechnology* 17, 1125–1129.

Hasenkampf, C.A. and Menzel, M.Y. (1980) Incipient genome differentiation in Gossypium. II. Comparison of 12 chromosomes in *G. hirsutum*, *G. mustelinum*, and *G. tomentosum* using heterozygous translocations. *Genetics* 95, 971–983.

Haygood, R., Ives, A.R. and Andow, D.A. (2003) Consequences of recurrent gene flow from crops to wild relatives. *Proceedings of the Royal Society of London, Series B* 270, 1879–1886.

Hutchinson, J.B. (1951) Intra-specific differentiation in *Gossypium hirsutum*. *Heredity* 5, 161–193.

Hutchinson, J.B., Silow, R.A. and Stephens, S.G. (1947) *The Evolution of Gossypium*. Oxford University Press, London.

Johnston, J., Blancas, L. and Borem, A. (2004) Gene flow and its consequences: a case study of Bt maize in Kenya. In: Hilbeck, A. and Andow, D.A. (eds) *Environmental Risk Assessment of Genetically Modified Organisms*, Volume 1: *A Case Study of Bt Maize in Kenya*. CAB International, Wallingford, UK.

Kearney, T.H. and Harrison, G.J. (1932) Pollen antagonism in cotton. *Journal of Agricultural Resources* 44, 191–226.

Mauer, F.M. (1930) The cottons of Mexico, Guatemala and Colombia. *Bulletin of Applied Botanical Genetics and Plant Breeding Supplement*, 543–553.

Moreira, J.D.N., Freire, E.C., Dossantos, J.W. and Vieira, R.M. (1995) Use of numerical taxonomy to compare mocó cotton with other cotton species and races. *Revista Brasileira de Genetica* 18, 99–103.

Nagylaki, T. (1977) *Selection in One- and Two-Locus Systems*. Springer-Verlag, Berlin.

Percy, R.G. and Wendel, J.F. (1990) Allozyme evidence for the origin and diversification of *Gossypium barbadense* L. *Theoretical and Applied Genetics* 79, 529–542.

Phillips, L.L. and Strickland, M.A. (1966) The cytology of a hybrid between *Gossypium hirsutum* and *G. longicalyx*. *Canadian Journal of Genetics and Cytology* 8, 91–95.

Seelanan, T., Schnabel, A. and Wendel, J.F. (1997) Congruence and consensus in the cotton tribe (Malvaceae). *Systematic Botany* 22, 259–290.

Snow, A.A. and Palma, P.M. (1997) Commercialization of transgenic plants: potential ecological risks. *Bioscience* 47, 86–96.

Snow, A.A., Pilson, D., Rieseberg, L.H., Paulsen, M.J., Pleskac, N., Reagon, M.R., Wolf, D.E. and Selbo, S.M. (2003) A Bt transgene reduces herbivory and enhances fecundity in wild sunflowers. *Ecological Applications* 13, 279–286.

Stephens, S.G. (1946) The genetics of 'corky' I. The New World alleles and their possible role as an interspecific isolating mechanism. *Journal of Genetics* 47, 150–161.

Stephens, S.G. (1950) The genetics of 'corky' II. Further studies on its genetic basis in relation to the general problem of interspecific isolating mechanisms. *Journal of Genetics* 50, 9–20.

Stephens, S.G. (1967) Evolution under domestication of the New World cottons (*Gossypium* spp.). *Ciencia e Cultura* 19, 118–134.

Stephens, S.G. and Phillips, L.L. (1972) The history and geographical distribution of a polymorphic system in New World cottons. *Biotropica* 4, 49–60.

Stewart, J.M. (1995) Potential for crop improvement with exotic germplasm and genetic engineering. In: Constable, G.A. and Forrester, N.W. (eds) *Challenging the Future: Proceedings of the World Cotton Research Conference 1*. CSIRO, Melbourne, Australia.

Turcotte, E.L. and Percy, R.G. (1990) Genetics of kidney seed in *Gossypium barbadense* L. *Crop Science* 30, 384–386.

Umbeck, P.F., Barton, K.A., Nordhaim, E.V., McCarty, J.C., Parrott, W.L. and Jenkins, J.N. (1991) Degree of pollen dispersal by insects from a field test of genetically engineered cotton. *Journal of Economic Entomology* 84, 1943–1950.

Watt, G. (1907) *The Wild and Cultivated Cotton Plants of the World*. Longmans, Green & Co, London.

Wendel, J.F. (1989) New World tetraploid cottons contain Old World cytoplasm. *Proceedings of the National Academy of Sciences USA* 86, 4132–4136.

Wendel, J.F., Rowley, R. and Stewart, J.M. (1994) Genetic diversity in and phylogenetic relationships of the Brazilian endemic cotton, *Gossypium mustelinum* (Malvaceae). *Plant Systematics and Evolution* 192, 49–59.

Glossary of Terms

Agricultural environment: Areas under active or recent cultivation and those areas directly impacted by agriculture.

Allotetraploid: A plant whose nuclear genome comprises two divergent diploid genomes (see also polyploid).

Centre of diversity: The geographical area of divergence for a group of related taxa.

Centre of origin: The location of the original gene pool that has diverged into a group of related taxa.

Corky: A phenotype observed in some interspecific hybrids between *G. barbadense* and *G. hirsutum* that is characterized by mosaic yellowing of the leaf mesophyll and excessive cork production on stems, leaf petioles and midribs. Because this cork production begins very early, growth is stunted. 'Corky' plants will flower and the pollen is viable, but female fertility is very low. The corky phenotype arises from the interaction of two of three codominant alleles at the 'corky' locus. These alleles are differentially distributed between *G. barbadense* and *G. hirsutum*: ck^0 is common in both species, ck^x is restricted to Caribbean and northern South American *G. hirsutum* and ck^y is found only in *G. barbadense*. Five of the six allelic combinations at this locus produce normal phenotypes, but interspecific hybrids that combine ck^x and ck^y express the 'corky' phenotype.

Cultigen: A domesticated form of a wild species. It is usually applied to indigenous forms that have undergone minimal human selection.

Day-length neutral: The characteristic of some cotton cultivars (and other cultivated plants) to flower regardless of day length. Primitive forms of the cultivated cottons are day-length sensitive and in temperate latitudes fail to initiate flowering early enough in the growing season for the bolls (capsules) to mature before the first killing frost. The ability to cultivate cotton in the higher latitudes was only possible with the discovery and selection of day-length neutral genotypes that would initiate flowering as early as physiologically possible after germination and without reference to day length.

Disturbed environment: Areas that are currently or recently impacted by human activity, including roadsides and areas indirectly affected by agriculture.

Dooryard cotton: Cotton that is grown near homes with little or no active cultivation.

Egyptian cotton: Extra-long staple cottons developed in Egypt, reportedly from a cross between a Sea Island cotton and an agronomically primitive *G. barbadense* tree cotton, Jumal's tree cotton. The Egyptian cottons were introduced into the south-western USA around 1900, where they were used as the primary genetic basis of the Pima cultivars.

Elite variety: Commercial variety that is the result of a formal breeding programme, which may be a protected variety.

Epicalyx: A whorl of bracts that subtend the flowers of the cotton species and their relatives. The epicalyx of the four cotton species consists of three large leaf-like (foliaceous) bracts that completely surround the flower in the early stages of floral development; together they are referred to as the square.

Feral populations: Sexual or vegetative progeny that are derived from crop plants and have established long-lived populations.

Fitness: The reproductive success of an entity; the contribution of a genotype to successive generations.

Fuzz (linters): The layer of short, unconvoluted, tightly adherent seed hairs that underlay the longer lint fibres of the cultivated cottons. Linters are unspinnable, but are processed as a source of cellulose.

Gene flow: The incorporation of genes into the gene pool of one population from one or more other populations.

Genetic diversity: Genetic differentiation among individuals and among populations.

Genome group: Related species whose nuclear genomes are cytogenetically more similar to each other than they are to other groups of species. Genome groups are most often recognized on the basis of comparative analyses of chromosome pairing behaviour in interspecific hybrids. Hybrids between members of the same genome group typically exhibit normal or near-normal meiotic configurations and partial to complete fertility, whereas interspecific hybrids between members from different genome groups show various degrees of abnormal pairing and complete sterility.

Introgression: The transfer, via hybridization and genetic recombination, of genes between gene pools.

Isozyme: Multiple molecular forms of proteins. These are often visualized using electrophoresis and histochemical staining techniques, and provide a set of readily scored genetic markers.

Kidney cotton: A group of *G. barbadense* indigenous cultigens from northeastern South America in which the seeds of each locule are fused into a solid kidney-shaped mass. Although they are not genetically distinct from the other indigenous forms of *G. barbadense*, they are sometimes recognized taxonomically as *G. barbadense* var. *braziliensis*. The kidney-seed phenotype is monogenic and is encoded by a recessive allele.

Landrace: Traditional varieties of cotton that are cultivated by small farmers; the seeds are saved for planting in subsequent cropping seasons.

Natural environment: Undisturbed ecosystems or areas of low disturbance.

New World cottons: A term that refers to the two cultivated cotton species that evolved and were domesticated in the New World, namely, *G. barbadense* and *G. hirsutum*.

Old World cottons: A term that refers to the two cultivated cotton species that evolved and were domesticated in the Old World, namely, *G. arboreum* and *G. herbaceum*.

Pollen gene flow: Movement of genetic material by pollen from one plant into another including the same or different species.

Polyploidy: The presence of multiple sets of chromosomes within a single nucleus. Allopolyploids recombine divergent sets of chromosomes. Autopolyploids contain multiple copies of a single set of chromosomes.

Race: A group of cultigens that have a common geographical or genetic point of origin. The term usually is applied to groups of indigenous cultigens.

Sea Island cotton: Day-length neutral *G. barbadense* cultivars grown on the coastal plains and islands of Georgia, South Carolina and Florida and in the West Indies that were developed from introduced South American *G. barbadense* cultigens. They were the first commercially successful extra-long staple cultivars. The Sea Island industry collapsed under boll weevil pressure by 1920, but the Sea Island gene pool contributed to the next generation of extra-long staple cottons developed in Egypt from which the Pima cottons were later developed.

Seed gene flow: Movement of genetic material from one site to another through the movement of seeds.

Sympatry: The occurrence of populations of more than one species in one geographical region.

Tree cotton: A term used to designate agronomically primitive arborescent forms of the four cultivated cotton species.

Tufted seed: A characteristic of some agronomically primitive *G. hirsutum* cultigens, in which the seed fuzz layer is missing, save a small 'tuft' around the micropyle. It is thought that tufted-seeded cultigens may have been preferred by aboriginal groups who ginned their cottonseed by hand.

Upland cotton: Modern, highly improved cotton cultivars and their derivatives that were developed in the southern USA cotton belt from introduced indigenous *G. hirsutum* cultigens. The term originally arose to discriminate between the *G. barbadense* cultivars, known as Sea Island cottons, grown on the coastal plains and islands of Georgia, South Carolina and Florida, from the mostly *G. hirsutum* cottons grown in inland (upland) regions. They are also distinct from the Cambodian cottons developed in Asia from a different set of indigenous introductions (mostly belong to the race 'punctatum').

Volunteer: Progeny from crop plants that survive to the next growing season, either as seeds or vegetative propagules.

Wild species: A plant species that occurs only in natural environments.

12 Resistance Risks of Bt Cotton and their Management in Brazil

G.P. Fitt, C. Omoto, A.H. Maia, J.M. Waquil, M. Caprio, M.A. Okech, E. Cia, Nguyen Huu Huan and D.A. Andow

Corresponding author: Gary Fitt, CSIRO Entomology, Australian Cotton CRC, Australia. Fax: +61 7 32142881, e-mail: Gary.Fitt@csiro.au

Introduction

This chapter addresses the risk that insect pests associated with *Bacillus thuringiensis* (Bt) cotton may evolve resistance to Bt proteins in Brazil. Insecticide resistance is a common response among insects to the selection pressure imposed by insecticides. The framework and concepts developed here are designed to be applicable to diploid arthropods evolving resistance to insecticidal transgenes expressed in transgenic plants. With some modification, they should also be relevant to assessing and managing resistance risk in haplodiploid and parthenogenetic arthropods, nematodes, virus, fungi or bacteria to nematicidal, viral, fungicidal or bactericidal transgenes, as well as to herbicide resistance. This framework follows that in Fitt *et al.* (2004), which was developed for the case of Bt maize in Kenya. In this chapter, we more formally integrate the framework into the risk-analysis process, considering briefly how the analysis should be staged to correspond with the development of the transgenic crop (see Andow *et al.*, Chapter 1, this volume).

We have established a series of informational needs that are essential to completing an assessment of risk and the development of a practical risk-management plan. In Fitt *et al.* (2004), we established a series of questions that should be addressed in this risk analysis. Here we integrate these questions under the informational needs to indicate clearly how each contributes to the risk analysis. This chapter focuses primarily on risk analysis prior to any field release of the transgenic crop plant. It concentrates on a comprehensive assessment of the pest/plant system and ecological attributes of the pests that help to define the risk of resistance and indicate possible resistance-management approaches. Additional research during field testing should be used to address key assumptions and develop an effective, workable and acceptable resistance-management plan and to establish details of the monitoring and response system.

It should be assumed that resistance is a real risk. Experience with insecticide use and basic consideration of evolutionary theory implies that if the Bt crop is used extensively without any resistance-management intervention, resistance should be considered inevitable. The real issue is how to delay its onset. For any given crop there are usually multiple pest species that require control, and any given pest control tactic usually affects multiple pest species. In cotton there are many insect pest species, so it is important to assess which species are at risk of resistance and which is most at risk. In this chapter, we use risk assessment to identify the species most at risk of resistance and then devise risk-management practices that could delay the onset of resistance in this species and all of the others.

To assess the relative resistance risk of a Bt crop, the following issues should be addressed prior to field release:

- Identification of the pest species that are at risk of evolving resistance to the transgene (species, geographic distribution, history of resistance).
- Determination of potential exposure of each species to selection.
- Determination of the likely 'dose' of the transgene toxin to which each species is likely to be exposed.

Dose is a simple measure of 'hazard' projecting the likely effect of exposure in relation to the time to resistance failures. Risk assessment then involves combining information on dose with an estimate of potential exposure. With this information, we can assess the relative resistance risk of the various pest species and identify the species that is most likely to evolve resistance before the others – which might be called the weak link.

Resistance management first focuses around the biological attributes of this weak-link species. We then assess whether the resistance-management strategy constructed around the weak link would also delay the evolution of resistance in other species at risk. While doing this, it is essential that the resistance-management plan be practicable, that is, growers can actually implement it. The resistance-management plan builds on the information from the previous risk assessment, using the following three steps:

- Determination of the likely requirements for resistance management, including refuges.
- Development of the components of a potentially workable resistance-management plan.
- Specification of monitoring needs and development of potential contingency responses.

We then discuss some issues to be considered after field release, but before commercial release. Field and laboratory experiments will be needed to develop an effective, feasible, acceptable resistance-management plan and monitoring and response system. The information presented in this chapter reflects the complexity and dynamic nature of Brazilian cropping systems where numerous uncertainties exist about the potential deployment of Bt crops and there is limited quantitative understanding of pest ecology. Despite these uncertainties, it is possible to assess resistance risk to various Bt transgenic cottons and to develop a reasonable management plan to delay resistance evolution.

Resistance Risk Assessment

Operational definition of resistance

Resistance is caused by genes that reduce susceptibility to a toxin, and is a trait of an individual. However, it will often happen that resistance is not yet known in a target species at the prerelease stage of development of the transgenic crop. Thus, it is important to define *resistance* operationally, so that resistance can be looked for in advance. This definition will, by necessity, be modified as information becomes available about the expression and inheritance of resistance. This definition is discussed in the section on 'potential exposure of target pests to *Bacillus thuringiensis* (Bt) cotton' below.

In addition, resistance occurs in a field population when there are enough resistant individuals to cause economic damage to the target crop. Hence, it is also necessary that we have an operational definition of *control failure from resistance*; this will be a characteristic of a population and should be easily and unambiguously implemented. An operational definition of control failure from resistance is necessary so that we know what we want to avoid during resistance management and know when we should admit failure and move on. Operationally, a control failure from resistance occurs when the pest causes significant economic damage to the crop. There are several alternative ways to implement this concept. For example, a control failure could be defined as occurring when the pest causes detectable economic damage to the crop, when the pest causes economic damage that is similar to that caused by susceptible insects on a non-resistant crop variety or when the economic damage is considered unacceptable to the grower.

Identification of pest species at risk

Identification of key pest species that could evolve resistance to Bt cotton involves identifying the key target pests first in each of the major geographic regions and then evaluating the resistance history of each species. In some cases, identification of the key target pest species can be difficult because the transgenic crop has not been tested against all relevant species. For Bt cotton in Brazil, there is about the target species but almost no information about the ability of Bt cotton to control them. The history of resistance is also important for determining resistance risk. This can be clearly illustrated by two Australian pest species, *Helicoverpa armigera* and *Helicoverpa punctigera*, which are significant pests of cotton and often controlled by a range of insecticides. *H. armigera* has historically evolved resistance to all major classes of insecticides deployed against it (Forrester *et al.*, 1993), while *H. punctigera* has not developed field resistance to any insecticide despite being exposed to the same selection in cotton as *H. armigera*. This difference reflects the differing host range and mobility of the two species (Fitt, 1989; Fitt and Daly, 1990; Gregg *et al.*, 1995), and results in a substantial proportion of the *H. punctigera* population avoiding selection in unsprayed crops and non-crop plants. When Bt cotton was introduced into Australia, this history of insecticide resistance clearly identified *H. armigera* as a resistance risk and hence the target for a pre-emptive resistance-management strategy (Fitt, 1997).

We have identified four key target pests for deployment of Bt cotton in Brazil. These are cotton budworm, *Heliothis virescens* (Fabricius); pink bollworm, *Pectinophora gossypiella* (Saunders); cotton leafworm, *Alabama argillacea* (Hübner) and fall armyworm, *Spodoptera frugiperda* (J.E. Smith). These species are all regarded as significant pests of cotton and, historically, *H. virescens* and *A. argillacea* have been regarded as key pests. *P. gossypiella* and *A. argillacea* are specialists, largely restricted to cotton or closely related Malvaceae. *H. virescens* and *S. frugiperda*, in contrast, are polyphagous species with a wide range of unrelated host plants. *H. virescens* appears to be somewhat more specialized in Brazil than in the USA (Gallo *et al.*, 2002). However, with the expansion of agriculture in the Midwest region in Brazil, *H. virescens* also started infesting soybean fields in this region.

Since the introduction of boll weevil in 1983 and of virus-susceptible US varieties in the last 10 years, both the pest spectrum and the distribution of cotton in Brazil have changed dramatically (Ramalho, 1994; Fontes *et al.*, Chapter 2, this volume). About 70% of cotton area is now grown in the Midwest where large-scale developments have occurred only in the last 5–10 years. The infestation of boll weevil is not very critical in the Midwest yet; however, viral diseases (e.g. blue disease) transmitted by the cotton aphid have become extremely important. Maize production has also changed with a diversification of cropping to include extensive areas of autumn and irrigated winter production. As a consequence, *S. frugiperda* has emerged as one of the most significant threats to cotton production. It is regarded as an induced pest in the system. We are thus dealing with relatively new production systems and pest dynamics in environments where relatively little historical data are available.

A number of other Lepidoptera may occur on cotton but all are regarded as too minor to be considered here. Surprisingly, maize earworm, *Helicoverpa zea* (Boddie), a regular pest on US cotton, appears not to infest cotton in Brazil (Degrande, 1998; Gallo *et al.*, 2002). Given that, it could not be considered a resistance risk, although its host range in North America and Central America does include cotton. Research on the ecological genetics of host use in Brazilian *H. zea* may well generate interesting comparisons with its behaviour elsewhere in the Americas.

The geographic distribution of these species varies somewhat among the three major cotton agroecological regions in Brazil (Table 12.1). In the major production area, the Midwest, all of the Lepidoptera are important pests, although boll weevil is a significant and increasing pest and, together with

S. frugiperda, represents the main threat to production in the Midwest. In the Meridian region, *S. frugiperda* is a minor pest of cotton. In the North-east region, where the poorest cotton farmers are found, boll weevil and the specialized Lepidoptera, *A. argillacea* and *P. gossypiella*, are the most important pests. With the expansion of upland cotton in the North-east, *S. frugiperda* is also becoming an important cotton pest in this region. Some characteristics of the farming systems are given in Table 12.2 (see also Fontes *et al.*, Chapter 2, this volume). These data will be important to design a practical resistance-management strategy.

History of resistance

Species with a history of resistance should be prioritized because their *a priori* risk of evolving resistance to Bt cotton is high.

An extensive range of pesticides is registered for use in cotton and maize crops. The organophosphates (OPs), synthetic pyrethroids (SPs) and insect growth regulators (IGRs) are used predominantly (see Table 2.3 in Fontes *et al.*, Chapter 2, this volume). In extensive cotton production of varieties largely derived from US germplasm, the key targets for pesticides are now early-season aphids, which are vectors for viral diseases, and boll weevils that occur in most regions of Brazil. Spraying for these pests suppresses many of the Lepidopteran pests, although three to four insecticide applications are likely to target Lepidoptera specifically. The average frequency of insecticide applications in the three main cotton-production regions varies from 4 to 20 applications per crop (Table 12.2). The Midwest, which produces 86% of Brazilian cotton fibre, is an intensive production system with the highest pesticide input with an average of 16–20 applications per growing season.

Table 12.1. Regional differences in severity of arthropod pests of Brazilian cotton.

Region		Midwest	Meridian	North-east	Resistance history[a]
	% total cotton area[b]	76.5	12	11	
Target pests	*Heliothis virescens*	***	***	*	OCl, OPs
	Alabama argillacea	***	***	***	SPs, OPs
	Pectinophora gossypiella	***	***	***	None
	Spodoptera frugiperda	***	*	–	Widespread
Other key pests	*Anthonomus grandis*	**	***	***	None
	Aphis gossypii	***	**	*	OPs

*Severity: *** = most severe.
[a]Insecticide class: OCl = organochlorine; OP = organophosphate; SP = synthetic pyrethroid.
[b]Remaining area not included here is in the northern region (State of Tocantins).

Table 12.2. Regional differences in farm size, inputs and anticipated impact of Bt cotton on insecticide use in Brazilian cotton.

Region	Midwest	Meridian	North-east
% of cotton production[a]	87	10	2.5
Farm size	Large	Intermediate to large	Small
Inputs	Very high	Intermediate to high	Very low
Average yield (kg/ha)[b]	3357	2518	1770
Insecticide applications	12–20	6–12	4–8
Anticipated reduction in insecticide use from Bt cotton	2–4 applications	2 applications	2 applications

[a]Remaining production not included here is in the northern region (State of Tocantins).
[b]Source: CONAB (2003).

Most of the applications are targeted to control aphids in the virus-susceptible cotton varieties (six to eight applications), followed by Lepidopteran pests (three to four), boll weevil (two to four) and stink bugs (one to two).

Limited information is available to assess the past history of resistance in Lepidoptera pests of cotton in Brazil (Table 12.1) because there has been no regular monitoring programme in place. Recent baseline assessments and monitoring have demonstrated high frequency of resistance in *S. frugiperda* to OPs and SPs in cotton-producing regions (Diez-Rodriguez and Omoto, 2001). There is also evidence of incipient resistance to IGRs in this species (C. Omoto, Piracicaba, 2004, personal communication). For *A. argillacea*, there is a high probability of resistance to OPs and SPs based on frequent field failures for controlling this pest; however, there is no evidence of pesticide resistance in *P. gossypiella* in Brazil, although measurable background frequency of Bt resistance has been found in field populations in the USA (Tabashnik *et al.*, 2000a). Despite the widespread deployment of Bt cotton in that environment, Bt resistance has not increased in frequency due to other factors (Carrière *et al.*, 2002).

During the 1970s, there are accounts of field failures of organochlorines (OCls) and OPs against *H. virescens* to the extent that cotton production declined in Brazil. Whether this was due to resistance was not documented at that time. During the early 1980s, prior to introduction of the boll weevil, significant natural enemy populations had developed following reduced use of pesticides. These included high populations of egg parasitoids as one important component of a successful integrated pest management (IPM) approach. In 1983, with the introduction of the boll weevil, significantly increased use of insecticide sprays reduced beneficial organisms, complicating management of

H. virescens (Ramalho, 1994). Subsequently, during the mid 1990s, some field failures of insecticides were reported, but as before, there were no efforts to document that the failures were caused by resistance. These episodes none the less indicate that the Brazilian cotton-production systems are probably quite effective at selecting for resistance in *H. virescens*. We urge that appropriate baseline information on resistance frequencies be collected for the identified key pests prior to deployment of Bt cottons.

Based on the history of resistance to insecticides, we conclude that of the four target species, *H. virescens*, *S. frugiperda* and *A. argillacea* have the highest risk of resistance in Brazil. In addition, based on recent reports of resistance to Bt cotton in the USA (Tabashnik *et al.*, 2000a), *P. gossypiella* also poses a significant resistance risk.

Potential exposure of target pests to Bt cotton

Association with Bt cotton

The association of the target species with Bt cotton is the *maximum period of overlap* of the species on the target crop, in terms of area, spatial distribution and seasonal availability of the crop. Overlap can be evaluated on the basis of presence and absence and general knowledge about the species. More precise, quantitative evaluations will become necessary to develop realistic resistance-management plans (see next section).

The four target species differ markedly in host range and association with cotton (Table 12.3). *P. gossypiella* is a specialist on Malvaceae, particularly on *Gossypium* species. It is thought to feed only on cotton in Brazil. *A. argillacea* is also restricted to cotton. No alternative hosts are known. *H. virescens* has a wide range of recorded hosts, including crops and wild hosts, although in Brazil it is thought to be closely associated with cotton and rarely occurs on other crops (Gallo *et al.*, 2002). Likewise, *S. frugiperda* has an extremely broad range of potential hosts (Table 12.4), though actual usage may be much more constrained.

Table 12.3. Association and fecundity of target Lepidopteran species in Brazilian cotton.

Species	Number of generations in cotton	Number of generations per year	Adult dispersiveness	Cannibalism	Fecundity
Alabama argillacea	3	3	Very high	Low	500–800
Pectinophora gossypiella	3–5	3–5	Very low	Very low	250–500
Heliothis virescens	2–3	2–3	Low	Very high	800–1000
Spodoptera frugiperda	2	6	High	Very high	750–1250

Figure 12.1 illustrates the main periods of cotton production in the three main cotton agroclimatic zones of Brazil. Planting of cotton occurs progressively later from south to north, again reflecting that most cotton is rainfed and rains commence earlier in the south. With the expansion of the second cotton season (planting in January, Fig. 12.1) in the Midwest region, the pest pressure and consequently the pest exposure to cotton will increase in this region. All the four target species have at least two complete generations in cotton. For *P. gossypiella, A. argillacea* and *H. virescens*, all generations are associated with cotton (Table 12.3).

S. frugiperda displays a very wide host range (Table 12.4), but clearly prefers grasses. This species is comprised of two genetically different, but

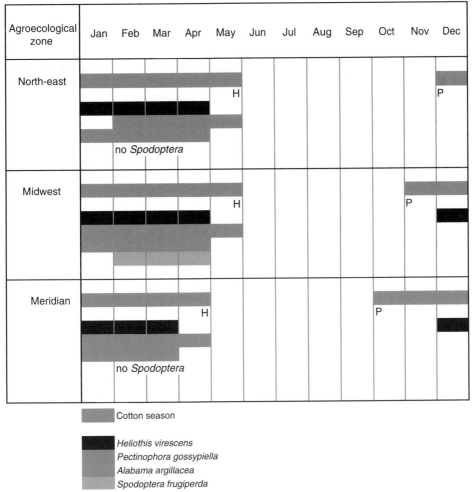

Fig. 12.1. Main cotton-production season and occurrence of key Lepidopteran pests across three main regions. 'P' indicates the typical planting date and 'H' indicates the typical harvest date.

morphologically identical host strains (Pashley, 1986), a 'maize' strain and a 'rice' strain. The 'maize' strain feeds principally on maize, sorghum and cotton, and will feed on a few other hosts when they grow near the primary hosts. In Brazil, maize and cotton are important hosts, as are a number of other crops (Cruz, 1995; Degrande, 1998). Maize appears to be preferred over other hosts. The extent to which populations occur in wild host plants in Brazil is unknown. Non-crop hosts are abundant only during the summer

Table 12.4. List of host plants for fall armyworm, *Spodoptera frugiperda* (Smith), adapted from the *Spodoptera* database (Lepidoptera: Noctuidae) (Pogue, 1995).

Scientific name	Common name	Botanical order	Botanical family
Agrostis alba (L.)		Graminales	Poaceae
Agrostis hyemals (Walt.)		Graminales	Poaceae
Allium cepa (L.)	Onion	Liliales	Liliaceae
Althaea rosea (Cav.)	Hollyhock	Malvales	Malvaceae
Amaranthus sp.	Pigweed	Caryophyllales	Amaranthaceae
Andropogon virginicus (L.)		Graminales	Poaceae
Arachis hypogaea (L.)	Groundnut	Rosales	Fabaceae
Atropa belladonna (L.)	Deadly nightshade	Polemoniales	Solanaceae
Avena sativa (L.)	Oats	Graminales	Poaceae
Beta vulgaris (L.)	Mangold, beet, sugarbeet	Caryophyllales	Chenopodiaceae
Brassica napus var. napobrassica (L.) Rchb.	Rutabaga	Papaverales	Brassicaceae
Brassica oleracea (L.) var. *viridis* (L.)	Kale, collards	Papaverales	Brassicaceae
Brassica oleracea (L.)	Kale	Papaverales	Brassicaceae
Capsicum annuum (L.) var. *annuum*	Green, bell, sweet, red pepper, chilli	Polemoniales	Solanaceae
Carex sp.	Sedges	Graminales	Cyperaceae
Carya illinoinensis (Wangenh.) K. Koch	Pecan	Juglandales	Juglandaceae
Carya sp.	Hickories	Juglandales	Juglandaceae
Cenchrus pauciflorus (Benth.)		Graminales	Poaceae
Chenopodium quinoa (Willd.)	Quinoa	Caryophyllales	Chenopodiaceae
Chloris gayana (Kunth)	Rhodes grass	Graminales	Poaceae

Table 12.4. List of host plants for fall armyworm, *Spodoptera frugiperda* (Smith), adapted from the *Spodoptera* database (Lepidoptera: Noctuidae) (Pogue, 1995) – cont'd.

Scientific name	Common name	Botanical order	Botanical family
Chrysanthemum sp.	Chrysanthemum	Asterales	Asteraceae
Cicer arietinum (L.)	Garbanzo, chick-pea	Rosales	Fabaceae
Citrullus lanatus var. *lanatus* (Thumb.) Matsum. & Nakai	Watermelon	Loasales	Cucurbitaceae
Citrus unshiu (Marcow)	Satsuma orange	Sapindales	Rutaceae
Convolvulus sp.		Polemoniales	Convolvulaceae
Corchorus capsularis (L.)	White jute	Malvales	Tiliaceae
Corchorus olitorius (L.)	Jute	Malvales	Tiliaceae
Croton capitatus (Michx.)	Woolly croton	Euphorbiales	Euphorbiaceae
Cucumis sativus (L.)	Cucumber	Loasales	Cucurbitaceae
Cynodon dactylon (L.) Pers.	Bermuda grass	Graminales	Poaceae
Cyperus rotundus (L.)	Nut sedge	Graminales	Cyperaceae
Dactyloctenium aegyptium (L.) Willd.	Crowfoot grass	Graminales	Poaceae
Dahlia hybrid	Dahlia	Asterales	Asteraceae
Digitaria eriantha (Steud.)	Pangola grass	Graminales	Poaceae
Digitaria ischaemum (Schreb.) (Schreb. ex. Muhl.)	Smooth crabgrass	Graminales	Poaceae
Digitaria sanguinalis (L.) Scop.	Large crabgrass	Graminales	Poaceae
Echinochloa colona (L.) Link	Jungle rice	Graminales	Poaceae
Eleusine indica (L.) Gaertn.	Goose grass	Graminales	Poaceae
Eriochloa polysta-chya (Kunth.)	Carib grass	Graminales	Poaceae
Fragaria chiloensis (L.) Duchesne	Beach strawberry, Chilean strawberry	Rosales	Rosaceae
Geranium sp.	Geranium	Graminales	Poaceae
Gladiolus glandavensis	Gladiolus	Liliales	Iridaceae
Glycine max (L.) Merrill	Soybean	Rosales	Fabaceae
Gossypium herbacium (L.)	Tree cotton	Malvales	Malvaceae

Continued

Table 12.4. List of host plants for fall armyworm, *Spodoptera frugiperda* (Smith), adapted from the *Spodoptera* database (Lepidoptera: Noctuidae) (Pogue, 1995) – *cont'd.*

Scientific name	Common name	Botanical order	Botanical family
Gossypium hirsutum	Upland cotton	Malvales	Malvaceae
Hevea brasiliensis (Muell. -Arg.)	Pokok getah, rubber tree	Euphorbiales	Euphorbiaceae
Hordeum vulgare (L.)	Barley	Graminales	Poaceae
Ipomoea batatas (L.) Lam.	Sweet potato	Polemoniales	Convolvulaceae
Ipomoea purpurea (L.) Roth	Common morn-ing-glory	Polemoniales	Convolvulaceae
Linum usitatissi-mum (L.)	Flax, linseed	Geraniales	Linaceae
Lolium perenne (L.)	Ryegrass	Graminales	Poaceae
Lycopersicon esculentum	Tomato	Polemoniales	Solanaceae
Malus domestica (Borkh.)	Apple	Rosales	Rosaceae
Manihot esculenta (Crantz)	Ubi kayu, tapioca plant, Cassava	Euphorbiales	Euphorbiaceae
Medicago sativa (L.)	Lucerne	Rosales	Fabaceae
Mucuna pruiens (Wall. ex Wight) Baker ex Burck	Velvet bean	Rosales	Fabaceae
Musa x (L.)	Banana	Musales	Musaceae
Nicotiana tabacum (L.)	Tobacco	Polemoniales	Solanaceae
Onosmodium virginianum (L.)		Polemoniales	Boraginaceae
Oryza sativa (L.)	Rice	Graminales	Poaceae
Panicum maximum (Jacq.)	Guinea grass	Graminales	Poaceae
Panicum miliaceum (L.)	Common millet	Graminales	Poaceae
Panicum purpur-ascens (Raddi)		Graminales	Poaceae
Panicum texanum (Buckley)		Graminales	Poaceae
Paspalum conjug-atum (P.J. Bergius)	Sour paspalum	Graminales	*Poaceae*
Pennisetum clandestinum (Hochst. ex Chiov.)	Kikuyu grass	Graminales	Poaceae
Pennisetum glaucum (L.) R. Br.	Pearl millet, bulrush millet	Graminales	Poaceae
Phleum pratense (L.)	Pearl millet, bulrush millet	Graminales	Poaceae

Table 12.4. List of host plants for fall armyworm, *Spodoptera frugiperda* (Smith), adapted from the *Spodoptera* database (Lepidoptera: Noctuidae) (Pogue, 1995) – *cont'd.*

Scientific name	Common name	Botanical order	Botanical family
Phytolacca americana (L.)	Pokeweed, poke, pigeonberry	Caryophyllales	Phytolaccaceae
Pisum sativum (L.)	Pea	Rosales	Fabaceae
Platanus occidentalis (L.)	Sycamore	Hamamelidales	Platanaceae
Plumeria rubra (L.)	Frangipani	Gentianales	Apocynaceae
Poa annua (L.)	Annual bluegrass, annual meadow grass	Graminales	Poaceae
Poa pratensis (L.)	Kentucky bluegrass	Graminales	Poaceae
Poa spp.	Pasture grass	Graminales	Poaceae
Polytrias praemorsa (Nees) Hack.		Graminales	Poaceae
Portulaca oleracea (L.)	Purslane	Caryophyllales	Portulacaceae
Prunus persica (L.) Batsch	Peach	Rosales	Rosaceae
Saccharum officinarum (L.)	Sugarcane	Graminales	Poaceae
Secale cereale (L.)	Rye	Graminales	Poaceae
Setaria italica (L.) P. Beauv.	Foxtail millet, Italian millet, German millet, Hungarian millet	Graminales	Poaceae
Solanum dulcamara (L.)	Bittersweet, bitter nightshade	Polemoniales	Solanaceae
Solanum melongena (L.)	Aubergine	Polemoniales	Solanaceae
Solanum tuberosum (L.)	Potato	Polemoniales	Solanaceae
Sorghum bicolor (L.) Moench	Grain sorghum, sorghum, sweet sorghum, milo	Graminales	Poaceae
Sorghum halpense (L.) Pers.	Johnsongrass	Graminales	Poaceae
Spinacia aleracea (L.)	Spinach	Caryophyllales	Chenopodiaceae
Trifolium pratense (L.)	Red clover, purple clover, peavine clover	Rosales	Fabaceae
Triticum aestivum (L.), nom. cons.	Bread wheat, wheat	Graminales	Poaceae

Continued

Table 12.4. List of host plants for fall armyworm, *Spodoptera frugiperda* (Smith), adapted from the *Spodoptera* database (Lepidoptera: Noctuidae) (Pogue, 1995) – *cont'd.*

Scientific name	Common name	Botanical order	Botanical family
Urochloa decumbens (Stapf) R. D. Webster	Signal grass	Graminales	Poaceae
Urochloa mutica (Forssk.) T.Q. Nguyen	Mauritius grass, paragrass	Graminales	Poaceae
Vaccinium macro-carpon (Aiton)	Cranberry, large cranberry, American cranberry	Ericales	Ericaceae
Vigna unguiculata ssp. *unguiculata* (L.) Walp.	Southern-pea, cowpea, black-eyed pea, crowder-pea	Rosales	Fabaceae
Viola sp.	Violets	Violales	Violaceae
Vitus sp.	Grapes	Rhamnales	Vitaceae
Vitus vinifera (L.)	Grape vine	Rhamnales	Vitaceae
Wisteria hispida (Maxium)	—	*Rosale*	Fabaceae
Xanthium strumarium (L.)	Common cocklebur, California-bur	Asterales	Asteraceae
Zea mays (L.)	Maize	Graminales	Poaceae

rainy period and it may be unlikely that much of the population occurs outside maize and cotton during the autumn and winter periods. The 'rice' strain feeds principally on rice, Bermuda grass and Johnson grass, and will also feed on several other species of grass when they are available. When the larvae are very numerous they defoliate the preferred plants and disperse in large numbers, consuming nearly all vegetation in their path. Many host records of this species reflect such periods of abundance, and are not truly indicative of oviposition and feeding behaviour under normal conditions. Meagher and Nagoshi (2004) discuss techniques to monitor the spatial abundance of the host races in Florida and identify differences in distribution of the races in relation to agroecosystem features, which may also apply in Brazil. Nagoshi and Meagher (2003) provide molecular techniques to distinguish the host races, which may be useful in Brazil to untangle the interaction between wild hosts and maize.

Based on their association with cotton, *P. gossypiella*, *A. argillacea* and *H. virescens* are likely to be exposed to more intense selection in Bt cotton than *S. frugiperda*.

Association on other plants with Bt toxin

S. frugiperda, and to a lesser extent *H. virescens*, are the only species that are likely to occur outside of cotton in the field. *H. virescens* will occasionally occur in crops such as soybean, where it may be exposed to Bt sprays, but we regard this exposure as insignificant.

In relation to *S. frugiperda*, some similar Bt transgenes are proposed for use in maize crops. Maize production has diversified in recent years in Brazil, with the main crop grown during summer and a smaller crop of hybrid varieties in autumn and an even smaller winter crop of hybrid varieties in some regions. It is anticipated that 50–100% of the autumn and winter hybrids may eventually be planted to Bt maize that could express Cry1Ab or Cry1F, which provide the best control of *S. frugiperda* (Waquil *et al.*, 2002). By contrast, Cry9C has no effect on this species (Waquil *et al.*, 2002). Under laboratory selection, increased resistance (fivefold based on median lethal concentration (LC_{50})) of *S. frugiperda* to Cry1Ab was observed after four generations (Vilela *et al.*, 2002).

S. frugiperda represents a great risk of resistance evolution should both Cry1Ab maize and Cry1Ac/Cry 2Ab cotton be deployed. This would represent a mosaic of Bt-protein exposure and selection in the two main host plants of this species, hosts which overlap extensively in time and space. Considerable thought will be required to balance the possible introduction of Bt maize into this system, alongside Bt cotton. Below, in the section on 'resistance-risk management', we discuss some ways to balance these needs.

Exposure affected by farming system

The key pests of cotton in the main production regions are currently aphids and boll weevil. Insecticides applied for these pests provide considerable control of the Lepidopteran species and appears to mask the damaging potential of those species. The strength of selection of Bt cotton on the four target Lepidoptera varies directly with the amount of control that Bt cotton will exert above and beyond the control exerted by the insecticides. While there is some disagreement as to how much additional control Bt cotton may exert, it is possible that it will provide little additional control, although the selection pressure for resistance may still be significant (M. Caprio, Mississippi, 2004, personal communication).

Scale of adult movement

The scale of adult movement determines how much mixing and mating can occur between individuals emerging from different fields. For the purposes of relative resistance-risk assessment of the target species, it is not necessary to have precise quantitative data on the species. In general, the less dispersive a species is the greater the risk for resistance evolution (Carrière *et al.*, 2004a). This occurs because sedentary species will be more likely to mate with individuals from the field in which they emerged, and to oviposit in the same fields, which is likely to lead to greater selection pressure on that local part of the population. Hence in assessing the resistance risk, it can suffice to rank the dispersiveness of the target species.

There is little information specific to Brazil on adult movement of the four target species, although *H. virescens*, *P. gossypiella* and *S. frugiperda* have been studied extensively elsewhere. *H. virescens* has been found to be relatively sedentary, although capable of extensive local movement (Schneider, 1999, 2004). *S. frugiperda* is capable of extensive local and interregional movement (Pashley *et al.*, 1985), while *P. gossypiella* is probably the most sedentary of them all (Tabashnik *et al.*, 1999; Carrière *et al.*, 2001, 2004a,b). For *P. gossypiella* in Arizona, it was determined that refuges should not be further away than 0.75 km from Bt-cotton fields (Carrière *et al.*, 2004a,b). *A. argillacea* is believed to undertake extensive long-range movements from Argentina and Central South America through southern and central Brazil to north-eastern Brazil (Medeiros *et al.*, 2003). This movement largely follows the seasonal front of onset of the rainy season. Resident populations undoubtedly remain in all regions. There is no clear evidence of return movements from north to south.

Given this and knowledge from elsewhere, it seems reasonable to rank the dispersiveness as: *P. gossypiella* < *H. virescens* < *S. frugiperda* < *A. argillacea* (Table 12.3).

Likely dose of Cry toxins in Bt cotton

The dose of insecticidal toxin in Bt cotton will be a major factor determining the level of resistance risk. Dose is a measure of 'hazard', which is one of the two components that enter into risk assessment. Dose is defined by both the concentration of the Cry toxin in the Bt plant and the genetic characteristics of the target pest. Is the Bt crop a 'high-dose' or a 'low-dose' plant? A 'high-dose' is defined as one that kills a high proportion (> 95%) of heterozygous resistance genotypes similar to homozygous susceptible genotypes (Tabashnik, 1994a; Roush, 1997; Andow and Hutchison, 1998; Gould, 1998). A 'low-dose' is anything that is not a high-dose.

Resistance management will differ for high-dose versus low-dose plants. Simulation models clearly show that a high dose can delay the evolution of resistance more effectively than a low dose (Roush, 1994; Alstad and Andow, 1995; Gould, 1998; Tabashnik *et al.*, 2003a). A high dose may also allow greater options for resistance management with less restrictions on how non-transgenic refuges are managed (Carrière and Tabashnik, 2001; Ives and Andow, 2002; Onstad *et al.*, 2002; Storer *et al.*, 2003), and so may be more readily implemented compared to low-dose events. Low-dose events will require larger non-transgenic refuges and/or restrictions on the management of these refuges (Ives and Andow, 2002). Indeed, in Australia, growers agreed to cap the area of single gene Bt cotton (low dose for *H. armigera*) to 30% of the total crop in addition to the requirement for refuges (50% sprayed cotton refuge or 10% unsprayed cotton refuge) (Fitt, 2004). In the US, it has been argued that a 50% refuge may be needed for low-dose plants (Gould and Tabashnik, 1998). Simulations conducted during the Brazil workshop also indicated that a 50% refuge was needed for low-dose plants (Fig. 12.2).

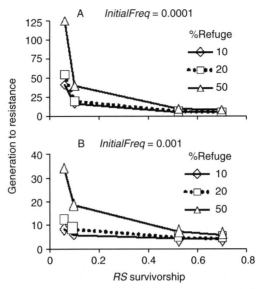

Fig. 12.2. Relationship between Bt cotton efficacy and time to resistance (details in the text under resistance-management section).

To evaluate the 'dose' it is essential to have insects resistant to the Bt crop that can be crossed to create heterozygous individuals that can be challenged with the Bt plant. However, in most cases prior to field release, resistant insects will not have been discovered. When resistance in a target species is not yet known it is not possible to evaluate heterozygous genotypes, so it is impossible to determine if a transgenic plant is high-dose or not. When this occurs, a temporary, provisional, operational definition of 'high-dose' must be used. One such definition for a provisionally 'high-dose' is: a plant that expresses toxin at a concentration that is 25 times the lethal concentration (LC_{99}) of the target pest (Gould and Tabashnik, 1998). This operational definition has been accepted for use by the US-Environmental Protection Agency (US-EPA). One alternative definition is a high dose produces at least 99.99% mortality of homozygote susceptibles relative to a non-Bt control (ILSI, 1999).

As discussed in more detail in Grossi *et al.* (Chapter 4, this volume), several Bt toxins have been incorporated into cotton that have been commercialized outside of Brazil, and many more toxins are under commercial development. Cry1Ac was the first to be commercialized as Mon531 (Bollgard®, INGARD® in Australia). Cry2Ab has been recently commercialized together with Mon531 (Bollgard II®). Cry1F has been combined with Cry1Ac in varieties that are under development (Widestrike®), and Vip3A may be combined with a Cry1A toxin.

The four target pests differ in tolerance to Bt insecticidal proteins (Table 12.5). These evaluations are based on preliminary bioassay and field-performance data, and could be revised as rigorous evaluations are completed. Cry1Ac and Cry2Ab are present at high enough concentrations in the present

Table 12.5. Estimated effective 'dose' provided by different Bt toxins for target
Lepidopteran pests of Brazilian cotton.

Insect	Cry1Ac	Cry2Ab	Cry1Ab	Cry1F	Vip3A
Heliothis virescens	High	High	High	Low	High
Alabama argillacea	High	High	High	Low	?
Pectinophora gossypiella	High	High	High	Low	?
Spodoptera frugiperda	Low	Low	Low, 689.81 ng/cm^{2a}	High, 36.46 ng/cm^2	High

[a]Surface treatment of artificial diet with pure toxin using laboratory colony of *S. frugiperda* in
USA. Source: Waquil *et al.* (2004).

Bt cottons to express a high dose for *H. virescens*, *A. argillacea* and *P. gossypiella*. However, toxin expression in these same cotton varieties is a low dose for *S. frugiperda*. Cry1F exhibits the opposite pattern, showing the characteristics of a high-dose event for *S. frugiperda* and low dose for the other three species. The toxicity of Vip3A against the target pests in Brazil is not known to us, but evidence of its high efficacy against *H. armigera* (G. Fitt, Brisbane, 2004, personal communication; Llewellyn *et al.*, 2005, in preparation) suggest it may be high dose for *H. virescens* and possibly for *S. frugiperda*.

Comprehensive laboratory or field information is not available for any combination of pests and specific Bt varieties in Brazil. Based on experience of some of these pests elsewhere, and some preliminary work for some species, we believe that Cry1Ac/Cry2Ab cotton expresses a high dose for *H. virescens*, *P. gossypiella* and *A. argillacea*. For *S. frugiperda*, we consider Cry1Ac/Cry2Ab cotton as a low-dose event. Cry1F/Cry1Ac cotton may be a high dose for all four species because Cry1F is high dose for *S. frugiperda*, while Cry1Ac is high dose for the other three species. We had insufficient information about Vip3A or any Vip3A/Cry1A cottons to evaluate their dose.

Bioassays for estimating LC_{99}

Bioassays estimating LC_{50} or LC_{99} or sublethal effects that have been previously correlated with LC_{50} or LC_{99} are recommended (Sims *et al.*, 1996). It will be most convenient to conduct the bioassays with purified toxin equivalent to that produced by the transgenic plant. The use of purified toxin allows experiments to evaluate the effects of toxin concentrations many times higher than that present in the transgenic plant.

There are many ways to conduct bioassays. First, the carrier of the toxin should be selected. This can be a natural food source (plant tissue) or artificial diet. Generally, the plant tissue is treated with the toxin by surface application with a series of toxin dilutions. Using artificial diet, the toxin can be provided as a mixture (Gould *et al.*, 1997; Hilbeck *et al.*, 1998) or surface treatment. The surface treatment of artificial diet can be done by applying each dilution to the diet surface in a 128-well bioassay tray (Marçon *et al.*, 1999; Waquil *et al.*, 2004). This method conserves toxin, and is acceptable only when small amounts of toxin are available. This method underexposes larvae that bore into the diet (Bolin *et al.*, 1999). In all cases, neonate larvae from the F_2 or F_3 generation can be used. In some cases, older larvae can be used. Normally, the trays are incubated for at least 7 days at 27°C, 80% RH and 24 h scotophase or photophase, and mortality and larval biomass are measured to estimate the LC and the growth inhibition (GI) (Marçon *et al.*, 1999).

Transgenic plants can also be used to create a series of toxin concentrations by diluting the tissue into an artificial diet (e.g. Olsen and Daly, 2000). This is advantageous because the toxins are in the same form as expressed in the plant, but is disadvantageous because the maximum toxin concentration that can be evaluated is less than what actually occurs in the plant (Andow and Hilbeck, 2004). When using plant material the specific testing should include:

1. Plant tissues that express the highest concentrations of toxins because they will allow a greater range of toxin concentrations to be tested. However, if the plant produces other secondary plant compounds that adversely affect the target insect pests, tissues that express lower concentrations of these chemicals would be favourable to avoid potential confounding mortality in the assays.
2. Quantification of the actual amount of transgene expression in the tissue that is used.

Plant tissue used without dilution does not allow estimation of an LC_{99}. However, this tissue can, in some circumstances, be used as a discriminating concentration to separate resistant and susceptible phenotypes. This method has been questioned recently by Zhao *et al.* (2002), who showed that the transgenic plant may be less accurate as a discriminating concentration than toxin incorporated into an artificial diet. Thus, the transgenic plant should not be used as a discriminating concentration unless it has been experimentally demonstrated to be an accurate method.

Need to find resistance
Because the actual dose expressed by the Bt plant cannot be determined until resistance genes are recovered in natural populations, assessments of risk without this information should be regarded as preliminary. We have used the limited information available to estimate dose, and we made the precautionary assumption that unless there is evidence that the Bt plant expresses a high dose indicated by consistently high efficacy against a range of field colonies of the pest, then the plant expresses a low dose.

Hence, it is of considerable importance to identify resistance genes in field populations and test their inheritance in the laboratory on Bt plants. Such tests should provide definitive evidence that the Bt plant is a high- or low-dose plant.

For potentially low-dose species, mass selection on laboratory colonies derived from recently collected individuals from the field should be initiated (Akhurst *et al.*, 2003). For potential high-dose target species, mass selection may be less likely to recover resistance (e.g. Bolin *et al.*, 1999; Huang *et al.*, 1999), but in some cases it can be successful (Gould *et al.*, 1997; Tabashnik *et al.*, 2000a, 2002, 2003b; Morin *et al.*, 2003). For high-dose species, additional methods include F_2 screens (Andow and Alstad, 1998; Genissel *et al.*, 2003), in-field screens (Tabashnik *et al.*, 2000a; Venette *et al.*, 2000) and any other approach that can maximize the probability of finding resistant individuals.

If relevant resistance genes have already been recovered in one of the target species in another region or country, a collaboration may be advisable both to use previous data as well as to access the resistant colony for future research. Note, however, that the genetic composition of insect populations varies geographically. Thus, the genetic basis of resistance could differ from one region to the next.

Risk assessment

Potential adverse consequences of resistance

The main potentially adverse consequences of resistance are: control failures; yield loss and economic hardship, when the pest is otherwise difficult to control; increased use of pest-management tactics, such as pesticides, that may be a significant human health or environmental risk; and reduced management options for growers that can increase production costs.

Key pests in Brazil are boll weevil and aphids (as virus vectors) in addition to the four target Lepidopteran species. Insecticides are the primary management tool used presently, with significant numbers of applications in all regions (Table 12.2). IPM systems to reduce insecticide applications have been developed and demonstrated but are not widely adopted (J.M. Waquil, Sete Lagoas, 2004, personal communication). As the four target Lepidoptera are not the key pests, we consider that two to four pesticide applications are the maximum that could be saved by Bt cotton (Table 12.2). Bt cotton is unlikely to reduce insecticide use for boll weevil, but may assist with aphid management if the pesticides currently applied for Lepidoptera control induce outbreaks of aphids. Applications to control boll weevil and aphids also suppress the Lepidoptera. Hence, under the present management practices, control failures and increased pesticide use are unlikely to occur if resistance arises.

Bt cotton may provide more impetus to adopt IPM approaches and so increase pesticide savings and consequent environmental benefits (Fitt, 2002). Resistance evolution would jeopardize these advances, limiting management options for growers. It is difficult to conclude that resistance evolution to the present Bt-cotton varieties will of itself cause substantial harm, because the major problems for farmers are currently boll weevil and aphids. None the less, any reduction in pesticide use will bring environmental benefits. Outside the Midwest region a reduction of two sprays could represent a 30–50% reduction in pesticide use, which is significant and would be threatened if field resistance were to occur. However, this assessment would change substantially if new, environmentally friendly control tactics for boll weevil and aphids become available.

Ranked resistance risk

A history of resistance to pesticides, low mobility of adults, high expected exposure of the population to Bt cotton and a low dose together imply a high resistance risk. Because resistance management usually relies on changing the exposure rate, the greatest risk will usually be a low-dose species and secondarily species with low dispersal distances. Of course, a history of resistance is a good indicator that resistance to Bt cotton might also occur readily. All of the four target species in cotton have a history of resistance evolution in Brazil or elsewhere, so none can be inferred to have a low resistance risk.

In the Midwest region, all four target species are important pests. The weakest link is *S. frugiperda*, except in locations where considerable areas of non-Bt maize is grown. If Bt maize is allowed or little maize is grown, the resistance risk in *S. frugiperda* is very high (Table 12.6). A low-dose strategy for *S. frugiperda* will be necessary for Cry1Ac/Cry2Ab cottons in this region.

In the Meridian region, *S. frugiperda* is not a major pest in cotton, and all of the other three species are weak links (Table 12.6). Although *P. gossypiella* has the lowest adult mobility of the three species, both *H. virescens* and *A. argillacea* have a great potential for resistance in Brazil and should be taken seriously. A high-dose strategy can be used in this region.

In the North-east region, the key Lepidopteran pests are *A. argillacea* and *P. gossypiella*. Both are weak links for resistance management for reasons similar to the Meridian region (Table 12.6). A high-dose strategy can be used in this region. However, with the expansion of upland cotton in part of this region, *S. frugiperda* is becoming an important pest. In this situation, the same considerations from the Midwest region should be followed.

Resistance Risk Management

Resistance-management requirements

The requirements for resistance management focus firstly around the weak link for each of the three cotton-producing regions of Brazil, then consider

Table 12.6. Summary assessment of the relative resistance risk of the four target Lepidopteran pests of Brazilian cotton. 1 indicates a very high risk and 4 indicates a low risk. All species are likely to be at risk of resistance evolution.

Insect	Region	History of resistance	Movement and risk	Exposure	Hazard (dose)	Risk
Heliothis virescens	Midwest, Meridian	1	1/2	1	2/3	2
Alabama argillacea	Midwest, Meridian, North-east	1	3	1	2/3	2
Pectinophora gossypiella	Midwest, Meridian, North-east	2	1	1	2/3	2
Spodoptera frugiperda	Midwest	1	2	1/2	1/3	1/3

requirements for the other species at risk. In this way the minimum essential requirements for resistance management can be developed.

Seed mixtures should not be used

A seed mixture is often considered as a possible resistance-management tactic. It involves mixing the seeds of Bt and non-Bt cotton in the seed bags or planters so that a mixture of Bt and non-Bt plants occurs in each field. The idea is that the non-Bt plants provide an effective refuge from selection on the Bt plants. While it is true that seed mixtures delay resistance evolution compared to having no refuge at all (Tabashnik, 1994a), they can be seriously compromised by the movement of larvae between plants (Mallet and Porter, 1992). The worst case occurs when Bt resistant heterozygotes can survive the Bt plant long enough to move to a neighbouring non-Bt plant, where they can complete development and vice versa, where susceptible larvae and resistant heterozygotes feeding on non-Bt plants move to Bt plants where susceptibles are killed and heterozygotes survive, so reducing the real value of the refuge.

   Our understanding of the behaviour of larvae of *S. frugiperda*, *H. virescens* and *A. argillacea* suggest that interplant movement of larvae is significant and would compromise the effectiveness of seed mixtures as refuges. *S. frugiperda* deposits egg masses from which larvae feed communally, and larvae may move considerable distances from fields or patches of host plants when plant quality declines or the crop is destroyed (Degrande, 1998). *A. argillacea* and *H. virescens* lay eggs singly on certain plant parts. The larvae of these three species move from plant to plant. Larvae of *P. gossypiella*, on the other hand, are very sedentary and rarely move between bolls on a plant. If this species were the only pest of cotton, seed mixtures might be a feasible tactic. However, there is no region in Brazil where this species is the only Lepidopteran pest of cotton. Consequently, we conclude that seed mixtures should not be used in Brazilian cotton (Table 12.8).

Kinds of refuges

A refuge is a habitat in which the target pest can maintain a viable population in the presence of Bt-cotton fields, where there is no additional selection for resistance to Bt toxins and insects occur at the same time as in the Bt fields (Ives and Andow, 2002). The refuge can be managed to control pest damage as long as the control methods do not reduce the population to such low levels that susceptible populations are driven to extinction (Ives and Andow, 2002). Because current cotton pest-management practices in Brazil have not come close to eliminating Lepidopteran pest populations (Ramalho, 1994), it is possible that continuing normal pest management on refuges would not jeopardize resistance management. However, this possibility should be investigated experimentally. Sprayed refuges are likely to be required to be larger in extent than unsprayed ones.

   Preliminary data suggest that few wild hosts exist for the four target species during the cotton-growing season. Should the presence of such hosts be proposed, such as for *S. frugiperda*, it will be necessary to provide scientific data to prove the suitability of such hosts. Specific data requirements are the net population replacement rate, the area of the refuge, the density of adults produced, the relative fitness of these adults (relative to non-Bt cotton or non-Bt maize), temporal synchrony in moth production between the crops and wild hosts and consistency (production year after year).

For *P. gossypiella*, *A. argillacea* and *H. virescens*, which are found mainly on cotton, the only refuge habitat possible at this time is cotton. Cotton is also a suitable refuge for *S. frugiperda*. This means that in the North-east and Meridian regions cotton is the only suitable refuge for resistance management (Table 12.8).

When the same or similar Bt toxins are not used in maize, maize can be a suitable refuge for *S. frugiperda*. *S. frugiperda* is an important pest in the Midwest region, and can be found on cotton from February to April (Fig. 12.1). Maize is produced year round in the Midwest, with significant areas grown at the same time as cotton (Table 12.7). Sorghum is not a suitable refuge for *S. frugiperda* because it is not available at the same time as the *S. frugiperda* is infesting cotton (Table 12.7). Figure 12.3 indicates when during the year each planting of maize is available to *S. frugiperda* in each of the cotton-growing regions of Brazil. The autumn maize crop overlaps with the time during the cotton growth season when *S. frugiperda* is attacking cotton. Thus, the rainfed autumn maize crop can act as a refuge for Bt cotton in those areas of the Midwest where there is significant autumn maize. In addition, millet, which is commonly used to produce biomass in no-tillage systems, may be a suitable refuge for *S. frugiperda*.

Agroecological zone	Jan	Feb	Mar	Apr	May	Jun	Jul	Aug	Sep	Oct	Nov	Dec
North-east												
Summer (rainfed)		H							P			
Autumn (rainfed)												
Winter (irrigated)					P					H		
Midwest												
Summer (rainfed)		H							P			
Autumn (rainfed)	P					H						
Winter (irrigated)					P					H		
Meridian												
Summer (rainfed)		H							P			
Autumn (rainfed)	P					H						
Winter (irrigated)					P					H		

Fig. 12.3. Phenology of maize production in the three cotton-production regions of Brazil. H = harvesting, P = planting

Table 12.7. Cultivated area (10³ haᵃ) of *Spodoptera frugiperda*-susceptible crops for each season in the Midwest region of Brazil. Figures in parentheses are % of total susceptible crop area.

Crop	Summer	Autumn	Winter[d]	Total
Cotton[b]	545.6 (8.6)	60.0[c] (0.9)	0	605.6 (9.6)
Maize	777.1 (12.3)	1165.3 (18.4)	370.0 (5.8)	2312.4 (36.5)
Sorghum	0	517.6 (8.2)	0	517.6 (8.2)
Millet[b]	768 (12.1)	664 (10.5)	0	1432 (22.6)
Rice[b]	848.5 (13.4)	0	0	848.5 (13.4)
Wheat	0	122.0 (1.9)	0	122.0 (1.9)
Sugarcane	245.0 (3.9)	245.0 (3.9)	0	490.0 (7.7)
Total	2969 (47.0)	2773.5 (43.8)	370.0 (5.8)	6328.1 (100)

[a]Source: CONAB (2003), average of the previous 2 years, except for cotton and rice (just the last season) which had more than 38% increase last year.
[b]Cultivated area in 2003/2004 season.
[c]Source: ORO Consultoria, Rio Verde, GO and Barreiras, BH, Brazil.
[d]Irrigated area.

When the same or similar Bt toxins are used in both cotton and maize, then the value of maize refuges could be compromised depending on the area of Bt maize grown. In the section on 'workable resistance-management plans' below, we address this possibility in detail.

Distance to refuge

ADULT MOVEMENT. The necessary maximum distance from Bt cotton to the refuge depends on the frequency and distance that adults disperse. Although detailed dispersal data do not exist for any of the target Lepidopteran species in Brazil, based on observations on these species in Brazil and elsewhere in the world, it may be reasonable to assume that all species undertake sufficient movement at a local (farm) scale of 1–2 km area from their emergence site. For the four target species, it will be necessary to structure a resistance-management strategy to take account of the least mobile of the four (*P. gossypiella*) where adult movements would likely be < 1 km (Table 12.8; Tabashnik *et al.*, 1999).

Table 12.8. Necessary characteristics for resistance management for Bt cotton in Brazil.

Region	Seed mixture	Refuge types	Maximum distance to refuge
Meridian	No	Cotton and maize[a]	2 km
Midwest	No	Cotton and maize[a]	2 km
North-east	No	Cotton	2 km

[a]Maize is a suitable refuge provided Bt maize is not planted in the region and the maize is available to the pest Lepidoptera at the same time as cotton.

Methodologies for study of adult movement could involve mark-recapture techniques (Fitt *et al.*, 1995; Tabashnik *et al.*, 1999; Carrière *et al.*, 2001; Cameron *et al.*, 2002; Kfir *et al.*, 2002) or quantification of various polymorphisms within and among populations to estimate gene flow (Han and Caprio, 2002). Flight mill or wind tunnel methods may indicate the propensity for flight but do not indicate the likelihood of long distance movement, and are not recommended. Methods based on genetic polymorphism could be appropriate to estimate gene flow in any of the four target species in Brazil, because they have been long-standing components of the Brazilian environment. It is possible, however, that recent changes in cropping systems, from conventional to no-tillage, during the last 10 years may have dramatically influenced population structure and invalidate this method. Mark-recapture methods would be more suitable than methods based on genetic polymorphism if there is significant variation in dispersal among generations.

It is questionable, however, that such studies would dramatically change the spatial scale of movement assumed in the first paragraph of this section or change the relative rankings among the species. Hence, we do not recommend considerable research on this problem.

MATING BEHAVIOUR. An understanding of adult mating behaviour in relation to the movement of each sex can be crucial to the design of a high dose-management strategy. It is less clear that these details are important for low-dose strategies. Identifying when mating occurs in relation to the site and time of emergence can be critical for high-dose strategies. If all mating occurs soon after emergence and in the natal patch, then intermating among refuge and Bt crop populations may be compromised. Key questions include: When do moths emerge? When and where do they mate after emergence? How far do they move before first mating and subsequently?

In Brazil, these questions will be significant only for the North-east and the Meridian regions, because these are regions where the high-dose strategy can be implemented (at locations where *S. frugiperda* is not a problem in cotton). Very little information is available on mating dynamics of any of the species in Brazil. Despite this lack of knowledge, we believe that additional studies of mating behaviour in relation to movement of each sex are not necessary during the prerelease period. Simulation models can address the various mating scenarios and allow uncertainty to be managed in the initial design of the resistance-management strategy. More specific experimental information will become necessary as the response systems are designed and verified. This should occur when Bt varieties are being tested in the field prior to commercial release.

Workable resistance-management plans

In this section, we propose workable resistance-management plans that incorporate the known scientific information to delay resistance evolution for at least 20 years. These plans are necessarily preliminary, because several critical pieces of information are missing. Specifically, we consider a high-dose plan for the North-east and Meridian regions (Scenario 1) and two low-dose plans for the Midwest region, one where there is no Bt maize (Scenario 2) and one where Bt maize occurs (Scenario 3). In addition, we outline a high-dose plan for the Midwest region if, sometime in the future, a Bt cotton that expresses a high dose against all four target Lepidopteran pests becomes available (Scenario 4).

Key assumptions underlying the development of these resistance-management plans are given in Box 12.1. We strongly recommend that Bt cotton be integrated into a comprehensive IPM approach that manages aphids and boll weevil effectively and may maximize efficacy of Lepidopteran control by Bt cotton. Research results from Brazil indicate that an IPM approach could have significant value in reducing pesticide applications from the current average of about 15 applications (F.S. Ramalho, Campina Grande, 2003, personal communication). It is entirely feasible that Bt cotton could be included as part of an IPM approach to further reduce pesticide applications in the Midwest region. Unfortunately, cotton growers have not yet adopted IPM approaches, despite extensive efforts to demonstrate on-farm benefits. It appears that many cotton producers no longer fully bury or destroy cotton residues. This undoubtedly exacerbates boll weevil problems. Cultural control of both pink bollworm and boll weevil will be enhanced substantially if the present low-tillage practices are accompanied by adequate stalk destruction (Ramalho, 1994; Degrande, 1998). More extensive field trials are essential to fully demonstrate the value of IPM to growers. For both resistance management of Bt cotton and IPM more broadly, there are distinct advantages in having a contained planting window for cotton, hence we suggest that the second planting season should be avoided even with non-Bt cotton.

Box 12.1. Key assumptions about IPM with resistance management.

Resistance management should be implemented within the framework of a multi-tactic IPM system. For cotton in Brazil, this should include the following elements:

• Appropriate emphasis on monitoring and thresholds for deciding when to make pesticide application, and on conservation of natural enemies.
• Strong emphasis on cultural control of boll weevil. This may include planting windows, crop-residue destruction and available control tactics that conserve natural enemies.
• Development and use of cotton varieties with resistance to aphid-vectored viruses.
• Establishment of regional working groups for providing on-going oversight of resistance-management recommendations, refuge requirements and remedial action.
• Funding of laboratories to conduct baseline susceptibility studies, sustain on-going resistance monitoring and respond to putative resistance events.

Four approaches can be used to delay resistance evolution. The approach most widely used is to reduce the exposure of the pests to Bt cotton by planting refuges. Specific issues include size, placement, timing of planting and management of refuges. A second approach is to reduce the selective differential between resistant and susceptible insects. The selective differential is the fitness advantage of resistant phenotypes over susceptible phenotypes when both are exposed to the transgenic plant. This can be accomplished by suppressing pests emerging from the transgenic crop with insecticides, physical controls, biological control, etc. A third approach is to reduce heterozygote fitness. A heterozygote has one susceptible allele and one resistant allele. Heterozygotes may have a susceptible or a resistant phenotype. If they are phenotypically susceptible, then they have low fitness on the Bt plant (resistance is recessive), and the rate of resistance evolution is slow. It is possible that natural enemies can alter heterozygote fitness. Little is known about potential selective feeding by natural enemies. If IPM for aphids and boll weevil can be implemented, there would be considerable potential for egg parasitoids and egg predators to reduce exposure of larvae to Bt proteins. The fourth approach can be used only with high-dose strategies. For some species it may be possible to manage the sex-specific movement and mating frequencies to delay resistance evolution (Andow and Ives, 2002). By using chemical and environmental attractants, it may be possible to enhance the movement of males and simultaneously reduce the movement of females from refuges to transgenic fields.

In what follows, we rely on refuges to reduce the exposure of pests to selection. None of the other approaches have been developed sufficiently to incorporate into a scientifically justified resistance-management plan. Refuges should be provided in a structured way with specific areas of non-Bt cotton or non-Bt maize. Such refuges need to be planted in association with Bt cotton in an appropriate spatial arrangement and to provide insects that overlap temporally with those from the Bt-cotton crop.

Given this general outline and what is known of the biology and ecology of the target species in Brazil, we distinguish three scenarios: Scenario 1, where high-dose Bt cotton is introduced and no alternative hosts are present; Scenario 2, where low-dose Bt cotton is introduced with non-Bt maize as an alternative crop; and Scenario 3, where low-dose Bt cotton is introduced and Bt and non-Bt maize co-occur. Scenarios 2 and 3 change the risks and management options for *S. frugiperda*. The other species are not affected, because they are not associated with maize. In addition, we describe briefly a case where high-dose Bt cotton is introduced with Bt maize. This may correspond to future Bt cottons that have yet to be commercialized anywhere.

Refuge options will be needed to cover the spectrum of target pests. In the case of *S. frugiperda*, non-Bt maize (Scenario 2) could be a significant refuge host. Few wild hosts for *S. frugiperda* are known to occur during this period in the growing season. In the case where Bt-maize is introduced (Scenario 3), it is recommended to limit the exposure of *S. frugiperda* by limiting the entire Bt-crop area to 50% of the total area (Fig. 12.2). It is necessary to limit exposure because Bt-cotton does not express a high-dose relative to *S. frugiperda*, and capping the entire Bt-crop area at 50% is the only way to meaningfully delay resistance evolution and increase the probability that the event will last more than 20 years.

A deterministic simulation model (Caprio, 1998a) was used to assess the requirements for refuge area for Bt cotton in Brazil. The simulations assume monogenic inheritance of a recessive trait and assumed that *H. virescens*, *P. gossypiella* and *A. argillacea* experience a high-dose Bt cotton, whilst *S. frugiperda* experiences a low-dose Bt cotton. The simulations were performed using the deterministic module of the software RRiskBt (Maia and Dourado-Neto, 2003), based on the deterministic version of Caprio's model (Caprio, 1998a). Two simulation-modelling approaches are available in RRiskBt:

• Deterministic approach – prediction of the number of pest generations until the *R* allele frequency (*RFreq*) exceeds a critical value using a deterministic model (Caprio, 1998b).
• Probabilistic approach – the probability of *RFreq* exceeding a critical frequency is predicted using a probabilistic risk-assessment model (Maia and Dourado Neto, 2003).

RRiskBt was developed using the Visual Basic language. A sensitivity analysis module allows us to investigate the influence of *R* allele initial frequency and/or the functional dominance of resistance on the model outputs. Uncertainty analysis tools were incorporated to obtain resistance-risk estimates.

Deterministic approach

The deterministic version of Caprio's model is a simplified version of the stochastic model developed by the same author (Caprio, 1998a). The events of interest are modelled at the time scale of a pest generation instead of a daily scale. Aspects related to refuge layout are not explicitly considered, but incorporated into pre- and postmating pest-dispersal indexes. It is a biological model in which the economics of the refuge are not taken into account. Using this model, the resistance-allele frequency in the target pest population can be projected as a function of pest generation. Such estimates are based on genetic and biological parameters of the target pest and operational factors related to the transgenic crop (Table 12.9). A description of both the deterministic and stochastic versions of the 'Caprio' model can be found in Caprio (2001).

We adopted the deterministic approach here because the studies required for characterization of parameter uncertainty are still in a preliminary stage. We chose sets of parameter values for simulations taking into account the characteristics of the three broad scenarios outlined above for Bt-cotton systems in Brazil (Table 12.10). The simulations were set by varying initial *R* allele frequency, proportion of refuge and survivorship of pest subpopulations corresponding to the genotypes *SS*, *RS* and *RR* in the transgenic cotton (Table 12.11). Low- (LD) or high-dose (HD) scenarios were represented by changing survival rates of susceptible (*SS*) and/or heterozygote (*RS*) genotypes. The *CriticalFreq* was set to 0.50, the *EndCoef* was assumed to be zero and

Table 12.9. Input parameters of the deterministic Caprio model (Caprio, 1998b).

Parameter	Description[a]
InitialFreq	*R* allele initial frequency in the target pest population
CriticalFreq	*R* allele critical frequency
RefSurv	Target pest immature-stage survivorship rate from an insecticide applied in the refuge
EndCoef	Endogamy coefficient in the target pest population
SurvSS	Immature stage survivorship of the target pest *SS* (homozygous susceptible) subpopulation in the refuge
SurvRR	Immature stage survivorship of the target pest *RR* (homozygous resistant) subpopulation in the refuge
SurvRS	Immature stage survivorship of the target pest *RS* (heterozygous) subpopulation in the refuge
PreDisp	Premating dispersal index
PosDisp	Postmating dispersal index

[a]The range for all the parameters is the interval [0,1].

Table 12.10. Characterization of the three broad scenarios for potential Bt cotton-growing in regions of Brazil.

Scenario characterization	Scenario 1	Scenario 2	Scenario 3
Crops	Bt and non-Bt cotton (only)	Bt and non-Bt cotton, non-Bt maize (only)	Bt and non-Bt cotton, Bt and non-Bt maize
Bt genes in cotton	*cry*1Ac + *cry*2Ab	*cry*1Ac + *cry*2Ab	*cry*1Ac + *cry*2Ab
Bt genes in maize	None	None	*cry*1Ab[a], *cry*1F[b] or *vip*3A[c]
Bt-cotton toxin(s) potency	HD against *H. virescens*, *P. gossypiella* and *A. argillacea*; LD against *S. frugiperda*	HD against *H. virescens*, *P. gossypiella* and *A. argillacea*; LD against *S. frugiperda*	HD against *H. virescens*, *P. gossypiella* and *A. argillacea*; LD for *S. frugiperda*
Bt-maize toxin(s) potency	None	None	[a]LD against *S. frugiperda* [b]HD against *S. frugiperda* [c]Unknown potency against *S. frugiperda*

HD = high dose, LD = low dose.
[a]Potency of *cry*1Ab.
[b]Potency of *cry*1F.
[c]Potency of *vip*3A.

Table 12.11. Number of pest generations until resistance (N^*) for several high-dose (HD) scenarios (> 95% mortality of heterozygotes relative to susceptibles) resulting from combination of partial recessiveness levels (expressed by SS and SR survival indexes), refuge size and R allele initial frequency (*InitialFreq*). Number of years until resistance for each pest species was calculated by dividing N^* by the number of generations in cotton (Table 12.3).

Scenario	InitialFreq	Survivorship			Refuge area (%)	N^*	Number of years		
		SS	RS	RR			A. argillacea	P. gossypiella	H. virescens
HD1	0.0001	0.01	0.02	1	10	22	7.3	4.4–7.3	7.3–11
HD2	0.0001	0.01	0.02	1	20	35	11.7	7–11.7	11.7–17.5
HD3	0.0001	0.01	0.02	1	50	106	35.3	21.2–35.3	35.5–53
HD4	0.0001	0.01	0.06	1	10	6	2.0	1.2–2	2.0–3
HD5	0.0001	0.01	0.06	1	20	9	3.0	1.8–3	3–4.5
HD6	0.0001	0.01	0.06	1	50	25	8.3	5–8.3	8.3–12.5
HD7	0.0001	0.001	0.002	1	10	63	21.0	12.6–21	21–31.5
HD8	0.0001	0.001	0.002	1	20	131	43.7	26.2–43.7	43.7–65.5
HD9	0.0001	0.001	0.002	1	50	492	164.0	98.4–164	164–246
HD10	0.0001	0.001	0.006	1	10	8	2.7	1.6–2.7	2.7–4
HD11	0.0001	0.001	0.006	1	20	18	6.0	3.6–6	6.0–9
HD12	0.0001	0.001	0.006	1	50	98	32.7	19.6–32.7	32.7–49
HD1	0.001	0.01	0.02	1	10	4	1.3	0.8–1.3	1.3–2
HD2	0.001	0.01	0.02	1	20	7	2.3	1.4–2.3	2.3–3.5
HD3	0.001	0.01	0.02	1	50	27	9.0	5.4–9	9–13.5
HD4	0.001	0.01	0.06	1	10	4	1.3	0.8–1.3	1.3–2
HD5	0.001	0.01	0.06	1	20	5	1.7	1–1.7	1.7–2.5
HD6	0.001	0.01	0.06	1	50	15	5.0	3.0–5	5–7.5
HD7	0.001	0.001	0.002	1	10	4	1.3	0.8–1.3	1.3–2
HD8	0.001	0.001	0.002	1	20	6	2.0	1.2–2	2.0–3
HD9	0.001	0.001	0.002	1	50	38	12.7	7.6–12.7	12.7–19
HD10	0.001	0.001	0.006	1	10	4	1.3	0.8–1.3	1.3–2
HD11	0.001	0.001	0.006	1	20	6	2.0	1.2–2	2.0–3
HD12	0.001	0.001	0.006	1	50	31	10.3	6.2–10.3	10.3–15.5

the dispersal indexes, *PreDisp, PosDisp* were set to their maximum values (1.0), which assume that the refuge layout allows complete mixing of adults coming from refuge and Bt-crop areas before mating and complete postmating dispersal of females for oviposition. We assumed that the refuge would be sprayed with non-Bt insecticides with efficacy of 80% (*RefSurv* = 0.20). The alternative refuge sizes considered for simulation were 10%, 20% or 50%, based on refuge recommendations adopted in the USA and Australia. The number of pest generations until resistance for high- (Table 12.11) and low-dose scenarios (Table 12.12) was estimated. In some of the high-dose simulations (Table 12.11), resistance evolved in less than 20 years. For these cases, the genetics of resistance of these hypothetical toxins suggests that they would be poorly suited for use in Bt cotton, because it would be difficult to manage resistance.

The correspondence between these simulations and the broad scenarios above (1,2) is determined by the potency of Bt toxin(s) expressed in the cotton/maize system, the target pests that are at resistance risk and the number of generations of these pests in cotton and/or maize. For example, for the high-dose species (*A. argillacea, P. gossypiella* and *H. virescens*), scenarios 1, 2 and 3 are equivalent and a 20% refuge is projected to give > 20 years of durability under HD8 conditions (Table 12.11), namely the initial frequency of resistance is < 0.0001, *SS* survival is < 0.001 and *RS* survival is < 0.002 (nearly completely recessive). These survival rates might be expected for effective high-dose events. A 50% refuge gives > 20 years of durability under wider conditions, *SS* survival < 0.01 and *RS* survival < 0.02. Both require that the initial frequency of resistance be low (< 0.0001).

The low-dose cases correspond to *S. frugiperda* (Table 12.12). For those scenarios, *SurvSR* was set to values corresponding to different functional-dominance levels ranging from partial recessiveness to partial dominance (0.06, 0.10, 0.525 or 0.70), keeping *SurvSS* = 0.05 and *SurvRR* = 1.00. The model gives similar results for Scenarios 1 and 2 (Table 12.12) because they differ only in that maize can be a refuge under Scenario 2, while only cotton is the refuge under Scenario 1. Thus, larger refuges will be more readily obtained under Scenario 2 compared with Scenario 1. Under these simulations, the 50% refuge would likely provide about 20 years of durability when initial resistance frequency < 0.0001, for *SS* survival < 0.05 and *RS* survival < 0.10. These might be reasonable survival levels for an efficacious low-dose event. Scenario 3 is more complicated. If Bt maize is produced with a Cry1Ab event, it will represent a low dose for *S. frugiperda* and there will likely be cross-resistance to Cry1Ac (Tabashnik *et al.*, 2000b). In this case, selection for similar resistance alleles will occur in both cotton and maize, and the time to resistance will be reduced according to the number of generations of selection in maize. If Bt maize is produced with a Cry1F event, it will likely be high dose for *S. frugiperda* and there may be no cross-resistance. In this case, the maize refuge for Cry1F and the cotton refuge for Cry1Ac could function together as a refuge for both toxins, and selection for resistance to Cry1Ac would only occur during the two generations in cotton. If Bt maize is a Cry1F event and there is cross-resistance between Cry1Ac and Cry1F, then the evolutionary dynamics will be more complex but faster than the case where

Table 12.12. Number of pest generations until resistance (N^*) for several low-dose (LD) scenarios resulting from combination of R allele initial frequency (*InitialFreq*), refuge size and functional dominance (*DFRes*) levels (PR, partial recessiveness; CD, codominance; PD, partial dominance) expressed by SS and SR survival indexes. Number of years until resistance is calculated for *Spodoptera frugiperda* by dividing N^* by the number of generations on cotton (Table 12.3; no Bt maize and high-dose (HD) Bt maize) or cotton and maize (LD Bt maize).

Scenario	Initial Freq	DFRes level	Survivorship			Refuge area (%)	N^*	Number of years			
			SS	RS	RR			No maize	No Bt maize	LD Bt maize[a]	HD Bt maize[b]
LD1	0.0001	PR	0.05	0.06	1	10	41	20.5	20.5	10.3	20.5
LD2	0.0001	PR	0.05	0.06	1	20	54	27.0	27.0	13.5	27.0
LD3	0.0001	PR	0.05	0.06	1	50	125	62.5	62.5	31.3	62.5
LD4	0.0001	PR	0.05	0.10	1	10	16	8.0	8.0	4.0	8.0
LD5	0.0001	PR	0.05	0.10	1	20	19	9.5	9.5	4.8	9.5
LD6	0.0001	PR	0.05	0.10	1	50	39	19.5	19.5	9.8	19.5
LD7	0.0001	CD	0.05	0.525	1	10	5	2.5	2.5	1.3	2.5
LD8	0.0001	CD	0.05	0.525	1	20	6	3.0	3.0	1.5	3.0
LD9	0.0001	CD	0.05	0.525	1	50	9	4.5	4.5	2.3	4.5
LD10	0.0001	PD	0.05	0.700	1	10	5	2.5	2.5	1.3	2.5
LD11	0.0001	PD	0.05	0.700	1	20	5	2.5	2.5	1.3	2.5
LD12	0.0001	PD	0.05	0.700	1	50	8	4.0	4.0	2.0	4.0
LD1	0.001	PR	0.05	0.06	1	10	8	4.0	4.0	2.0	4.0
LD2	0.001	PR	0.05	0.06	1	20	12	6.0	6.0	3.0	6.0
LD3	0.001	PR	0.05	0.06	1	50	34	17.0	17.0	8.5	17.0
LD4	0.001	PR	0.05	0.10	1	10	6	3.0	3.0	1.5	3.0
LD5	0.001	PR	0.05	0.10	1	20	8	4.0	4.0	2.0	4.0
LD6	0.001	PR	0.05	0.10	1	50	18	9.0	9.0	4.5	9.0
LD7	0.001	CD	0.05	0.525	1	10	4	2.0	2.0	1.0	2.0
LD8	0.001	CD	0.05	0.525	1	20	5	2.5	2.5	1.3	2.5
LD9	0.001	CD	0.05	0.525	1	50	7	3.5	3.5	1.8	3.5
LD10	0.001	PD	0.05	0.700	1	10	4	2.0	2.0	1.0	2.0
LD11	0.001	PD	0.05	0.700	1	20	4	2.0	2.0	1.0	2.0
LD12	0.001	PD	0.05	0.700	1	50	6	3.0	3.0	1.5	3.0

[a]Assumes cross-resistance between Cry1Ac and Cry1Ab, and two generations per year in maize.

[b]Assumes no cross-resistance between Cry1Ac and Cry1F. If there is cross-resistance, then resistance evolution will be faster and determined by evolution in both maize and cotton.

there is no cross-resistance. In this case, it will be essential to consider resistance both to Cry1Ac and Cry1F.

For both the high-dose and the low-dose simulations (Tables 12.11 and 12.12), we have assumed that all adults disperse maximally. As discussed below and indicated in previous modelling efforts (Comins, 1977; Caprio, 2001; Ives and Andow, 2002), reduced movement can significantly extend the time to resistance. Several of the target pests are unlikely to disperse maximally during all generations, so these simulation results are likely to represent the worst possible case.

Probabilistic approach

The probabilistic model developed by Maia and Dourado Neto (2003) was based on Caprio's deterministic model. Uncertainty associated with the initial frequency of resistance (*InitialFreq*) was incorporated using Monte Carlo methods (Hoffman and Kaplan, 1999; Hayse, 2000; Abrahamssom, 2002). Initial resistance frequency can be characterized by a probability-distribution function (e.g. truncated normal, uniform, triangular), referred to as the input-parameter distribution. Using Monte Carlo methods, a sample of *InitialFreq* values is taken from the input distribution. We will then run the deterministic model for each *InitialFreq* sampled value, producing in this way an output distribution for *RFreq* at the end of each pest generation. Such a probabilistic approach allows prediction of resistance risk (probability of *FreqR* exceeding a critical value) over time. For details, see Maia (2003) and Maia and Dourado-Neto (2004).

Scenario 1. High-dose Bt cotton, no alternative hosts (Box 12.2)

This strategy is appropriate for the North-east and Meridian regions, where *S. frugiperda* is not an important pest (Table 12.1), and the presently available

---

Box 12.2. Scenario 1. High-dose Bt cotton, no alternative hosts for target pests.

Suitable for North-east and Meridian regions.
Non-Bt cotton is the only refuge.

Provisional requirements:

- 20% of total cotton area to be planted to non-Bt cotton varieties.
- Bt cotton fields must be located within 1.5 km of their corresponding non-Bt cotton refuge fields.
- Refuge fields must be at least 60 rows wide.
- Refuges can be sprayed with insecticides (cannot be sprayed with Bt products).

Convene regional working groups to formulate appropriate new refuge requirements based on:

1. *G. barbadense* and *G. hirsutum* var. *marie-galante* as potential refuges, including cultivated, feral, volunteer, dooryard and landrace populations. In addition, they should consider how gene flow would affect these refuges (see Johnston *et al.*, Chapter 11, this volume).
2. Geography and extent of use of Bt cotton.

Bt cottons express a high dose for the remaining target Lepidopteran pests (Table 12.5). In these regions, we consider non-Bt cotton as the only suitable refuge.

There are four key provisional requirements for this strategy. First, 20% of the total cotton area should be planted to non-Bt cotton varieties (Table 12.11). This 20% refuge recommendation is consistent with other model results (Andow and Hutchison, 1998; Onstad and Gould, 1998; Roush, 1998; Caprio, 2001; Ives and Andow, 2002; Tabashnik *et al.*, 2003b).

Second, every Bt-cotton field should be located within 2 km of a non-Bt cotton refuge field, which was discussed in the section on 'resistance-management practices' above. Third, the refuge fields should be at least 60 rows wide. The purpose of this recommendation is to reduce movement of females from the refuge to the Bt fields, so that females are more likely to oviposit in the refuge. While this increases the likelihood that adults emerging from the Bt field will mate with each other, this is more than offset by the reduction in selection caused by the non-random oviposition (Ives and Andow, 2002). For example, for the high-dose example HD2 (Table 12.11), resistance occurred in 35 generations. If we assume that 80% of the adults that emerged from the refuges mated globally at random, and of the 20% that mated locally at random, 50% of their eggs were laid in the refuge; then 56 generations were required before resistance occurred. When the non-random oviposition parameter was increased to 80%, then the time to resistance was slowed to 77 generations. Increasing the minimal width of the refuges increases non-random oviposition, which is predicted to decrease the rate of resistance evolution. Fourth, refuges can be sprayed with insecticides but cannot be sprayed with Bt products, as was described in the section on 'resistance-management practices' above.

It will be important to convene working groups in the North-east and Meridian regions to formulate new refuge requirements based on *Gossypium hirsutum* var. *latifolium*, *Gossypium barbadense* and *G. hirsutum* var. *marie-galante*. Especially in the North-east, *G. barbadense* and *G. hirsutum* var. *marie-galante* occur as cultivated, feral, volunteer, dooryard and landrace populations (see Johnston *et al.*, Chapter 11, this volume). The potential effect of gene flow to the feral, volunteer, dooryard and landrace populations should also be considered. In addition, it may be possible to use information on the geography and extent of use of Bt and non-Bt cotton to modify requirements. For example, if the technology is not used very much by growers and refuges occur by happenstance, then it might become possible to relax some refuge requirements.

Scenario 2. High/low-dose Bt cotton, only non-Bt maize (Box 12.3)

This strategy is appropriate for the Midwest region, where *S. frugiperda* is an important pest (Table 12.1). The presently available Bt cottons express a low dose for *S. frugiperda* and a high dose for the remaining target Lepidopteran pests (*H. virescens*, *P. gossypiella* and *A. argillacea*; Table 12.5). Here, non-Bt cotton is the only suitable refuge for the other species, but maize could be an effective refuge for *S. frugiperda*.

> **Box 12.3.** Scenario 2. Bt and non-Bt cotton, non-Bt maize (only).
>
> Potentially suitable for Midwest region.
> Non-Bt cotton and non-Bt maize can be used as refuges.
> *Assumptions*
> Maize contributes substantially as a refuge for susceptible *S. frugiperda*.
>
> *Provisional requirements*:
>
> - Two Bt toxins in cotton: Cry1Ac/Cry2Ab.
> - 20% of total cotton area must be planted to non-Bt cotton varieties.
> - No Bt plants should be more than 1.5 km from their refuge area.
> - Refuge areas must be at least 60 rows wide.
> - Refuges can be sprayed with insecticides (cannot be sprayed with Bt products).

There are five key provisional requirements for this strategy. First, only two-gene Bt cotton with both Cry1Ac and Cry2Ab should be used. The one-gene Cry1Ac cotton should not be used in Brazil. The two-gene Bt cotton ensures that Bt cotton will act as a high dose against *H. virescens*, *P. gossypiella* and *A. argillacea*, and provides added resilience to the resistance-management strategy (Roush, 1998; Zhao *et al.*, 2003). Unfortunately, this two-gene cotton still expresses a low dose for *S. frugiperda*, so resistance management must be aimed at this weak link.

Second, 20% of the total cotton area should be planted to non-Bt cotton varieties. This is appropriate for the three high-dose species (*H. virescens*, *P. gossypiella* and *A. argillacea*) as discussed in Scenario 1 above (Table 12.11). For low-dose situations, such as *S. frugiperda*, a larger refuge is required. Specifically, our model results suggest that a 50% refuge is needed to ensure a sufficient delay in the time to resistance in this species (Table 12.12). Because autumn maize is an effective refuge for *S. frugiperda*, and we have assumed that none of it is Bt maize, the entire autumn maize crop supplements the 20% non-Bt cotton refuge. It is expected that maize will make up the difference in the needed refuge for *S. frugiperda*.

The remaining conditions are similar to those for Scenario 1. Every Bt-cotton field should be located within 2 km of a non-Bt cotton refuge field; the cotton refuge fields should be at least 60 rows wide; and cotton and maize refuges can be sprayed with insecticides, but cannot be sprayed with Bt products. In addition, we suggest that refuges should preferentially be within Bt-cotton fields. By using embedded refuges, growers may be more likely to manage the refuge in a similar way to the rest of the cotton crop. They may be less likely to spray it unless they are also treating the Bt-crop, and irrigation, crop scouting, etc., are all likely to be similar. By treating them the same, it is more likely that the refuge will function effectively.

Scenario 3. High/low-dose Bt cotton with Bt and non-Bt maize (Box 12.4)

This strategy is appropriate for the Midwest region, where *S. frugiperda* is an important pest (Table 12.1). This is similar to the previous scenario, except that some of the maize refuge for *S. frugiperda* is likely to be ineffective

---

Box 12.4. Scenario 3. High/low-dose Bt cotton with Bt and non-Bt maize.

Potentially suitable for Midwest region.
Non-Bt cotton and non-Bt maize can be used as refuges.

Assumptions

We are unable to predict the impact of Bt maize on resistance evolution of *S. frugiperda* at this time. Thus, a trigger has been established for refining refuge requirements in cotton on a regional basis in cases where Bt crops gain a 50% market share.

Provisional requirements

• Two Bt genes in cotton – Cry1Ac + Cry2Ab.
• 20% of total cotton area must be planted to non-Bt cotton varieties – halt expansion of Bt crops and refine refuge requirements for cotton if the total percentage of Bt crops in any state equals or exceeds 50% of the planted crops during any month.
• Bt fields must be located within 1.5 km of their corresponding refuge fields.
• Refuge fields must be at least 60 rows wide.
• Refuges can be sprayed with insecticides (cannot be sprayed with Bt products).

Convene regional working group to formulate appropriate new refuge requirements based on:

1. New information on production of susceptible *S. frugiperda* in refuges and wild hosts.
2. New information regarding survival of *S. frugiperda* in Bt crops.
3. Geography and extent of use of Bt cotton.

---

because it will be Bt maize and hence will not produce unselected moths. This means that additional requirements on refuge size must be considered.

At this time we are unable to predict the impact of Bt maize on resistance evolution of *S. frugiperda*. There are three Bt genes which may be deployed in maize and that vary in their dose against *S. frugiperda*: Cry1Ab – low dose; Cry1F – high dose (Waquil *et al.*, 2004); Vip3A – high dose. Resistance-management requirements in maize will depend on the relative use of these Bt genes. For example, if the low-dose maize becomes prevalent and cross-resistance is a possibility, it may be necessary to impose an area cap on the use of all Bt crops of say 50% of the crop area. If high-dose maize becomes prevalent, then the relative survival of *S. frugiperda* on all possible hosts may play an important role in determining refuge requirements. Thus, a trigger has been established for refining refuge requirements in cotton on a regional basis in cases where Bt maize gains greater than 50% market share.

The provisional requirements for this scenario are the same five requirements as for scenario 2, except that the refuge requirement is altered to take into account the use of Bt maize. The refuge should remain as conventional cotton at 20% of the total cotton area, which is appropriate as a high-dose strategy for the three species (*H. virescens*, *P. gossypiella* and *A. argillacea*) as discussed in Scenario 1 above. However, if the total area of Bt cotton and Bt maize in any state equals or exceeds 50% of the planted cotton and maize during any month, the expansion of Bt crops should be halted and the refuge requirements should be refined on the basis of the new information. This will ensure at least a 50% refuge for *S. frugiperda*.

It will be important to convene a representative working group for the Midwest region to formulate appropriate new refuge requirements based on research data on the production of *S. frugiperda* from refuges, wild hosts and Bt crops. In addition, it may be possible to use information on the geography and extent of use of Bt and non-Bt cotton and Bt and non-Bt maize to modify requirements.

New kinds of Bt cotton with Bt and non-Bt maize

In addition to the Cry1Ac and Cry1Ac/Cry2Ab Bt cottons, which are now commercially produced in some countries outside of Brazil, several new kinds of Bt cotton are under development. One is based on combining Cry1F and Cry1Ac (Widestrike™, from Dow AgroSciences), and another is based on combining Vip3A with a Cry1A toxin (Syngenta). If Cry1F and Vip3A are expressed at high enough concentrations that they work as high-dose toxins against *S. frugiperda*, then these new combinations would act as high-dose toxins against all four target Lepidopteran pests of Brazilian cotton. Resistance management for these events could follow Scenario 1 described above.

Methods to involve stakeholders, especially growers

It is vitally important that stakeholders, particularly growers, are intimately involved in the implementation of resistance-management strategies. The initial reaction of Brazilian cotton producers is often 'resistance management might work fine in the USA or Australia, but it can't be done here'. This position is understandable given the highly technical basis for the strategies discussed here, the uncertainties that surround them and the added costs in terms of time and money that they impose on growers. None the less, growers need to be convinced of the importance of rigorous management strategies if Bt cotton is to be used sustainably. To this end it will be crucial to involve growers in the regional working groups to develop alternative refuge approaches, and that significant investment is provided for effective extension and educational programmes to support implementation of a proactive resistance management for Bt cotton. All sectors of industry need to provide committed support to sustainable cotton-production systems that minimize environmental impact and optimize grower returns. Ongoing research and extension should support an IPM approach for cotton in which Bt cottons are one important component.

Design of transgenic plants for improved resistance management

We strongly recommend that Bt cotton be deployed as a pyramided two-gene product with both genes expressing a high dose. Pyramiding of additional transgenes can make heterozygotes phenotypically more like susceptible homozygotes and a pyramided plant provides additional safeguards, provided it is used with a refuge (Roush, 1998; Zhao *et al.*, 2003). In the absence of a refuge or when there is considerable cross-resistance, pyramiding is not of itself an effective resistance-management tool, but pyramided plants will still retain efficacy longer than single-gene Bt plants. The combination of Cry1Ac and

Cry2Ab is one example, providing a double high-dose against *H. virescens*, *P. gossypiella* and *A. argillacea*. The combination of Cry1Ac and Cry1F would not represent an acceptable pyramided variety because only one of the toxins provides a high-dose against each target pest. Cry1F appears to be highly active for *S. frugiperda*, but has little activity against the other species (Table 12.5).

Monitoring and response plans

The ultimate goal of monitoring is to obtain timely information that can be used to avoid or lessen the ramifications that pest resistance will have on the economics of agricultural production, and pesticide exposure of humans, wildlife and the environment. In the case of Bt cotton, monitoring information may be used to change the way that Bt cotton is deployed prior to widespread control failures due to resistance, or to justifying continuation of the current use strategy. Necessary steps in achieving this goal with Bt cotton in Brazil will include: (i) establishment of baseline susceptibility of target pests; (ii) detection and isolation of resistant phenotypes; (iii) investigation of putative field control failures; and (iv) documenting the use of Bt cotton and compliance with the resistance-management plan. These objectives necessitate the funding of centralized laboratory facilities in the major cotton-producing regions for the four target pests of Bt cotton. Additionally, Regional Bt-Cotton Resistance Working Groups should be convened annually in each region in order to evaluate the resistance-management strategy in light of new findings, to disseminate new research information and to identify the most critical regional research and education needs.

Monitoring for Bt resistance requires comparing the susceptibility of field-collected individuals with baseline susceptibility data and/or a susceptible laboratory colony. Centralized rearing and bioassay facilities should be identified, and appropriately funded and staffed for this purpose. All methodologies for testing should be standardized between laboratories and strictly adhered to. Monitoring should also involve some level of evaluation of compliance with refuge requirements in each state or region. Baseline susceptibility should be evaluated before the Bt crop is commercialized. Techniques for establishing baseline susceptibility to pesticides of various types are well established (Stone and Sims, 1993; Tabashnik, 1994; Robertson *et al.*, 1995; Sims *et al.*, 1996; Andow and Alstad, 1998; Gould, 1998; Marcon *et al.*, 1999).

Possible monitoring methods
Three kinds of monitoring are essential: crop-damage monitoring, resistance monitoring and compliance monitoring. Crop-damage monitoring involves observation by all parties involved (growers, consultants, extension staff, researchers) of increased damage to crops or numbers of insects. Larvae surviving in Bt-cotton fields should be collected and transported to a facility for bioassay by local farmers, NGOs or agricultural extension or research personnel. Resistance monitoring could involve one of several approaches. A signifi-

cant question is whether the monitoring programme can be sufficiently widespread and intensive to provide a realistic early warning of change in gene frequency for resistance.

All resistance monitoring requires collection of appropriate life stages from the field. In the case of Bt cotton, eggs, egg masses, larvae or adults (using light traps) could be collected and sent to a regional facility for bioassay using a phenotypic screen (Tabashnik *et al.*, 2000a) or for use in a more complicated, but more informative, F_2 screen (Andow *et al.*, 2000). Both methods require a discriminating dose methodology (Tabashnik *et al.*, 2000a) to distinguish resistant and susceptible phenotypes. If a resistant laboratory colony can be developed, then field-collected adults can be crossed with resistant individuals of known genotype and the progeny bioassayed using a discriminating dose to provide information on the genotype of the field-collected adults (Gould *et al.*, 1997). A field-based method could involve sentinel plots or samples could be taken from both refuge and Bt-cotton plots to compare larval densities as an indicator of emerging resistance (Venette *et al.*, 2000). All of these methods establish the frequency of resistance alleles. When conducted over successive years, they allow detection of regional change in resistance frequency. Andow and Ives (2002) compare the cost efficiency of these various methods for Bt maize; a similar comparison could be conducted for Bt cotton.

For refuge compliance (primarily large farms), producers should be required to keep detailed maps of the placement of Bt cotton and refuge crops. Compliance could be estimated by statistical sub-sampling of cotton fields to evaluate the degree of concordance between maps provided by the producer and the observed size and location of Bt and non-Bt fields. Collection of bolls could be made in Bt and non-Bt fields to corroborate designations or antibody tests could be conducted on plant tissue for the same purpose.

Possible responses

Provided it was possible to detect changes in resistance frequency early enough there could be three responses (see Andow and Ives, 2002):

1. Use of other control strategies that result in absolute or high mortality of a putative resistant population (e.g. pesticide overspray, inundative releases of parasitoids, destruction of crop).
2. Increase in size of structured refuge.
3. Eliminate planting of Bt crops in the affected area until susceptibility has returned. A decision on the area affected by resistance should be based on field surveys of damage in Bt crops coupled with knowledge of pest-dispersal propensity (Carrière *et al.*, 2001).

The capacity to respond will be highly dependent on rapid and accurate communication with all producers. It may be very difficult to withdraw Bt cotton from the market once released, particularly when in the hands of smallholders not subject to restrictions on seed availability. Mitigation plans should be formulated with and disseminated to growers through education channels as soon as possible, to illustrate the importance of following recommended resistance-management strategies.

Issues Addressed during Field Testing

Issues to initiate and complete prior to field testing:

- Determine dose of all Bt-cotton types against all four target Lepidoptera.

Issues to initiate prior to field release and continue during field testing:

- Search for resistance in natural populations.
- Conduct resistance monitoring.
- Develop and implement IPM methods (Box 12.1).
- Develop goals and strategies for stakeholder involvement.

Issues to initiate during field testing:

- Evaluate the utility of alternative refuges to *G. hirsutum* var. *latifolium* in the North-east region for *P. gossypiella* and *A. argillacea*, and in the Midwest for *S. frugiperda*, especially maize, sorghum and millet, and also wild hosts.
- Evaluate the distance requirement for *P. gossypiella* and examine pre- and postmating sex-specific movement for all high-dose species.
- Estimate the production of *S. frugiperda* from refuges, wild hosts and Bt crops in Midwest.
- Implement education programmes.
- Develop the monitoring programme.

   1. Specify the monitoring/auditing methods and reporting procedures for: (i) establishment of baseline tolerance to the target crop; (ii) early detection of resistance; (iii) control failures due to resistance; and (iv) compliance to resistance-management plan.
   2. How can these methods be integrated with other monitoring programmes used on the landscape?
   3. Have any discussions occurred with representatives of the farm industry? Has there been any discussion with other stakeholder groups, including the technology industries, and community and environmental groups?
   4. How have growers been integrated into monitoring?
   5. Quality control. The quality of the data generated in the monitoring efforts may degrade with time. Methods for ensuring that data quality is maintained need to be specified for: (i) early detection of resistance; (ii) control failures due to resistance; and (iii) compliance to resistance-management plan.
   6. Using monitoring information. Processing and reporting the monitoring results, and the linkage with the response strategies, need to be evaluated periodically for: (i) early detection of resistance; (ii) control failures due to resistance; and (iii) compliance to resistance-management plan.

- Develop the response plan.

   1. Specify the elements of a response plan that is triggered by some monitoring threshold associated with the methods specified above (I.D.1).
   2. How will these responses be integrated with the original proactive resistance-management plan?

3. Have any discussions occurred with farmers or other representatives of the farm industry?

4. Has there been any discussion with other stakeholder groups, including the technology industries and community and environmental groups?

Conclusions

There is no question that resistance is a potential risk to be considered with the deployment of Bt cotton. At the same time there is accumulating evidence to show that resistance risks can be managed (Tabashnik *et al.*, 2003b). Due to the complexity and dynamic nature of Brazilian cropping systems, appropriate methods for evaluating resistance risks and their management are necessary. For that reason we evaluated a range of different scenarios to accommodate the possible deployment of Bt cotton and Bt maize. The suite of pest Lepidoptera to consider and their differing sensitivities to Bt proteins further complicates the issues to be considered.

Our overview of the Brazilian agricultural system, the interactions of crops and pests and the conclusions reached here for the resistance-management needs of Bt-cotton deployment were the result of discussions among scientists from different countries, including Brazil. While we identified many gaps and research needs, we conclude that it is possible to formulate reasonable recommendations to minimize the resistance risk for Bt cotton in Brazil. These recommendations should serve as a starting point for a comprehensive research and extension effort to fill the many gaps and commence the critical task of educating growers about the serious need for resistance management, if the potential benefits of Bt cotton are to be realized and sustained. We reiterate that Bt cotton can provide an important component for IPM approaches to cotton production and, despite the added complexity this might impose on growers, the environmental and economic benefits will justify such investment.

For Brazil, the critical first steps are to critically evaluate the efficacy of all Bt-cotton types against all four target Lepidoptera through regulated small-scale trials. At the same time, research should establish the baseline tolerance of field populations to relevant Bt proteins using well-established techniques. This information will serve as the basis for future monitoring programmes and may also uncover resistant individuals that can form the basis for selection experiments to establish resistant strains in the laboratory.

References

Abrahamssom, M. (2002) Uncertainty in quantitative risk analysis – characterization and methods of treatment. *Report No.* 1024. Lund University, Lund, Denmark.

Akhurst, R.J., James, W., Bird, L.J. and Beard, C. (2003) Resistance to the Cry1Ac delta-endotoxin of *Bacillus thuringiensis* in the cotton bollworm, *Helicoverpa armigera* (Lepidoptera: Noctuidae). *Journal of Economic Entomology* 96, 1290–1299.

Alstad, D.N. and Andow, D.A. (1995) Managing the evolution of insect resistance to transgenic plants. *Science* 268, 1894–1896.

Andow, D.A. and Alstad, D.N. (1998) The F_2 screen for rare resistance alleles. *Journal of Economic Entomology* 91, 572–578.

Andow, D.A. and Hilbeck, A. (2004) Science-based risk assessment for non-target effects of transgenic crops. *BioScience* 54(7), 637–649.

Andow, D.A. and Hutchison, W.D. (1998) Bt-corn resistance management. In: Mellon, M. and Rissler, J. (eds) *Now or Never: Serious New Plans to Save a Natural Pest Control.* Union of Concerned Scientists, Cambridge, Massachussetts, pp. 19–66.

Andow, D.A. and Ives, A.R. (2002) Monitoring and adaptive resistance management. *Ecological Applications* 12, 1378–1390.

Andow, D.A., Olson, D.M., Hellmich, R.L., Alstad, D.N. and Hutchison, W.D. (2000) Frequency of resistance alleles to *Bacillus thuringiensis* toxin in an Iowa population of European corn borer. *Journal of Economic Entomology* 93, 26–30.

Bolin, P.A., Hutchison, W.D. and Andow, D.A. (1999) Long-term selection for resistance to *Bacillus thuringiensis* Cry1Ac endotoxin in a Minnesota population of European corn borer (Lepidoptera: Crambidae). *Journal of Economic Entomology* 92, 1021–1030.

Cameron, P.J., Walker, G.P., Penny, G.M. and Wisley, P.J. (2002) Movement of potato tuber worm (Lepidoptera: Gelechiidae) within and between crops, and some comparisons with diamondback moth (Lepidoptera: Plutellidae). *Environmental Entomology* 31, 462–468.

Caprio, M.A. (1998a) Non random mating model. Available at: http://www.msstate.edu/Entomology/PGjava/ILSI model.html (accessed September 2003).

Caprio, M.A. (1998b) Evaluating resistance management for multiple toxins in presence of external refuges. *Journal of Economic Entomology* 91, 1021–1031.

Caprio, M.A. (2001) Source-sink dynamics between transgenic and non-transgenic habitats and their role in the evolution of resistance. *Journal of Economic Entomology* 94, 698–705.

Carrière, Y. and Tabashnik, B.E. (2001) Reversing insect adaptation to transgenic insecticidal plants. *Proceedings of the Royal Society of London Series B* 268, 1475–1480.

Carrière, Y., Dennehy, T.J., Petersen, B., Haller, S., Ellers-Kirk, C., Antilla, L., Liu, Y.B., Willot, E. and Tabashnik, B.E. (2001) Large-scale management of insect resistance to transgenic cotton in Arizona: can transgenic insecticidal crops be sustained? *Journal of Economic Entomology* 94, 315–325.

Carrière, Y., Dennehy, T., Ellers-Kirk, C., Holley, D., Liu, Y.-B., Simms, M. and Tabashnik, B.E. (2002) Fitness costs, incomplete resistance, and management of resistance to Bt crops. In: Akhurst, R.J., Beard, C.E. and Hughes, P. (eds) *Biotechnology of* Bacillus thuringiensis *and its Environmental Impact.* Proceedings of the 4th Pacific Rim conference, Canberra, Australia, pp. 82–92.

Carrière, Y., Sisterson, M. and Tabashnik, B.E. (2004a) Resistance management for sustainable use of Bt crops in integrated pest management. In: Horowitz, A.R. and Ishaaya, I. (eds) *Insect Pest Management: Field and Protected Crops.* Springer, Berlin, pp. 65–95.

Carrière, Y., Dutilleul, P., Ellers-Kirk, C., Pedersen, B., Haller, S., Antilla, L., Dennehy, T.J. and Tabashnik, B.E. (2004b) Sources, sinks, and zone of influence of refuges for managing insect resistance to Bt crops. *Ecological Applications* 14(6), 1615–1623.

Comins, H.N. (1977) The development of insecticide resistance in the presence of migration. *Journal of Theoretical Biology* 64, 177–197.

CONAB (2003) *Algodão em pluma informativo especial Julho 2003*. Available at: http://www.conab.gov.br/download/safra/safra20032004Lev04.pdf (accessed July 2003).

Cruz, I. (1995) *A lagarta-do-cartucho na cultura do milho*. Circular Técncia, 21. EMBRAPA/CNPMS, Sete Lagoas, Brazil.

Degrande, P.E. (1998) *Guia prático de controle das pragas do algodoeiro*. UFMS, Dourados, MS, Brazil.

Diez-Rodriguez, G.I. and Omoto, C. (2001) Herança da Resistência de *Spodoptera frugiperda* (J.E. Smith) (Lepidoptera: Noctuidae) à Lambda-Cialotrina. *Neotropical Entomology* 30, 311–316.

Fitt, G.P. (1989) The ecology of *Heliothis* species in relation to agro-ecosystems. *Annual Review of Entomology* 34, 17–52.

Fitt, G.P. (1997) Risks, deployment and integration of insect resistant crops expressing genes from *Bacillus thuringiensis*. In: McLean, G.D., Waterhouse, P.M., Evans, G. and Gibbs, M.J. (eds) *Commercialisation of Transgenic Crops: Risk, Benefit and Trade Considerations*. CRC for Plant Science and Bureau of Resource Sciences, Canberra, Australia, pp. 273–284.

Fitt, G.P. (2002) Transgenic cotton as a foundation for integrated pest management – towards green cotton. *Australian Biologist* 15(2), 56–63.

Fitt, G.P. (2004) Implementation and impact of transgenic Bt cottons in Australia. In: *Cotton Production for the New Millennium. Proceedings of the Third World Cotton Research Conference*, 9–13 March, 2003, Cape Town, South Africa. Agricultural Research Council – Institute for Industrial Crops, Pretoria, South Africa, pp. 371–381.

Fitt, G.P. and Daly, J.C. (1990) Abundance of overwintering pupae and the spring generation of *Helicoverpa* spp. (Lepidoptera: Noctuidae) in New South Wales, Australia: consequences for pest management. *Journal of Economic Entomology* 83(5), 1827–1836.

Fitt, G.P., Dillon, M.L. and Hamilton, J.G. (1995) Spatial dynamics of *Helicoverpa* populations in Australia: simulation modelling and empirical studies of adult movement. *Computers and Electronics in Agriculture* 13, 177–192.

Fitt, G.P., Andow, D.A., Carrière, Y., Moar, W.A., Schuler, T., Omoto, C., Kanya, J., Okech, M., Arama, P. and Maniania, N.K. (2004) Resistance risks and management associated with Bt maize in Kenya. In: Hilbeck, A. and Andow, D.A. (eds) *Environmental Risk Assessment of Genetically Modified Organisms, Volume 1: A Case Study of Bt Maize in Kenya*. CAB International, Wallingford, UK, pp. 209–249.

Forrester, N.W., Cahill, M., Bird, L.J. and Layland, J.K. (1993) Management of pyrethroid and endosulfan resistance in *Helicoverpa armigera* (Hubner) in Australia. *Bulletin of Entomological Research Supplement Series No. 1*, 132 pp.

Gallo, D., Nakano, O., Silveira Neto, S., Carvalho, R.P.L., Baptista, G.C., Berti Filho, E., Parra, J.R.P., Zucchi, R.A., Alves, S.B., Vendramim, J.D., Marchini, L.C., Lopes, J.R.S. and Omoto, C. (2002) *Entomologia Agrícola*. FEALQ, Piracicaba, Brazil.

Genissel, A., Augustin, S., Courtin, C., Pilate, G., Lorme, P. and Bourguet, D. (2003) Initial frequency of alleles conferring resistance to *Bacillus thuringiensis* poplar in a field population of *Chrysomela tremulae*. *Proceedings of the Royal Society Biological Sciences Series B* 270(1517), 791–797.

Gould, F. (1998) Sustainability of transgenic insecticidal cultivars: integrating pest genetics and ecology. *Annual Review of Entomology* 43, 701–726.

Gould, F. and Tabashnik, B.E. (1998) Bt-cotton resistance management. In: Mellon, M. and Rissler, J. (eds) *Now or Never: Serious New Plans to Save a Natural Pest Control.* Union of Concerned Scientists, Cambridge, Massachussetts, pp. 67–106.

Gould, F., Anderson, A., Jones, A., Summerford, D., Heckel, G.G., Lopez, J., Micinski, S., Leanard, R. and Laster, M. (1997) Initial frequency of alleles for resistance to *Bacillus thuringiensis* toxins in field populations of *Heliothis virescens. Proceedings of the National Academy of Sciences USA* 94, 3519–3523.

Gregg, P.C., Fitt, G.P., Zalucki, M.P. and Murray, D.A.H. (1995) Insect migration in an arid continent II. *Helicoverpa* spp. in Australia. In: Drake, V.A. and Gatehouse, A.G. (eds) *Insect Migration: Physical Factors and Physiological Mechanisms.* Cambridge University Press, Cambridge, UK, pp. 151–172.

Han, Q.F. and Caprio, M.A. (2002) Temporal and spatial patterns of allelic frequency in cotton bollworm (Lepidoptera: Noctuidae). *Environmental Entomology* 31, 462–468.

Hayse, J.W. (2000) Using Monte Carlo analysis in ecological risk assessment. Available at: http://web.ead.anl.gov/ecorisk/issue/pdf/montecarlo.pdf (accessed 13 May 2004).

Hilbeck, A., Moar, W.J., Pusztai-Carey, M., Filippini, A. and Bigler, F. (1998) Toxicity of *Bacillus thuringiensis* Cry1Ab toxin to the predator *Chrysoperla carnea* (Neuroptera: Chrysopidae). *Environmental Entomology* 27(5), 1255–1263.

Hoffman, F.O. and Kaplan, S. (1999) Beyond the domain of direct observation: how to specify a probability distribution that represents the 'state of knowledge' about uncertainty inputs. *Risk Analysis* 19(1), 131–134.

Huang, F., Buschman, L.L., Higgins, R.A. and McGaughey, W.H. (1999) Inheritance of resistance to *Bacillus thuringiensis* toxin (Dipel ES) in the European corn borer. *Science* 284, 965–967.

ILSI (International Life Sciences Institute) (1999) *An evaluation of insect resistance management in Bt field corn: a science-based framework for risk assessment and risk management.* International Life Sciences Institute Health and Environmental Sciences Institute, Washington, DC.

Ives, A.R. and Andow, D.A. (2002) Evolution of resistance to Bt crops: directional selection in structured environments. *Ecology Letters* 5, 792–801.

Kfir, R., Overholt, W.A., Khan, Z.R. and Polaszek, A. (2002) Biology and management of economically important Lepidopteran cereal stem borers in Africa. *Annual Review of Entomology* 47, 701–731.

Maia, A.H.N. (2003) Modelagem da evolução da resistência de pragas a toxinas Bt expressas em culturas transgênicas: quantificação de risco utilizando análise de incertezas. PhD thesis, Universidade de São Paulo, Piracicaba, Brazil.

Maia, A.H.N. and Dourado-Neto, D. (2003) RRiskBt – um programa computacional para quantificar risco de resistência de pragas a toxinas Bt expressas em culturas transgênicas. In: *Proceedings: Congresso Brasileiro da Sociedade Brasileira de Informática Aplicada à Agropecuária e Agroindústria*, 4, Porto Seguro, BA, Brazil, pp. 571–573.

Maia, A.H.N. and Dourado-Neto, D. (2004) Probabilistic tools for assessment of pest resistance risk associated with insecticidal transgenic crops. *Scientia Agricola* 61(5), 481–485.

Mallet, J. and Porter, P. (1992) Preventing insect adaptation to insect-resistant crops: are seed mixtures or refugia the best strategy? *Proceedings of the Royal Society of London Series B* 250, 165–169.

Marçon, P.C., Yong, L.J., Steffely, K.L. and Siegfried, B.D. (1999) Baseline susceptibility of European corn borer (Lepidoptera: Crambidae) to *Bacillus thuringiensis* toxins. *Journal of Economic Entomology* 92, 279–285.

Meagher, R.L. and Nagoshi, R.N. (2004) Population dynamics and occurrence of *Spodoptera frugiperda* host strains in southern Florida. *Ecological Entomology* 29, 614–620.

Medeiros, R.S., Ramalho, F.S., Zanúncio, J.C. and Serrão, J.E. (2003) Estimate of *Alabama argillacea* (Hübner) (Lepidoptera: Noctuidae) development with nonlinear models. *Brazilian Journal of Biology* 63, 589–598.

Morin, S., Biggs, R.W., Sisterson, M.S., Shriver, L., Ellers-Kirk, C., Higginson, D., Holley, D., Gahan, L.J., Heckel, D.G., Carrière, Y., Dennehy, T.J., Brown, J.K. and Tabashnik, B.E. (2003) Three cadherin alleles associated with resistance to *Bacillus thuringiensis* in pink bollworm. *Proceedings of the National Academy of Sciences USA* 100, 5004–5009.

Nagoshi, R.N. and Meagher, R.L. (2003) FR tandem-repeat sequence in fall armyworm (Lepidoptera: Noctuidae) host strains. *Annals of the Entomological Society of America* 96, 329–335.

Olsen, K.M. and Daly, J.C. (2000) Plant-toxin interactions in transgenic Bt cotton and their effect on mortality of *Helicoverpa armigera* (Lepidoptera: Noctuidae). *Journal of Economic Entomology* 93, 1293–1299.

Onstad, D.W. and Gould, F. (1998) Modelling the dynamics of adaptation to transgenic maize by European corn borer (Lepidoptera: Pyralidae). *Journal of Economic Entomology* 91, 585–593.

Onstad, D.W., Guse, C.A., Porter, P., Buschman, L.L., Higgins, R.A., Sloderbeck, P.E., Peairs, F.B. and Gronholm, G.B. (2002) Modelling the development of resistance by stalk-boring lepidopteran insects (Crambidae) in areas with transgenic corn and frequent insecticide use. *Journal of Economic Entomology* 95, 1033–1043.

Pashley, D.P. (1986) Host-associated genetic differentiation in fall armyworm (Lepidoptera: Noctuidae): a sibling species complex? *Annals of the Entomological Society of America* 79, 898–904.

Pashley, D.P., Johnson, S.J. and Sparks, A.N. (1985) Genetic population structure of migratory moths: the fall armyworm (Lepidoptera: Noctuidae). *Annals of the Entomological Society of America* 78, 756–762.

Pogue, M.G. (1995) World Spodoptera Database (Lepidoptera: Noctuidae). Available at: http://www.sel.barc.usda.gov/lep/spodoptera/spodoptera.html (accessed July 2004).

Ramalho, F.S. (1994) Cotton pest management: Part 4. A Brazilian perspective. *Annual Review of Entomology* 39, 563–578.

Robertson, J.L., Preisler, H.K., Ng, S.S., Hickle, L.A. and Gelernter, W.D. (1995) Natural variation – a complicating factor in bioassays with chemical and microbial pesticides. *Journal of Economic Entomology* 88, 1–10.

Roush, R.T. (1994) Managing pests and their resistance to *Bacillus thuringiensis*: can transgenics be better than sprays? *Biocontrol Science and Technology* 4, 501–516.

Roush, R.T. (1997) Managing resistance to transgenic crops. In: Carozzi, N. and Koziel, M. (eds) *Advances in Insect Control: the Role of Transgenic Plants*. Taylor & Francis, London, pp. 271–294.

Roush, R.T. (1998) Two-toxin strategies for management of insect resistant transgenic crops: can pyramiding succeed where pesticide mixtures have not? *Philosophical Transactions of the Royal Society of London B* 353, 1777–1786.

Schneider, J.C. (1999) Dispersal of a highly vagile insect in a heterogeneous environment. *Ecology* 80, 2740–2749.

Schneider, J.C. (2004) Overwintering of *Heliothis virescens* (F.) and *Helicoverpa zea* (Boddie) (Lepidoptera: Noctuidae) in cotton fields of North-east Mississippi. *Journal of Economic Entomology* 96, 1433–1447.

Sims, S.B., Greenplate, J.T., Stone, T.B., Caprio, M.A. and Gould, F. (1996) Monitoring strategies for early detection of Lepidoptera resistance to *Bacillus thuringiensis* insecticidal proteins. In: Brown, T.M. (ed.) *Molecular Genetics and Evolution of Pesticide Resistance.* ACS Symposium Series No. 645, Washington, DC, pp. 229–242.

Stone, T.B. and Sims, S.R. (1993) Geographic susceptibility of *Heliothis virescens* and *Helicoverpa zea* (Lepidoptera: Noctuidae) to *Bacillus thuringiensis. Journal of Economic Entomology* 86(4), 989–994.

Storer, N.P., Peck, S.L., Gould, F., Van Duyn, J.W. and Kennedy, G.G. (2003) Spatial processes in the evolution of resistance in *Helicoverpa zea* (Lepidoptera: Noctuidae) to Bt transgenic corn and cotton in a mixed agro-ecosystem: a biology-rich stochastic simulation model. *Journal of Economic Entomology* 96, 156–172.

Tabashnik, B.E. (1994a) Delaying insect adaptation to transgenic plants: seed mixtures and refugia considered. *Proceedings of the Royal Society of London Series B* 255, 7–12.

Tabashnik, B.E. (1994b) Evolution of resistance to *Bacillus thuringiensis. Annual Review of Entomology* 39, 47–79.

Tabashnik, B.E., Patin, A.L., Dennehy, T.J., Liu, Y.-B., Miller, E. and Staten, R.T. (1999) Dispersal of pink bollworm (Lepidoptera: Gelechiidae) males in transgenic cotton that produces a *Bacillus thuringiensis* toxin. *Journal of Economic Entomology* 92, 772–780.

Tabashnik, B.E., Patin, A.L., Dennehy, T.J., Liu, Y.-B., Carrière, Y. and Antilla, L. (2000a) Frequency of resistance to *Bacillus thuringiensis* in field populations of pink bollworm. *Proceedings of the National Academy of Sciences USA* 21, 12980–12984.

Tabashnik, B.E., Liu, Y.-B., de Maagd, R.A. and Dennehy, T.J. (2000b) Cross-resistance of pink bollworm (*Pectinophora gossypiella*) to *Bacillus thuringiensis* toxins. *Applied Environmental Microbiology* 66, 4582–4584.

Tabashnik, B.E., Liu, Y.-B., Dennehy, T.J., Sims, M.A., Sisterson, M., Biggs, R. and Carrière, Y. (2002) Inheritance of resistance to Bt toxin Cry1Ac in a field-derived strain of pink bollworm (Lepidoptera: Gelechiidae). *Journal of Economic Entomology* 95, 1018–1026.

Tabashnik, B.E., Gould, F. and Carrière, Y. (2003a) Delaying evolution of insect resistance to transgenic crops by decreasing dominance and heritability. *Journal of Evolutionary Biology* 17, 904–912.

Tabashnik, B.E., Carrière, Y., Dennehy, T.J., Morin, S., Sisterson, M., Roush, R.T., Shelton, A.M. and Zhao, J.-Z. (2003b) Insect resistance to transgenic Bt crops: lessons from the laboratory and field. *Journal of Economic Entomology* (Forum) 96, 1031–1038.

Venette, R.C., Hutchison, W.D. and Andow, D.A. (2000) An in-field screen for early detection and monitoring of insect resistance to *Bacillus thuringiensis* in transgenic crops. *Journal of Economic Entomology* 93, 1055–1064.

Vilela, F.M.F., Waquil, J.M., Vilela, E.F., Siegfried, B.D. and Foster, J.E. (2002) Selection of the fall armyworm, *Spodoptera frugiperda* (Smith) (Lepidoptera: Noctuidae) for survival on Cry1A(b) Bt toxin. *Revista Brasileira de Milho e Sorgo* 1, 12–17.

Waquil, J.M., Vilela, F.M.F. and Foster, J.E. (2002) Resistência do milho (*Zea mays* L.) transgênico (Bt) à lagarta-do-cartucho, *Spodoptera frugiperda* (Smith) (Lepidoptera: Noctuidae). *Revista Brasileira de Milho e Sorgo* 1, 1–11.

Waquil, J.M., Vilela, F.M.F., Siegfried, B.D. and Foster, J.E. (2004) Actividade biológica das toxinas do *Bt*, Cry1A(b) e Cry1F em *Spodoptera frugiperda* (Smith) (Lepidoptera: Noctuidae). *Revista Brasileira de Milho e Sorgo* 3(2), 153–163.

Zhao, J.-Z., Li, Y.X., Collins, H.L. and Shelton, A.M. (2002) Examination of the F_2 screen for rare resistance alleles to *Bacillus thuringiensis* toxins in the diamondback moth (Lepidoptera: Plutellidae). *Journal of Economic Entomology* 95, 14–21.

Zhao, J.-Z., Cao, J., Li, Y.X., Collins, H.L., Roush, R.T., Earle, E.D. and Shelton, A.M. (2003) Transgenic plants expressing two *Bacillus thuringiensis* toxins delay insect resistance evolution. *Nature Biotechnology* 21, 1493–1497.

13 Supporting Risk Assessment of Bt Cotton in Brazil: Synthesis and Recommendations

D.A. Andow, E.M.G. Fontes, A. Hilbeck, J. Johnston, D.M.F. Capalbo, K.C. Nelson, E. Underwood, G.P. Fitt, E.R. Sujii, S. Arpaia, A.N.E. Birch, A. Pallini and R.E. Wheatley

Corresponding author: Dr D.A. Andow, Professor of Insect Ecology, University of Minnesota, 219 Hodson Hall, 1980 Folwell Avenue, St Paul, MN 55108, USA. Fax: +1 612 6255299, e-mail: dandow@umn.edu

Transgenic crops have been promoted vigorously throughout the world as the most significant recent development of agricultural technology. They have, however, engendered greater controversy than any previous agricultural technology. This controversy has focused on two related issues: (i) how can they be used to help humankind and (ii) how can they be evaluated to avoid harming biodiversity and humankind? These controversies are acute in Brazil, where impassioned discussions have continued for over 10 years (Capalbo *et al.*, Chapter 3, this volume).

In September 2003, the Cartagena Biosafety Protocol to the Convention on Biological Diversity entered into force. Among its several provisions, the protocol requires the use of a case-specific risk assessment to evaluate the environmental safety of a genetically engineered crop. Brazil is a party to the protocol, and has expressed considerable interest in developing the necessary expertise to conduct appropriate risk assessments.

This book focuses on the methods for conducting the science that supports environmental risk assessment of transgenic crops that are in the process of commercial development, using *Bacillus thuringiensis* (Bt) cotton as a case study, and expands on results reported in Capalbo and Fontes (2004). Throughout this book, we have relied on publicly available data to instantiate the evaluations and methodologies. While there is considerably more data in the private sector, by relying on the publicly available data we could ensure that our evaluations were transparent and would reinforce the fact that we had not conducted an actual risk assessment of Bt cotton. Moreover, it should be appreciated that the validity of the methodologies to support risk assessment is not influenced by the absence of private-sector data. Several kinds of Bt cotton were considered, but the case study concentrated on two private-sector events – a Cry1Ac event and a combined Cry1Ac/Cry2Ab event, targeted to control Lepidopteran pests (see Grossi *et al.*, Chapter 4, this volume). We recognize that there are several other private-sector events being developed that Brazil may consider, and that Brazilian Agricultural Research Corporation (Embrapa) is developing their own Bt cotton events; we suggest throughout the book that the methodologies should readily generalize to these other events. In this chapter, we will briefly summarize the agricultural context for the possible use of Bt cotton in Brazil and some of the main findings of the previous chapters. We will finish the chapter by describing some of the broader implications from the case study and suggest activities to further develop the scientific capacities for environmental risk assessment of transgenic crops in Brazil.

Cotton in Brazil and the Lepidopteran Pests

Cotton is an important and expanding cash crop in Brazil, and is grown in a wide range of climates, soils and crop-production systems (Fontes *et al.*, Chapter 2, this volume). There are three main production regions: the North-east, the Meridian and the Midwest. Cotton production in Brazil has over 100 years of history, but in the last 5–10 years large areas of new agricultural land have been brought into production in the Midwest, where about 76% of the area of Brazilian cotton is now grown. This shift to the Midwest has moved cotton production to poorer soils with associated changes in the cropping system and environment, including higher fertilizer use, higher altitudes, double-cropping with millet and direct drilling of seeds under minimal tillage. New high-yielding varieties are used, but these tend to be susceptible to viral diseases. Production is predominantly under very large-scale mechanized monoculture, with high use of pesticides and growth regulators. It is one of the most productive rainfed cotton areas in the world, and accounts for 85% of Brazilian fibre production. At the same time, socio-economic factors and the severe damage of the recent boll weevil invasion since the mid 1980s have contributed to the decline in cotton production by small-scale farmers in the North-east and Meridian regions.

Damage caused by pests and diseases is one of the main constraints to cotton productivity. The boll weevil and the cotton aphid are the most significant arthropod pests in all cotton-growing regions of Brazil. Unchecked, the boll weevil can cause total crop failure by killing or damaging the cotton bolls. The cotton aphid causes direct damage to cotton, but is a key limiting factor to cotton production as a vector of viral diseases, particularly blue disease caused by a luteovirus (Corrêa *et al.*, 2005). Particularly in the Midwest, where virus-susceptible varieties are widely grown, the risk of viral disease is prompting farmers to apply six to eight insecticide sprays to control cotton aphids from early in the growing season.

The most important Lepidopteran pests on cotton are cotton leafworm (*Alabama argillacea* (Hübner)), pink bollworm (*Pectinophora gossypiella* (Saunders)), tobacco budworm (*Heliothis virescens* (Fabricius)) and fall armyworm (*Spodoptera frugiperda* (J.E. Smith)). In the North-east, the first two species are the main Lepidopteran pests. In the Meridian region, the tobacco budworm is also important, and in the Midwest all four species are important. Cotton leafworm and pink bollworm are found feeding only on cotton, and in Brazil tobacco budworm is mainly restricted to cotton. Fall armyworm is highly polyphagous and is also a pest of maize and sorghum, which are often grown near cotton. More up-to-date information on pest status, damage and yield loss are needed, and there is clearly the need for a concentrated effort to develop an integrated pest management (IPM) programme to address the entire pest complex in an integrated way, emphasizing farmer education and extension.

Accurate predictions on how much farmers will reduce insecticide applications on Lepidopteran-active Bt cotton are difficult to make due to the complexity and dynamic nature of Brazilian cropping systems, and the range and differing sensitivity to Bt of the pest Lepidoptera (Fitt *et al.*, Chapter 12, this volume). The four target species differ markedly in host range and association with cotton. Bt cotton will not reduce insecticide use for boll weevil, but is likely to generate savings of ~2 pesticide applications for the target Lepidoptera (Fontes *et al.*, Chapter 2, this volume; Fitt *et al.*, Chapter 12, this volume). This would halve insecticide use in the North-east but only reduce it by ~10% in the Meridian and Midwest regions, where large amounts of insecticide are applied to control boll weevil and cotton aphid. Under conditions of decreasing or low pesticide use, natural enemies of pests may become more abundant, possibly enabling integration with biological controls. This will be possible only on virus-resistant varieties and when there is no need to apply insecticides to control other early-season pests. Other options are the application of cultural control methods, the selection of appropriate cultivars and the enhancement of biological control. In particular, new approaches for the integrated control of the boll weevil and the aphid-transmitted blue disease are urgently needed.

Problem Formulation and Options Assessment

Problem formulation and options assessment (PFOA) serves as a cornerstone in support of environmental risk assessment for transgenic organisms. One of the greatest challenges is to understand the conflicted nature of the discussion about transgenic organisms in Brazilian society, while finding a common ground from which it is possible to move forward. PFOA enables this to occur and, in so doing, the authors (Capalbo *et al.*, Chapter 3, this volume) reached several conclusions and reinforced findings of previous authors (Nelson *et al.*, 2004). Perhaps the most robust conclusion about PFOA is that all of the authors, who represent all of the key decision-making authorities in Brazil, agreed that a PFOA would be useful for Brazil and that the process should be further designed to meet the needs of Brazil.

PFOA proposes a deliberative process with multi-stakeholder participation and allows members of society to participate in the evaluation of critical needs and risks associated with transgenic organisms. It is best served if it is driven by sound, scientifically guided assessment and review. The purpose of a PFOA is to define the problem situation that is addressed by the transgenic organism under consideration and describe realistic options for addressing the problem situation. Thus, a PFOA specifies the context and scope for subsequent risk assessment and suggests alternatives to which a transgenic organism could be compared. These requirements imply that conducting a successful PFOA is a complex process – but this should not be considered as a possible argument against its use. Most important societal decisions are complex, messy and controversial, and because they are complex, messy and controversial we should work to improve them with transparent, systematic and scientifically based discussion. By doing so, the decision-making process gains social legitimacy and society gains greater confidence in the decisions taken. A PFOA can be used to provide such a discussion, and can play a significant role in environmental risk assessment.

Brazil has been working for nearly a decade to enact regulatory legislation, in part to provide the society adequate assurance that care and precaution will be taken to warrant both environmental and human health safety of all food/feed produced from transgenic organisms (see Capalbo *et al.*, Chapter 3, this volume, for a full discussion). The PFOA would need to be situated within the Brazilian regulatory framework to be effective. At present, this is the new Brazilian Biosafety Law (*Law no. 11.105, of March 24, 2005*) for transgenic organisms and their by-products that provides across-the-board regulation in the areas of agriculture, health and environment. Its provisions cover both authorizations for laboratory, greenhouse and field experiments and applications in fields such as forestry, bioremediation, agriculture and medicine. Inspection agencies and entities under the Ministry of Agriculture Cattle Raising and Supply, the Ministry for the Environment and the Ministry of Health are involved in inspections and enforcing penalties.

The National Technical Committee for Biosafety (CTNBio), which was created by Provisional Measures (PM) in 1996 and 2001, is given responsibility for assessment of health and environmental risks posed by transgenic organisms. Decisions of the CTNBio are sent to the above agencies; these agencies grant registrations and are responsible for monitoring in the respective areas. CTNBio shall rule, at the last and final jurisdiction, on cases when the activity involves a potential or effective agent of environmental degradation, and on the need for environmental licensing (Capalbo *et al.*, Chapter 3, this volume). Despite the evolution of the regulatory system, the PFOA's charge may be best conducted under the umbrella of the CTNBio, mainly because Article 15 of the new Biosafety Law gives authority to CTNBio to carry out public hearings where civil society participation is assured under the terms of the law. The situation remains fluid because new regulations may modify the roles of the agencies involved, and it may be too early to try to link the PFOA to any particular agency.

Some modifications will be necessary to make the PFOA responsive to multiple stakeholders. The CTNBio expert committee would need to be diversified to include expertise in economics, social science and broader areas of environmental science outside of agriculture. As the legal authority, the CTNBio would be able to charge a multi-stakeholder group to initiate a PFOA as a first step in risk assessment to help understand the problem, provide a preliminary review of options and highlight critical societal concerns. Then studies can be conducted during the subsequent stages in the development of a transgenic plant (Andow *et al.*, Chapter 1, this volume). A second PFOA meeting should be conducted when laboratory and field environment biosafety test results can be included in the evaluation by the PFOA group. As a suggestion to those who will consider these ideas for implementation, the authors encourage the nation to develop systems for monitoring new technologies, and consider the PFOA methodology as an approach that can be adapted to postrelease monitoring as well as prerelease evaluation.

In general, the Brazilian evaluation of the PFOA methodology builds on the Kenyan case of Bt maize (Nelson *et al.*, 2004) and confirms that this methodology is essential for assessing proposed uses of transgenic organisms because it provides constructive arenas for dialogue and potential consensus. Within Brazil it would need to be solidly anchored in the regulatory framework and needs to be structured to have greater emphasis on environmental concerns outside of agriculture in order to function successfully. Even in the highly politicized debates about Brazil's use of transgenic organisms, the authors (a diverse group of ministry officials) see some promise in the PFOA's guided discussion over a range of issues that enables societal consideration of a new transgenic organism. The next steps for designing and implementing a Brazilian PFOA require developing specifics of the multi-stakeholder process and consultation with a broader group of stakeholders.

Transgene Expression and Locus Structure

Characterizing transgene expression and locus structure provides the basic information necessary to conduct a risk assessment. The results and recommendations for Bt cotton (Grossi *et al.*, Chapter 4, this volume) reinforce two key recommendations from a previous case study on Bt maize in Kenya (Andow *et al.*, 2004). It is recommended that the transgene locus is sequenced with flanking regions in the plant to reduce the likelihood that the transgenic plant might produce unexpected or inadvertent (unintended) gene products. DNA sequencing of the transgene locus in the plant provides definitive scientific evidence for reducing uncertainties associated with the transgene locus, including ectopic expression, spurious open reading frames (ORFs), homologous recombination and disturbance of a native plant gene. Although this is not a requirement now in several national risk-assessment processes for transgenic plants, it is a technically feasible standard practice, and would be the easiest and most economical means to meet many regulatory standards. However, while DNA sequencing is necessary for risk assessment, it is not sufficient. It is also necessary to characterize the transgene phenotype. To do this, transgene expression should be measured in whole plants in the relevant environments in which they will be grown, because expression can vary among plant tissues, plant life stages and environmental conditions (such as soil types).

In this case study, we have focused on two Bt-cotton events. Publicly available information on these events was insufficient for risk-assessment purposes, lacking DNA sequence of the transgene locus and sufficient expression data from relevant plant tissues, growth stages and environments. Consequently, we focused primarily on methods to characterize the transgene locus and phenotype (Grossi *et al.*, Chapter 4, this volume). We found that all of the aspects of characterizing the transgene (locus structure and expression) can be addressed with scientifically feasible experimental protocols. In addition, we suggested that while transformation purely for research purposes may use many marker genes and the entire Ti plasmid backbone, transformation for commercial use would be advised to eliminate marker genes and the Ti plasmid backbone and to select simple transformation events at single transgene loci for development.

Non-Target and Biodiversity Effects

Risks to non-target species and biodiversity are difficult to assess because there are many non-target species and they interact in complicated ways. In Chapters 5–10, we have elaborated a methodology to provide scientific data for non-target risk assessment (Andow and Hilbeck, 2004; Birch *et al.*, 2004) that focuses on the species that are most at risk in the recipient environment. The objective of the methodology is to identify possible adverse effects and characterize risks to non-targets species and ecosystem processes in the environment of interest (Hilbeck *et al.*, Chapter 5, this volume). This is done by choosing significant functional categories of species and ecosystem processes, screening species and ecological functions within those categories for possible risks, conducting an exposure assessment, identifying possible adverse effects, generating testable scientific hypotheses about how risks might occur and developing experiments to test these hypotheses.

For Bt cotton in Brazil, the authors chose to focus the assessment on non-target cotton arthropod pests (Sujii *et al.*, Chapter 6, this volume), pollinators, flower visitors, endangered Lepidoptera (Arpaia *et al.*, Chapter 7, this volume), arthropod predators in cotton (Faria *et al.*, Chapter 8, this volume), parasitoids of important cotton pests (Pallini *et al.*, Chapter 9, this volume) and soil ecosystem processes (Mendonça *et al.*, Chapter 10, this volume). These choices reflected an overall concern that Bt cotton could have adverse effects on cotton production, possibly offsetting its potential benefits. This could be particularly problematic for the small-scale farmers in the North-east and Meridian regions. Non-target pests might be released from insecticides, competition, predation or parasitism, pollination might be reduced and the incidence of aphid-vectored disease might be exacerbated. Soil health may be reduced. Except for endangered Lepidoptera, it was not possible to address adverse effects on surrounding natural areas, such as the Cerrado in the Midwest, systematically in this book. This needs to be revisited in the future.

Species that could be at risk were identified using a Species Selection Matrix and an Exposure Assessment Matrix, both of which were useful to screen species and processes to identify a short list of candidates for actual experimental testing. The first matrix screens on the basis of exposure to the crop and significance of the ecological function, and requires information only about the crop and the intended environment of introduction. The second matrix requires information about transgene expression in the plant. From a total of over 90 species, seven species and three species groups (ladybird beetles, green lacewings and earthworms) were prioritized as possible candidates for risk-assessment experiments in all three of the cotton-growing regions, and three species were considered important regionally (Table 13.1). Five soil ecosystem functions were prioritized (Table 13.1). Several chapters suggested that the process should give special weight to the diversity of feeding modes in prioritizing the species.

These procedures rely heavily on expert judgement that depends greatly on the working information and experiences of the experts. Because cotton has only recently expanded into the Midwest, there is relatively little information or experience on cotton ecology from this region. New baseline information from the Midwest is likely to expand the initial listing of species (Arpaia *et al.*, Chapter 7, this volume), which may result in more species or a greater diversity of species being prioritized (see VI.2. below). The selection matrix also highlighted gaps of knowledge on the ecology of cotton, which can be used to prioritize the needs for baseline research. A precautionary approach to considering uncertainty was used (Hilbeck *et al.*, Chapter 5, this volume). The soil micro-, meso- and macrofauna are so complex and little known that a species-based approach to risk assessment is infeasible. Although additional information about these soil organisms would be helpful, we avoided the problems inherent in this uncertainty by focusing on the ecosystem processes in relation to non-target risk assessment.

For the prioritized candidate species and functions, we identified potential adverse environmental effects and formulated experimentally testable hypotheses about potential exposure and adverse-effect scenarios. We designed experiments to refute or confirm the hypotheses. In most cases, experiments were aimed to estimate the effects of Bt cotton on fitness or a component of fitness prior to field release of the transgenic crop. For parasitoids we placed additional emphasis on potential alterations of the chemical ecology, affecting host finding and acceptance (Pallini *et al.*, Chapter 9, this volume). Plant inputs have a strong driving influence on soil ecosystem processes. Therefore, for these processes, exposure routes, fate and transport of Bt toxins in soils largely determined the kinds of experiments proposed (Mendonça *et al.*, Chapter 10, this volume). Exposure routes in the soil are complex: Bt toxins occur embedded within decaying and living plant material, in root exudates, leakage from broken living roots, and bound on clay and humic substances. A few experiments were proposed for small-scale or large-scale field trials (Sujii *et al.*, Chapter 6, this volume), but it will be necessary to

Table 13.1. The non-target species, species groups and functions prioritized for possible testing of non-target effects of Bt cotton.

Name		Source
Prioritized in all regions of Brazil		
Cotton aphid	*Aphis gossypii*	Sujii *et al.*, Chapter 6, this volume
White fly	*Bemisia tabaci*	Sujii *et al.*, Chapter 6, this volume
Boll weevil	*Anthonomus grandis*	Sujii *et al.*, Chapter 6, this volume
Fall armyworm	*Spodoptera frugiperda*	Sujii *et al.*, Chapter 6, this volume
Africanized honey bee	*Apis mellifera*	Arpaia *et al.*, Chapter 7, this volume
Bumblebee	*Bombus* sp.	Arpaia *et al.*, Chapter 7, this volume
Green lacewing	*Chrysoperla* spp. complex	Faria *et al.*, Chapter 8, this volume
Ladybird beetles	Coccinellidae[a]	Faria *et al.*, Chapter 8, this volume; Arpaia *et al.*, Chapter 7, this volume
Egg parasitoid	*Trichogramma pretiosum*	Pallini *et al.*, Chapter 9, this volume
Larval parasitoid	*Catolaccus grandis*	Pallini *et al.*, Chapter 9, this volume
Residue decomposition		Mendonça *et al.*, Chapter 10, this volume
Ammonification		Mendonça *et al.*, Chapter 10, this volume
Nitrification		Mendonça *et al.*, Chapter 10, this volume
Root herbivory		Mendonça *et al.*, Chapter 10, this volume
Disease transmission		Mendonça *et al.*, Chapter 10, this volume
Earthworms	Lumbricidae	Mendonça *et al.*, Chapter 10, this volume
High priority regionally		
	Bracon vulgaris	Pallini *et al.*, Chapter 9, this volume
	Netelia sp.	Pallini *et al.*, Chapter 9, this volume
	Lysiphlebus testaceipes	Pallini *et al.*, Chapter 9, this volume

[a]Including *Cycloneda sanguinea*, *Hippodamia* spp., *Coleomegilla maculata*, *Scymnus* spp., *Eriopsis* spp.

develop the non-target methodology further by specifying the transition from confined testing to field testing.

Gene Flow and Consequences

All of the gene-flow concerns for Bt cotton in Brazil centre around the fact that the three tetraploid *Gossypium* species, *G. hirsutum*, *G. barbadense* and *G. mustelinum*, do not have substantial reproductive barriers between them, so that all crosses have a high likelihood of producing fertile F_1 progeny (Johnston *et al.*, Chapter 11, this volume). There are two major varieties of *G. hirsutum* in Brazil, *G. hirsutum* var. *latifolium* and *G. hirsutum* var. *marie-galante*. The first is the typical herbaceous cotton crop, and occurs as cultivated varieties and volunteers. The second is a perennial crop variety that occurs as commercial varieties, land races, volunteers, dooryard plants and in feral populations. There are two major varieties of *G. barbadense* in Brazil, *G. barbadense* var. *barbadense* and *G. barbadense* var. *braziliense*. Both occur as dooryard plants and in feral populations, but only *G. barbadense* var. *braziliense* occurs as land races and volunteers. *G. mustelinum* is a wild species. We concluded from the evidence presented that all of these indigenous Brazilian cottons belong to the primary germplasm pool for Bt cotton, and thus, that hybridization between Bt cotton and indigenous varieties will occur at some frequency. Specifically, there are important gaps in knowledge of the geographical distribution of *G. barbadense* var. *barbadense* that will be essential to understand before gene-flow risks can be assessed. It will be impossible to eliminate all risk of gene flow from genetically engineered cotton, but it may be possible to reduce the consequences of gene flow, for example, by restricting the area over which Bt cotton is planted, excluding potential gene-flow 'hotspots'. Whether and where to impose such restrictions in Brazil would depend on the potential consequences of gene flow.

There are many possible consequences of gene flow from Bt cotton to other *Gossypium* species in Brazil. The demographic and genetic changes that could occur in *G. mustelinum* populations as a result of gene flow from Bt cotton were identified as the most significant potential consequences of gene flow in this system. *G. mustelinum* is already recognized as rare and potentially endangered in Brazil. Exposure to gene flow might change the genetic composition of the species or create demographic fluctuations that could lead to extinction. While *G. mustelinum* is the least likely type of recipient population to overlap geographically with Bt-cotton cultivars, there are several ways that genetic bridges between Bt-cotton crops and these wild cotton populations could develop. As a worst-case scenario, repeated gene flow from Bt cotton (or other elite cotton varieties) could replace the unique genetic character of *G. mustelinum* with crop genes, essentially hybridizing the species out of existence. As a result, the unique germplasm of the species could be lost and the ecological communities in which the plant grows could be altered.

We were unable to thoroughly evaluate several of the other potential consequences of gene flow due to gaps in the knowledge of the biology and ecology of the traditionally cultivated or non-cultivated *Gossypium* species. Most importantly, we could not evaluate the fitness effect of the Bt gene on cotton fitness outside an agricultural setting because we were missing at least three types of information. First, we do not know if or how much the target insects (cotton leafworm, pink bollworm, cotton budworm and fall armyworm) are feeding on feral, dooryard, volunteer and wild cotton populations. Second, we do not know if plant fitness is affected by target insect herbivore pressure on cotton plants outside cultivation. And third, we do not know how the Bt gene will be expressed in a *G. barbadense* or *G. mustelinum* genetic background.

Resistance Risk and Management

Wherever Bt cotton is deployed, the evolution of resistance in target pests should be viewed as a risk that requires management (Fitt *et al.*, Chapter 12, this volume). Experience with insecticide use and basic consideration of evolutionary theory imply that if a Bt crop were used extensively without any resistance-management intervention, resistance would seem inevitable. We identified four key Lepidopteran targets for deployment of Bt cotton in Brazil: cotton budworm (*H. virescens*), pink bollworm (*P. gossypiella*), cotton leafworm (*A. argillacea*) and fall armyworm (*S. frugiperda*). The focus of our analysis for Bt cotton in Brazil is to identify pre-emptive strategies to delay the onset of resistance.

A critical factor in defining options for resistance management is the likely exposure of target pests to Bt cotton. Two of the four target species are specialists on cotton bollworm, pink bollworm and cotton leafworm, while cotton budworm is regarded as closely tied to cotton in Brazil. Only fall armyworm is highly polyphagous with a wide range of unrelated host plants. These host relationships would suggest that the three species with close associations with cotton are at greatest risk of resistance from Bt cotton. However, concomitant plans to introduce Bt maize into Brazil invalidate this conclusion, because maize is a significant host for fall armyworm. This species has evolved resistance to a number of insecticide groups and may represent a significant resistance threat across farming systems reliant on Bt crops.

One limitation in our assessment is that comprehensive information was not available on the dose of specific Bt-cotton varieties against any of the target pests in Brazil. A high dose of the Bt toxin, defined as one that kills nearly all susceptibles and > 95% of heterozygous resistance individuals, will help to delay the onset of resistance in combination with refuge crops. A low dose of the Bt toxin will require much larger refuges. The possible introduction of Bt maize in Brazil required an assessment of management options for different scenarios involving only Bt cotton or a combination of Bt cotton and Bt maize.

While there are many knowledge gaps and research needs, we conclude that it is possible to formulate reasonable recommendations to manage the resistance risk for Bt cotton in Brazil. Refuge crops are a key component of the recommended strategies for all scenarios. In addition, we strongly recommend that Bt cotton be deployed as a pyramided two-gene product with both genes expressing a high dose against the same target species. Pyramiding of two independently acting transgenes will provide additional safeguards against resistance, provided refuges are also used.

Overall, our recommendations should serve as a starting point for a comprehensive research and extension effort to fill the many gaps and commence the critical task of educating growers about the serious need for resistance management if the potential benefits of Bt cotton are to be realized and sustained.

Broad Implications of the Case Study

Conducting this case study provided the authors of the chapters of this book with the opportunity to understand the scientific methods that can be used to support a risk assessment while focusing on a topic of considerable importance and direct relevance to society. In addition, the case study approach was a test for the methodologies described in this book. While most of these methodologies have been proposed in whole or in part in other publications, assembling them all in this single case study enabled authors to gain better appreciation for how science can support risk assessment. This case study, similar to the previously published Kenya case study on Bt maize (Hilbeck and Andow, 2004), revealed many broad findings; we emphasize two of these.

The environmental risks associated with Bt cotton in Brazil will vary among the different regions of Brazil

Our preliminary assessment of the environmental risks of Bt cotton in Brazil has clarified that non-target gene flow and resistance risks will likely differ among the North-east, Midwest and Meridian regions within Brazil. This means that a risk assessment should be conducted to assess the risks specific to each region rather than being conducted as if the country were a homogeneous whole. Although this may increase the effort needed to complete a risk assessment, such an approach is case-specific as indicated by the Cartagena Protocol and relies on information about the transgene, the crop and the environment where the introduction is planned to occur (NRC, 1987, 2000, 2002; Tiedje et al., 1989; EU, 2001; Snow et al., 2005). This region-specificity also means that risk management can be tailored to the region, with different risk-management approaches for each region, allowing for more flexibility in management, as has been suggested for gene-flow risks (Johnston et al., Chapter 11, this volume) and resistance risks (Fitt et al., Chapter 12, this volume). This flexibility could result in a lower cost for risk management.

While there is also likely to be significant variation in environmental risk within each of these regions, conducting risk assessment at finer spatial scales may not improve environmental safety. Within each region, environmental risks may be related to variation in farm size and management, with large-scale monoculture having greater resistance risk than small-scale diversified production systems. However, stratifying risk assessment within regions, such as by farm size and management practice or by habitat heterogeneity would likely create an unduly complex risk-assessment process that could make risk management very difficult. Finding and implementing a socially desirable risk-management policy in a stratified social environment or a fine-scale spatial environment could be extremely complex.

NORTH-EAST REGION. Considerable environmental variability and physiological stress to cotton in this region may affect transgene expression. Gene flow to other *Gossypium* plants may be likely. *G. hirsutum* var. *marie-galante*, *G. barbadense* and *G. mustelinum* are present and could be affected. Farming is done on a small spatial scale and productivity is low for Brazil, but about half of the farmers and farming families in cotton in Brazil are in this region. Effective resistance management may be difficult because of the small scale of farming. Population changes to cotton boll weevil need careful assessment because it is a key pest and most farmers in the region cannot afford increased control costs should boll weevil become a more important pest. Other non-target species may be characteristic to this region and require their own assessment. Benefits of Bt cotton to small-scale farmers could be significant if pesticide applications can be reduced.

MIDWEST REGION. Large-scale production predominates and yields are high in this region. Production is new to this region in the past 5–10 years and continues to expand. Gene flow could occur to *G. barbadense* and mainly to other commercial cotton lines, and is therefore a less significant concern than in the North-east. Effective resistance management may be most easily realized in this region, because the scale of production is so large and farmers can internalize both the costs and benefits of resistance management. However, one of the key Lepidopteran pests is fall armyworm, which may require a large refuge of non-Bt cotton for effective resistance management. The simplified, large-scale agricultural production systems embedded in the unique and biologically diverse Cerrado landscapes result in a relatively unknown, but characteristic, non-target flora and fauna. It will be necessary to assess the possible risks of Bt cotton to this unique flora and fauna.

MERIDIAN REGION. Small- to large-scale production predominates, and cotton yields can be high, although not as high as in the Midwest. Gene flow may occur to other commercial cotton lines and possibly to *G. barbadense*, and is therefore a less significant concern than in the North-east and similar to the Midwest. Effective resistance management may be possible in this region, in part because of the diversified cropping systems. The scale of production is large enough that individual farmers may be able to implement effective resistance management. The landscapes are less uniform than in the Midwest, but the non-target fauna in cotton is better known than in the Midwest.

AMAZON REGION. At present, this region has little cotton production, and possible environmental risks in this region were not assessed in this case study. Cotton production is increasing in the borders of this region, so risk assessment may become necessary in the future. The Amazon region harbours wild relatives of *G. hirsutum* that could be adversely affected by gene flow.

We identified and prioritized the potential adverse effects to the environment and associated data gaps, and developed some protocols to address these issues. Our methods can be used to support environmental risk assessment of transgenic crops scientifically and transparently as required by the Protocol on Biosafety (Annex III).

Developing, adapting and using the methodologies in this book will enable a country such as Brazil to support a risk-assessment process that would meet its obligations under the Protocol on Biosafety. We have developed efficient and transparent scientific procedures to select from the high diversity of possible non-target species a much smaller number of species that could serve as test organisms to support non-target risk assessment (non-target selection matrices). We have developed straightforward methods for identifying potential recipients and adverse effects of gene flow, and for assessing resistance risk. All of these methodologies are initiated on a case-by-case basis as suggested by the protocol.

We recognize that the Cartagena Protocol allows for the use of a precautionary approach to decision making, but have focused our efforts on appropriate methods for addressing scientific uncertainty. We have developed scientific, transparent methods that can be layered on top of our main methodologies so that the influence of scientific uncertainty on the interpretation of results can be clarified. In general, we have relied on evaluating worst-case scenarios to clarify the influence of scientific uncertainty. We believe that it is important that scientific, transparent methods for addressing uncertainty are available to be incorporated, as appropriate, into the decision-support process.

We also developed a temporal staging of the scientific methodologies to correspond with the stages of development of a transgenic plant, starting from the transformation of cell lines to the release of a commercial product. Staging the scientific evaluations in this way allows the risk-assessment process to proceed in a timely manner, so that its quality is not compromised by unnecessarily shortened timelines, and it does not require more work than necessary. We suggest that during pre-commercial assessments, three levels of evaluation are probably needed:

- Laboratory and greenhouse
- Small-scale field trials
- Large-scale field trials

We suggest that the Brazilian regulatory system should be modified to allow large-scale field testing prior to commercial approval, because several elements of the risk-assessment process will require large-scale testing.

An additional significant result was the improved scientific capacity to conduct research on the environmental risks of transgenic crops in Brazil. By focusing on potential environmental risks of Bt cotton, a broad spectrum of Brazilian agricultural scientists now understand how to apply their research skills to conduct experiments to support the assessment of environmental risks of transgenic crops.

Some of the research gaps identified in this book have already been started to be filled in Brazil. Studies are being carried out to gather baseline data on flower-visiting insects in the Midwest and North-east, and have already revealed a strikingly higher diversity than was expected from the scientific literature, for instance five bee species had been identified from the literature, whereas 40 were sampled (E. Fontes, Brasília, Brazil, 2004, personal communication). Moreover, the survey found strong regional differences; in the

Midwest 23 species were surveyed and in the North-east 21 species were surveyed, but only four of these species were found at both sites. Scientists are investigating the *Spodoptera* biotypes on maize and cotton in the Midwest and have found them to be the same (C. Omoto, Piracicaba, Brazil, 2004, personal communication). A group is quantifying hybridization rates within and between cotton fields with certain separation distances and barrier crops, using glandless cotton as a marker, and also carrying out *in situ* characterization of wild, feral, dooryard and local varieties (Barroso *et al.*, 2004).

Recommendations and Future Outlook

The potential commercial release of Bt cotton in Brazil proved to be an instructive and challenging case, as illustrated by each chapter of this book. Cotton is an income source both for subsistence farmers of the North-east region and for large-scale agribusiness in the Midwest and Meridian regions. As highlighted by the PFOA (Capalbo *et al.*, Chapter 3, this volume), this provides vastly different contexts for understanding the potential utility and risks of Bt cotton. Brazil is one of the centres of diversification of cotton, and the cotton plant supports a rich fauna of herbivores and carnivores. This diversity of species, habitats and socio-economic conditions offered an array of options and a rich background for the evaluations evident in this book.

Consequently, this book offers the reader insights into how science can support environmental risk-assessment of one complex case of a transgenic crop. In so doing, it also offers the scientists, regulators and educators a rich source of information and methodologies to help identify possible adverse environmental effects, formulate hypotheses to evaluate the likely occurrence of possible environmental risks and plan appropriate scientific experiments to test these hypotheses. For regulatory personnel, it can be used as a tool to support decision making. While these methodologies can assist and support decision making, they will not replace the judgements and balancing of interests that decision makers must address.

It became clear that the baseline information needed to support environmental risk assessment in Brazil is in part unorganized or weak. Existing information needs to be systematically reviewed and made readily available, as demand requires. In addition, new data are needed to address important informational gaps. These data needs must be prioritized in funding programmes launched by the funding agencies. Moreover, as the assessment of pest resistance risk showed, there is a crucial and urgent need for farmer education and information programmes, particularly for insect-resistant transgenic crops.

It was also clear that scientific support of environmental risk assessment of transgenic crops in Brazil will require the involvement of a wide range of expertise. We identified a need to deepen the understanding of risk assessment and to engage more scientists in research in this area, so that scientists can more quickly provide targeted data needed to assess and manage risks. This can be encouraged by introducing the topic of environmental risk assessment in the programmes of scientific meetings, developing and conducting training courses and creating funding opportunities for risk assessment-related research. Although Brazil has the second largest agricultural research infrastructure in the world, there is a need to focus the research and development capacity in the country and to continue close collaboration and interchange with scientists from countries where risk-assessment research has a longer history.

We recommend that the methodologies developed in this book are consolidated in Brazil and expanded to other Latin American countries. This can be accomplished by developing course materials and offering training courses in these countries. The first audience of these training courses should be scientists or regulators who would be able to disseminate the methods to more people, i.e. priority should be given to training the trainers who can then teach others within their respective countries. Through these processes, we can face the challenges of how to characterize the risks and uncertainties to be useful for and understandable by decision makers.

References

Andow, D.A. and Hilbeck, A. (2004) Science-based risk assessment for non-target effects of transgenic crops. *BioScience* 54, 637–649.

Andow, D.A., Somers, D.A., Amugune, N., Aragão, F.J.L., Ghosh, K., Gudu, S., Magiri, E., Moar, W.J., Njihia, S. and Osir, E. (2004) Transgene locus structure and expression of Bt maize. In: Hilbeck, A. and Andow, D.A. (eds) *Environmental Risk Assessment of Transgenic Organisms: A Case Study of Bt Maize in Kenya*. CAB International, Wallingford, UK, pp. 83–116.

Barroso, P.A.V., Ciampi, A.Y., Hoffmann, L.V., Costa, J.N., Freire, E.C., Andrade, F.O., Carvalho, L.P. and Vidal Neto, F.C. (2004) Gene flow in cotton in Brazil. Poster shown at 8th International Symposium on the Biosafety of Genetically Modified Organisms, 26–30 September 2004, Montpellier, France.

Birch, A.N.E., Wheatley, R., Anyango, B., Arpaia, S., Capalbo, D., Getu Degaga, E., Fontes, E., Kalama, P., Lelmen, E., Lövei, G., Melo, I.S., Muyekho, F., Ngi-Song, A., Ochiendo, D., Ogwang, J., Pitelli, R., Sétamou, M., Sithanantham, S., Smith, J., Son, N.V., Songa, J., Sujii, E., Tan, T.Q., Wan, F.-H. and Hilbeck, A. (2004) Biodiversity and non-target impacts: a case study of Bt maize in Kenya. In: Hilbeck, A. and Andow, D.A. (eds) *Environmental Risk Assessment of Transgenic Organisms: A Case Study of Bt Maize in Kenya*. CAB International, Wallingford, UK.

Capalbo, D.M.F. and Fontes, E.M.G. (2004) *GMO Guidelines Project, Algodão Bt*. Embrapa Document 38, Embrapa Meio Ambiente, Jaguariúna, SP, Brazil. Available at: http://www.cnpma.embrapa.br/download/documentos_38.pdf

Corrêa, R.L., Silva, T.F., Araújo, J.L.S., Barroso, P.A.V., Vidal, M.S. and Vaslin, M.F.S. (2005) Molecular characterization of a virus from the family Luteoviridae associated with cotton blue disease. *Archives of Virology* 157(30), 1357–1367.

EU (European Union) (2001) Regulation of the European Parliament and of the Council on genetically modified food and feed. EC 2001/18. Available at: http://europa.eu.int/eur-lex/en/com/pdf/2001/en_501PC0425.pdf

Hilbeck, A. and Andow, D.A. (eds) (2004) *Environmental Risk Assessment of Transgenic Organisms: A Case Study of Bt Maize in Kenya*. CAB International, Wallingford, UK, 281 pp.

Nelson, K.C., Kibata, G., Lutta, M., Okuro, J.O., Muyekho, F., Odindo, M., Ely, A. and Waquil, J. (2004) Chapter 3: Problem formulation and options assessment (PFOA) for genetically modified organisms: the Kenya case study. In: Hilbeck, A. and Andow, D.A. (eds) *Risk Assessment of Transgenic Crops: A Case Study of Bt Maize in Kenya*. CAB International, Wallingford, UK, pp. 57–82.

NRC (National Research Council) (1987) *Field Testing Genetically Modified Organisms: Framework for Decisions*. National Academy Press, Washington, DC.

NRC (National Research Council) (2000) *Genetically Modified Pest-Protected Plants: Science and Regulation*. National Academy Press, Washington, DC.

NRC (National Research Council) (2002) *Environmental Effects of Transgenic Plants: the Scope and Adequacy of Regulation*. National Academy Press, Washington, DC.

Snow, A.A., Andow, D.A., Gepts, P., Hallerman, E.M., Power, A., Tiedje, J.M. and Wolfenbarger, L.L. (2005) Genetically modified organisms and the environment: current status and recommendation. *Ecological Applications* 15(2), 377–404.

Tiedje, J.M., Colwell, R.K., Grossman, Y.L., Hodson, R.E., Lenski, R.E., Mack, R.N. and Regal, P.J. (1989) The planned introduction of genetically engineered organisms: ecological considerations and recommendations. *Ecology* 70, 298–315.

Index

Page numbers in **bold** refer to illustrations and tables.

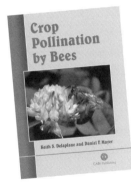
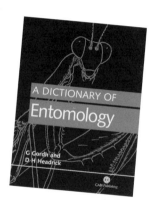
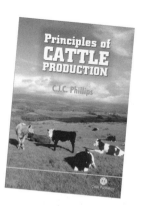